AF823103

REGENERATION OF STRIATED MUSCLE, AND MYOGENESIS

REGENERATION OF STRIATED MUSCLE, AND MYOGENESIS

PROCEEDINGS OF THE INTERNATIONAL CONFERENCE
CONVENED BY MUSCULAR DYSTROPHY
ASSOCIATIONS OF AMERICA AT THE
INSTITUTE FOR MUSCLE DISEASE, NEW YORK
MARCH 28-29 1969

EDITORS
ALEXANDER MAURO, SAIYID A. SHAFIQ
ADE T. MILHORAT

1970
EXCERPTA MEDICA, AMSTERDAM

1970, EXCERPTA MEDICA FOUNDATION

All rights reserved. No part of this publication may be reproduced or transmitted in any form or by any means, electronic or mechanical, including photocopying and recording or by any information storage and retrieval system, without permission in writing from the publishers.

INTERNATIONAL CONGRESS SERIES No. 218

Published December 1970

ISBN 90 219 0142 0

EXCERPTA MEDICA FOUNDATION OFFICES

Amsterdam	Herengracht 119-123
New York	New York Academy of Medicine Building, 2 East 103rd Street
London	Chandos House, 2 Queen Anne Street
Milan	Via San Senatore 6/2
Tokyo	Sankaido Building, 1-9-13 Akasaka, Minato-ku
Buenos Aires	Florida 165, Galeria Güemes, 4° Piso, Esc. 454/461

Printed in the Netherlands by Royal VanGorcum Ltd., Assen

FOREWORD

Striated muscle provides many of the conditions that give opportunity for the study of regeneration. It is an elaborately constructed tissue and a large number of its sub-units have been extensively investigated to the point where much now is known about their structure and function, thus, numerous parameters are available for observation and evaluation. The potential of striated muscle to regenerate generally is not appreciated by biologists and clinicians, although, under appropriate conditions, this regenerative capacity may be considerable. This phenomenon obviously is of much scientific and clinical interest. Many of the steps by which regeneration is accomplished resemble those seen in the embryogenesis of muscle, and the question whether damaged striated muscle can be restored to form and function resembling those of the original intact muscle has great significance to the hypotheses and practice of medicine.

Knowledge of regeneration has been increasing steadily. Biologists find of particular interest the origin and myogenic nature of the mononucleated cells found between the basement membrane and the plasmalemma, and the relation of these cells to myoblasts. Many investigators postulate that these mononucleated cells, so-called satellite cells, represent reserve myoblasts that could reproduce by mitotic division, the newly formed cells fusing to form myoblasts and finally multinucleated muscle fibers. Others believe that maturated multinucleated cells are themselves capable of producing mononucleated cells.

These hypotheses, and those concerned with sarcoplasmic sprouting and the myogenic role of invading cells may form the basis for elucidation of an important biological process and for the rational approach to the understanding and management of certain serious diseases of man. Probably most prominent among these disorders are the muscular dystrophies, a group of hereditary conditions in which muscle degeneration is the prominent and serious feature of the disease, with little evidence of regeneration. Regenerative attempts may be present but limited early in the course of the disorder, less evident or absent during the later stages, and never sufficient to alter the progressive pathological and clinical course. It is reasonable to speculate that an increase in our knowledge of the processes by which striated muscle regenerates could profoundly change our approach to these disorders.

To evaluate the evidence presently available and to indicate the paths that might be profitably followed in future investigations, Muscular Dystrophy Associations of America sponsored the international conference of which these proceedings are the published record. Thirty persons from this country and from abroad, all experts in this area of science, were invited to meet at the Institute for Muscle Disease in New York and discuss the present state of our knowledge of regeneration of striated muscle and the closely related subject of myogenesis. With the objective of stimulating active discussion and of giving the participants wide latitude of freedom, no formal papers were requested. The conference was planned to consist mainly of spontaneous presentations interspersed with numerous questions and answers in the hope thereby to achieve a content of information greater than could have been gained by a program of formal papers, some of which might have been written months before the conference. All discussions were recorded, and the participants had the opportunity of editing their transcripts and submitting photographs and figures for publication which they considered to be appropriate. In editing the transcripts every effort was made to preserve the spontaneous air of the presentations. A few of the speakers elected to supplement their contributions with short papers of a more formal nature.

I am happy that Dr. Mauro foresaw the value of the conference and was willing to serve as chairman.

I thank also all of the participants for their spirited cooperation in making the conference such a notable success. MDAA responded immediately to requests for financial support and Executive Director, Robert Ross, and his staff were most helpful. Excerpta Medica Foundation, the publisher, has our gratitude for the skills which give our publication the hoped for excellence.

ADE T. MILHORAT

CONTRIBUTORS

David Allbrook,
Department of Anatomy,
University of Western Australia,
Nedlands, Australia

Massimo Aloisi,
Institute of General Pathology,
University of Padua,
Padua, Italy

Michael E. Barberie,
Department of Anatomy,
University of Virginia School of Medicine,
Charlottesville, Va., U.S.A.

Richard Bischoff,
Department of Anatomy,
Washington University School of Medicine,
St. Louis, Mo., U.S.A.

Armin C. Braun,
Rockefeller University,
New York, N.Y., U.S.A.

Bruce M. Carlson,
Department of Anatomy,
University of Michigan,
Ann Arbor, Mich., U.S.A.

John C. T. Church,
Department of Anatomy,
Makerere University,
Kampala, Uganda, East Africa
present address:
Department of Surgery,
Faculty of Medicine,
Kenyatta National Hospital,
Nairobi, Kenya, East Africa

John R. Coleman,
Division of Biological and Medical Sciences,
Brown University,
Providence, R.I., U.S.A.

Patrick E. Conen,
Department of Pathology,
The Hospital for Sick Children,
Toronto, Ontario, Canada

James R. Florini,
Department of Zoology,
Syracuse University,
Syracuse, N.Y., U.S.A.

Roger K. Gilbert,
Department of Pathology,
Waterbury Hospital,
Waterbury, Conn., U.S.A.

Gabriel C. Godman,
Department of Microbiology,
College of Physicians and Surgeons,
Columbia University,
New York, N.Y., U.S.A.

Elizabeth D. Hay,
Department of Anatomy,
Harvard Medical School,
Boston, Mass., U.S.A.

Howard Holtzer,
Department of Anatomy,
University of Pennsylvania School of Medicine,
Philadelphia, Pa., U.S.A.

Harunori Ishikawa,
Department of Biochemistry,
University of Pennsylvania School of Medicine,
Philadelphia, Pa., U.S.A.
present address:
Department of Anatomy,
Faculty of Medicine,
University of Tokyo,
Bunkyo-ku Tokyo, Japan

M. Edward Kaighn,
Department of Virology,

The New York Blood Center,
New York, N.Y., U.S.A.

IRWIN R. KONIGSBERG,
Department of Biology,
University of Virginia School of Arts and Sciences,
Charlottesville, Va., U.S.A.

CHARLES P. LEBLOND,
Department of Anatomy,
McGill University,
Montreal, Quebec, Canada

ALEXANDER MAURO,
Rockefeller University,
New York, N.Y., U.S.A.

ADE T. MILHORAT,
Institute for Muscle Disease, Inc.,
New York, N.Y., U.S.A.

FRANCIS P. MOSS,
Department of Anatomy,
McGill University,
Montreal, Quebec, Canada

ALAN R. MUIR,
Department of Anatomy,
Royal (Dick) School of Veterinary Studies,
Edinburgh, Scotland, U.K.

MARK NAMEROFF,
Armed Forces Institute of Pathology,
Washington, D.C., U.S.A.

GEORGE PALADE,
Rockefeller University,
New York, N.Y., U.S.A.

MICHEL REZNIK,
Laboratoire d'Anatomie Pathologique,
Université de Liège,
Liège, Belgique

LEWIS P. ROWLAND,
Department of Neurology,
Hospital of the University of Pennsylvania,
Philadelphia, Pa., U.S.A.

ALEXANDER SANDOW,
Institute for Muscle Disease, Inc.,
New York, N.Y., U.S.A.

SAIYID A. SHAFIQ,
Institute for Muscle Disease, Inc.,
New York, N.Y., U.S.A.

JOHN C. SLOPER,
Department of Experimental Pathology,
Charing Cross Hospital Medical School,
London, England, U.K.

TRYGVE P. STEEN,
Department of Biology,
University of Virginia,
Charlottesville, Va., U.S.A.
present address:
Department of Biology,
Portland State University,
Portland, Ore., U.S.A.

JOHN H. VENABLE,
Department of Veterinary Anatomy,
College of Veterinary Medicine,
Oklahoma State University,
Stillwater, Okla., U.S.A.

ELEANOR ZAIMIS,
Department of Pharmacology,
Royal Free Hospital School of Medicine,
London, England, U.K.

JIŘINÁ ZELENÁ,
Institute for Muscle Disease, Inc.,
New York, N.Y., U.S.A.
present address:
Czechoslovak Academy of Sciences,
Institute of Physiology,
Prague, Czechoslovakia.

CONTENTS

INTRODUCTION

DR. MILHORAT: Ladies and gentlemen. On behalf of the trustees of the Institute for Muscle Disease and the directors of Muscular Dystrophy Associations, and my colleague Dr. Mauro of Rockefeller University, we want to thank you for coming today to discuss this most important problem of regeneration.

Dr. Mauro will say a few words about the agenda.

DR. MAURO: Thank you, Dr. Milhorat. Needless to say, Dr. Milhorat and I are very happy that you have agreed to come to the Conference on Regeneration of Striated Muscle, and Myogenesis. At this time I should like to extend my personal thanks to Dr. Milhorat and his colleagues, to Mr. Robert Ross, executive director of the Muscular Dystrophy Associations of America, and his many colleagues, especially Miss DeSola who took full responsibility for arranging travel and other incidentals. I must say you should thank her if your trip was facilitated. If you had any troubles, you can blame the troubles on me.

The strategy of our meeting is very simply this: We hope, because of your individual enthusiasm and interest for this subject, that we can carry on a rather unstructured program. Such a procedure is fraught with all sorts of negative possibilities which I think you can appreciate.

However, since we all are deeply interested in this fascinating phenomenon of cell biology, I think we can proceed without asking participants to make formal 1-hour presentations.

We thought that perhaps the way to start off the session would be to ask Dr. Elizabeth Hay and Dr. Bruce Carlson if they would discuss the two broad categories of regeneration in the skeletal muscles. Dr. Hay will review very briefly the phenomenon associated with the urodele amphibians, and Dr. Carlson will cover the regeneration phenomena in the so-called 'higher' vertebrates, *viz.* the rat and the frog.

They have urged me to urge you to break in during their presentation so that any points regarding terminology or simply the phenomenon that is being discussed may be clarified. I think this is also very important for all the speakers. During our discussions I think we should go out of our way to clarify points of nomenclature. I think nomenclature is one of the stumbling blocks in many fields; certainly it is in this field.

After their formal brief presentations, we can take a 5-minute break and try to develop an agenda. In other words, we will simply discuss among ourselves how we will proceed for the remainder of the morning, the afternoon session, and the two sessions tomorrow.

So with these brief remarks, I am very happy, again, to welcome you here to the Institute for Muscle Disease, and I hope we have a very productive and stimulating discussion. Dr. Hay, would you start off, please.

REGENERATION OF MUSCLE IN THE AMPUTATED AMPHIBIAN LIMB

ELIZABETH D. HAY

I think a little introduction is in order to the class of vertebrates. This is from Will Cuppy's book on the biology of the amphibians. I am sorry I was not able to get all of it on the slide: "As soon as writing was invented, everybody rushed into print with the statement that Salamanders live in the fire and are incombustible. The ancients used to notice Salamanders running out of bonfires built over their holes and drew the conclusion that they lived in the fire and were just going around the corner for a minute and would be right back. The Salamander could not make such a mistake about a fellow creature, since his brain is primitive and only equipped for ordinary common sense. To be as wrong as that, you have to have a cerebral cortex and other refinements peculiar to the higher vertebrates."

There are many myths about salamanders, but the fact that they regenerate their limbs is not a myth – it is a very dramatic property. The larval form of the salamander *Amblystoma punctatum* is a small fellow, about 2 centimeters long, a beautiful and convenient animal to work with, and much studied in the past. Not only is the tail fairly transparent, as in the case of its relative, the tadpole, but the limb is also relatively transparent at early stages. Figure 1 illustrates a study of regeneration of the finger of this animal in which we actually watched the appendage regenerate from day to day over a period of 2 to 3 weeks. This little animal can be strapped into a Speidel-type chamber, the arm placed up against the cover slip, and the cells actually observed with oil immersion. This is an invaluable way to study the origin of regenerating cells, since one does not have to reconstruct a series of static pictures from sections, which is the problem that plagues the morphologist in studying most regenerating systems.

The original work that led to the concept of dedifferentiation in amphibian regeneration was done on this same larva by Butler (1933, 1935). In sections of fixed material he noted that when the limb is amputated the skeleton (which is cartilaginous) disappears, releasing cells that appear to be undifferentiated; these cells then make up a small bud or blastema that grows out with contributions from other stump tissues and produces a new limb. He used the word 'dedifferentiation' to refer to the process by which differentiated cartilage cells lost the products that previously characterized them. In the living finger, it is seen that the blastema originates by a backward process of dedifferentiation rather than by emerging suddenly at the tip of the limb from a mythical reserve of undifferentiated cells. One can actually see the cartilage matrix gradually disappearing, and actually follow the release of each of the cells into the blastema, as *e.g.* cells b and c (2 and 3 days, Fig. 1) or cells v, y and z (2, 7 and 8 days, Fig. 1). In the finger illustrated, the regression of cartilage proceeded back to the level of cells x and x (10 days, Fig. 1) which is about the level of vessel B (2 days, Fig. 1). Then these blastema cells grow out and redifferentiate, in this case into cartilage only, for there is no muscle in the finger.

In 1938 one of Butler's students, Charles Thornton, published an elegant description of muscle dedifferentiation in this same larval system. Figure 2 depicts several of his original camera lucida drawings of sections of fixed tissue. It is unfortunate that the portion of the

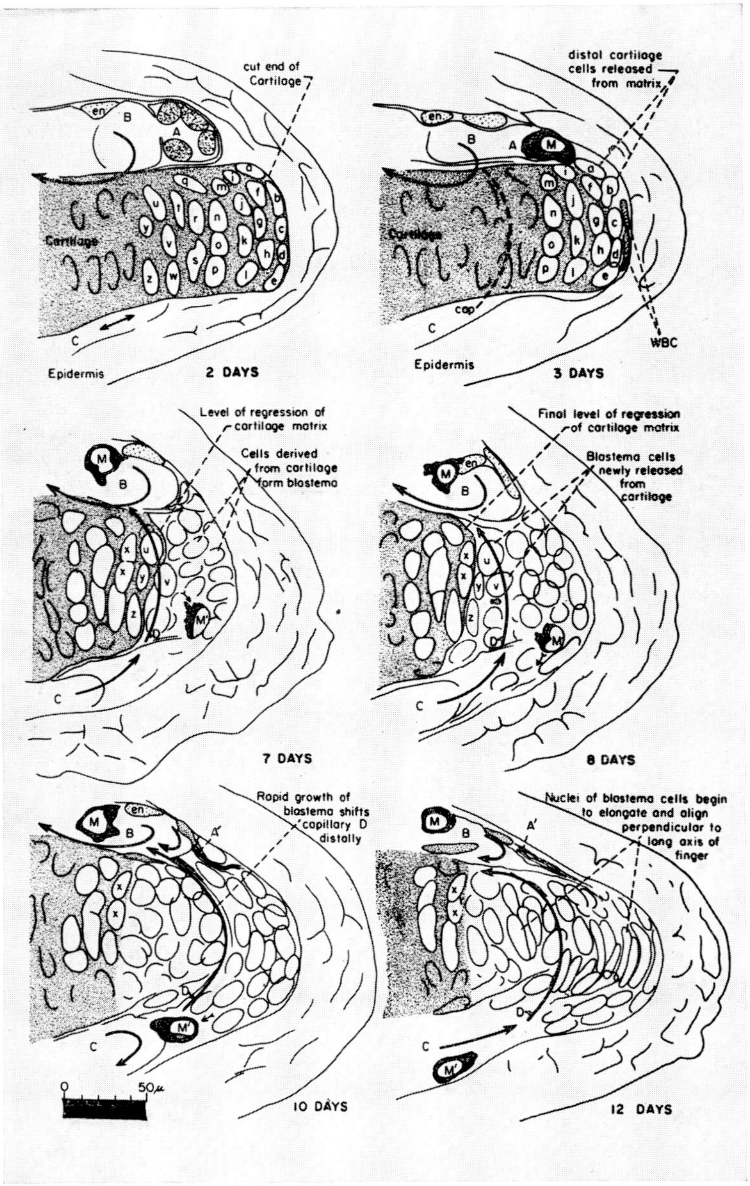

Fig. 1. Cartilage dedifferentiation, as observed in a regenerating finger of a living *Amblystoma punctatum* larva. At 2 days after amputation, the cut end of the finger is covered with epidermis, but no changes are detectable in the cartilage. Blood is circulating through the vessel B, but not through vessel A. An endothelial nucleus is labeled en. Vessel A will disappear during the period of dedifferentiation (3-8 days). The nuclei of most of the cartilage cells that were observed in this experiment are labeled in the 2 days illustration (a-z). At 3 days it can be seen that the distal cartilage cell nuclei (a-e) have rounded up and that the cells are becoming free of matrix (b, 3 days). At 7 days, most of the distal cartilage has dedifferentiated, forming a blastema at the tip of the limb. The cartilage cell labeled u in the drawing on the 7th day divided that afternoon, and on the 8th day, it was obser-

limb containing muscle is too thick to permit observation of living cells. Nevertheless, a convincing series of illustrations can be assembled which supports the idea that skeletal muscle gives rise to blastema cells. In areas in which muscle fibers are breaking up, there are a number of cells with scanty cytoplasm which would be basophilic in a properly stained section. These cells are closely associated with, and are believed to derive from, the muscle fibers (Fig. 2, B and C) and they are similar to the relatively undifferentiated mesenchymal cell type found in the blastema (Thornton, 1938, 1942). I should like to consider separately 2 different aspects of the process of muscle dedifferentiation: (1) Loss of myofibrils and other organelles occurs in all the muscle fibers near the injury, whether or not they give rise to mononucleated cells. (2) By a process as yet not fully understood, undifferentiated mononucleated cells derive from the muscle.

J. R. Coleman: How much cellular degeneration occurs in this blastema?

E. D. Hay: In these young larvae, there is but slight degeneration and it occurs before formation of blastema (Fig. 2, A). During the period of muscle dedifferentiation, there is no nuclear degeneration (Fig. 2, B). Some anucleate fragments of muscle cytoplasm may cleave off and they seem to degenerate and disappear, but there is not any massive degeneration, as is depicted in published studies of mammalian muscle regeneration. Incidentally, the nuclei of salamanders are enormous compared with the nuclei of mammalian or frog cells, and one often gets a false impression of the nucleocytoplasmic ratio in these fibers (Fig. 2). It is not that the fibers of the salamander muscle are that much smaller than the mammalian; it is that the nuclei are so much bigger.

We confirmed Thornton's light-microscopic observations of cytoplasmic dedifferentiation using the electron microscope (Hay, 1959). The process of dedifferentiation is characterized by the acquisition of 'embryonic' features and therefore, by our definition, is not degeneration. In electron micrographs (Fig. 3) it can be seen that areas of cytoplasm now devoid of myofibrils are filled with ribosomes and even have acquired granular endoplasmic reticulum (Hay, 1968; Lentz, 1969).

J. C. Sloper: Dr. Hay may I ask, as a higher mammalian physiologist or pathologist, what about the inflammatory reaction of the salamander?

E. D. Hay: There is hardly any in the salamander larva. One gets giant cells associated with the injured cartilage. I guess you will have to call them chondroclasts (they look like osteoclasts) and in the adult limb, which I shall show you in a minute, there are osteoclasts, for there is much bone there to be removed. In the first 4 or 5 days after amputation in the case of the adult salamander, a number of polymorphonuclear leucocytes also appear transiently, and mononuclear leucocytes give rise to the osteoclasts (Fischman and Hay, 1962). Moreover, the wound epithelium plays a role in removing debris (Singer and Salpeter, 1961; Thornton, 1968).

J. R. Coleman: *In vivo* then, you can actually see that chondrocytes which have lost their cartilaginous matrix eventually again become chondrocytes?

---

ved to become part of the blastema. The cells labeled v, y and z (2 days, 7 days) were also released from the regressing matrix on the 8th day. Due to continued proliferation of the blastema cells derived from the cartilage, the regenerate doubles in size by the 10th day. By the 12th day, it has begun to redifferentiate into cartilage and new vessels have appeared (A′, C, D). The extent of the regression of old cartilage is to the level of the pair of cells labeled x. Proximal cells not completely released from the matrix can be said to be partially dedifferentiated, whereas distal blastema cells have lost all the characteristics which identify them in their tissue of origin. Two macrophages (M, M′) wandered through the blastema during this period. The scale of the camera lucida drawings is indicated in the lower left hand corner. (*from*: Hay, 1962; copyright by The Ronald Press Co., New York, N.Y.)

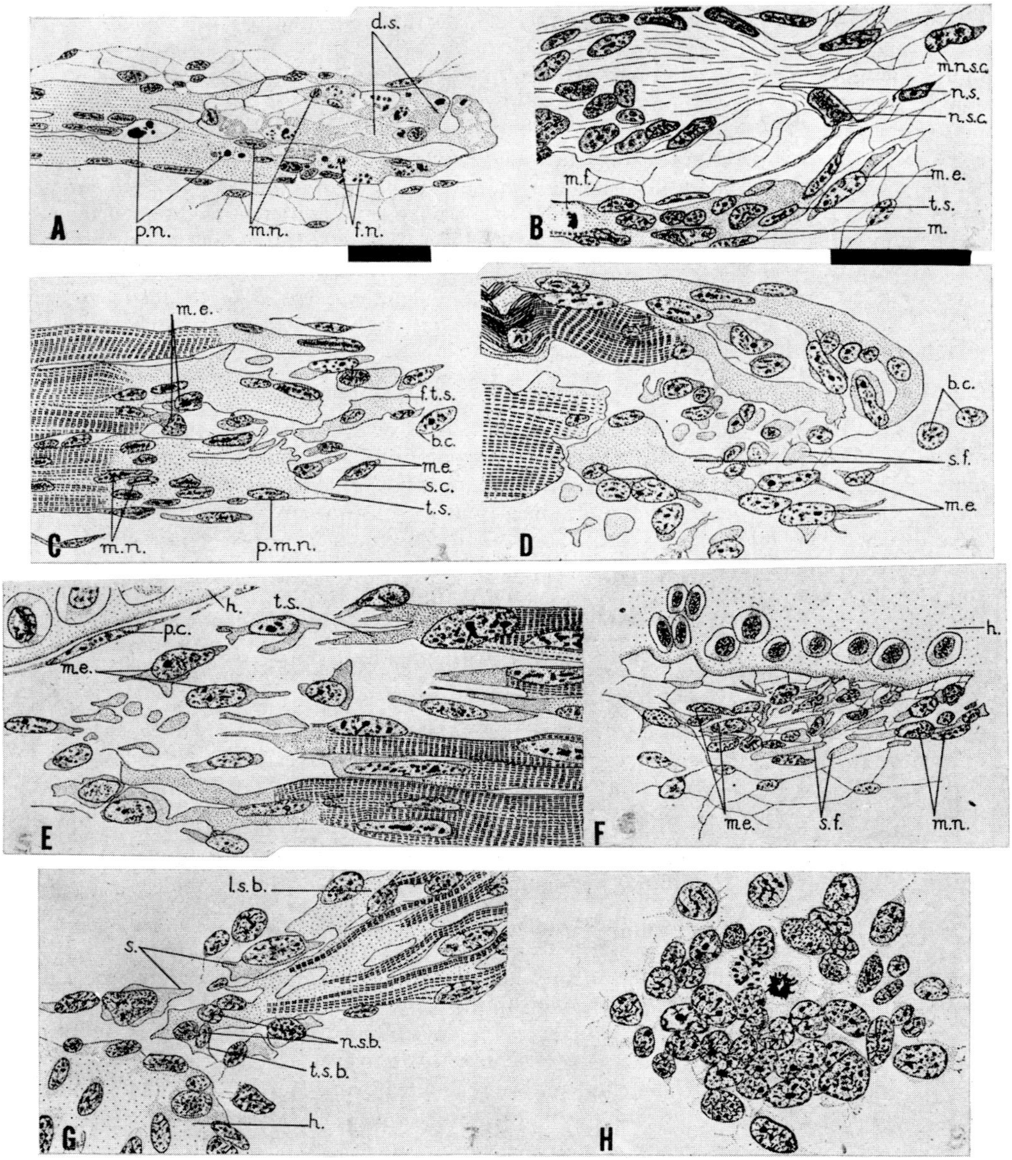

Fig. 2. Muscle dedifferentiation and redifferentiation during regeneration of the limb of *Amblystoma punctatum* larvae as originally illustrated by Thornton (1938). Amputation was through the proximal third of the humerus. A. One day after amputation, some degeneration of sarcoplasm can be detected at the tips of the cut muscle (d.s.). Several pycnotic (p.n.) and fragmented (f.m.) nuclei can be seen in the cut ends of the muscle fibers. B. Two days after amputation, dedifferentiation of the muscle into muscle elements (m.e.) has begun. Migrating nerve sheath cells (m.n.s.c.) are separating from the nerve sheath (n.s.). A mitotic figure (m.f.) can be seen in the dedifferentiating muscle (m.). The terminal sarcoplasm (t.s.) of the muscle fiber is becoming basophilic. C. Another illustration of the dedifferentiating end of a muscle 2 days after amputation, showing the steps in the separation of muscle elements (m.e.) from the terminal sarcoplasm (t.s.). They (m.e.) seem to give rise to blastema cells (labeled b.c., Fig. 2, C and D). D. Dedifferentiating muscle in a regenerating limb of a larva 3 days after amputation, showing fragmentation of the terminal sarcoplasm (s.f.) and the liberation of

E. D. HAY: Yes (Fig. 1, 12 days).

J. R. COLEMAN: The same with the muscle?

E. D. HAY: The muscle is too thick to study *in vivo* because I cannot see through that part of the limb. It is easy with the muscle to reconstruct the nuclear events in the sections, however. With the cartilage, it is more gratifying because you can see the live nuclei and watch them divide as the matrix regresses. In the case of the muscle, the nuclei round up (Fig. 2) like the dedifferentiating cartilage cell nuclei (Fig. 1) and become euchromatic. That these nuclei are metabolically active and not degenerating is evident not only from their morphology (Fig. 12) but also from the fact that they are all synthesizing RNA (unpublished autoradiographic observation). Some, but not all, of the nuclei associated with the dedifferentiating muscle fibers are making DNA (Hay and Fischman, 1961).

H. HOLTZER: You have made several statements of which I am not convinced. I see no critical evidence to back your assertions (1) that a nucleus, which at one moment was in a multinucleated muscle fiber, forms a progeny of mononucleated cells, and (2) that such mononucleated cells may then differentiate into cartilage, fibroblasts, or muscle cells.

E. D. HAY: Let me show you another view of the muscle breaking up in a larval limb (Fig. 4, A). This again is light microscopy. Here at the arrows are what we feel are individual cells with striations still present in the cytoplasm. The area labeled 'Mus' consists of what seem to be larger fragments, and if one looks at such an area in the electron microscope (Fig. 5), one finds that these fragments are limited by individual cell membranes. I call your attention to the fiber containing nucleus N and part of nucleus N′. Nucleus N is in a cell which seems to be splitting off the fiber (Fig. 5). If one were to look at the area in the rectangle under higher magnification, one would see a whole series of small vesicles (Fig. 6), which probably are fusing to form new cell membranes (Hay, 1959). One sees these images so commonly, that it is difficult to doubt that the muscle gives rise to numerous individual cells as Thornton (1938) illustrated so clearly (Fig. 2). The point as to whether or not the muscle fiber nuclei actually persist and are present in the mononucleated cells has to be answered indirectly, but there is simply no other source for the cells depicted (Fig. 2, D). Moreover, no muscle nuclei are dying (as I already indicated, they make RNA, DNA, or both during this period).

Let me show you how dramatic the replacement of muscle by mononucleated cells is in the limb of the adult newt. Figure 7 (A) shows a newt limb a few days after amputation to illustrate the enormous mass of muscle with which we begin. Approximately 2 weeks later, this area contains largely mononucleated cells (Fig. 7, B). The regeneration blastema of the limb (Fig. 7, C) arises 'backwards' as in the larval finger (Fig. 1), but this point can be

---

mononucleate cells from the muscle (m.e.). E. Dedifferentiating end of a shoulder muscle fiber in the regenerating limb of a larva 5 days after amputation, showing separation of mononucleate cells (m.e.) from the terminal sarcoplasm (t.s.). The perichondrium of the humerus (h.) has dedifferentiated into individual cell units (p.c.). F. Part of a dedifferentiating muscle area in the regenerating limb of a larva 6 days after amputation, showing the final fragmentation of the fiber into mononucleate cells (m.e.) with large nuclei (m.n.). The dedifferentiation of the muscle has progressed proximal to its origin on the humerus (h.). G. Following a period of growth, the regenerating cells redifferentiate. This drawing of a regenerating shoulder muscle 12 days after amputation shows formation of terminal (t.s.b.) and lateral (l.s.b.) sarcoplasmic buds which will attach the redifferentiating muscle fibers to the humerus (h.). H. Drawing showing redifferentiating muscle arising *de novo* in the blastema of a regenerating limb 12 days after amputation. Undifferentiated blastema cells are producing new muscle not in continuity with old muscle. Camera lucida drawings. Parts A, C, and F are magnified to the scale shown in A. The other figures are magnified to the scale shown in B. The bar equals 50 microns. (*from*: Thornton, 1938; copyright by The Wistar Institute Press, Philadelphia, Pa.)

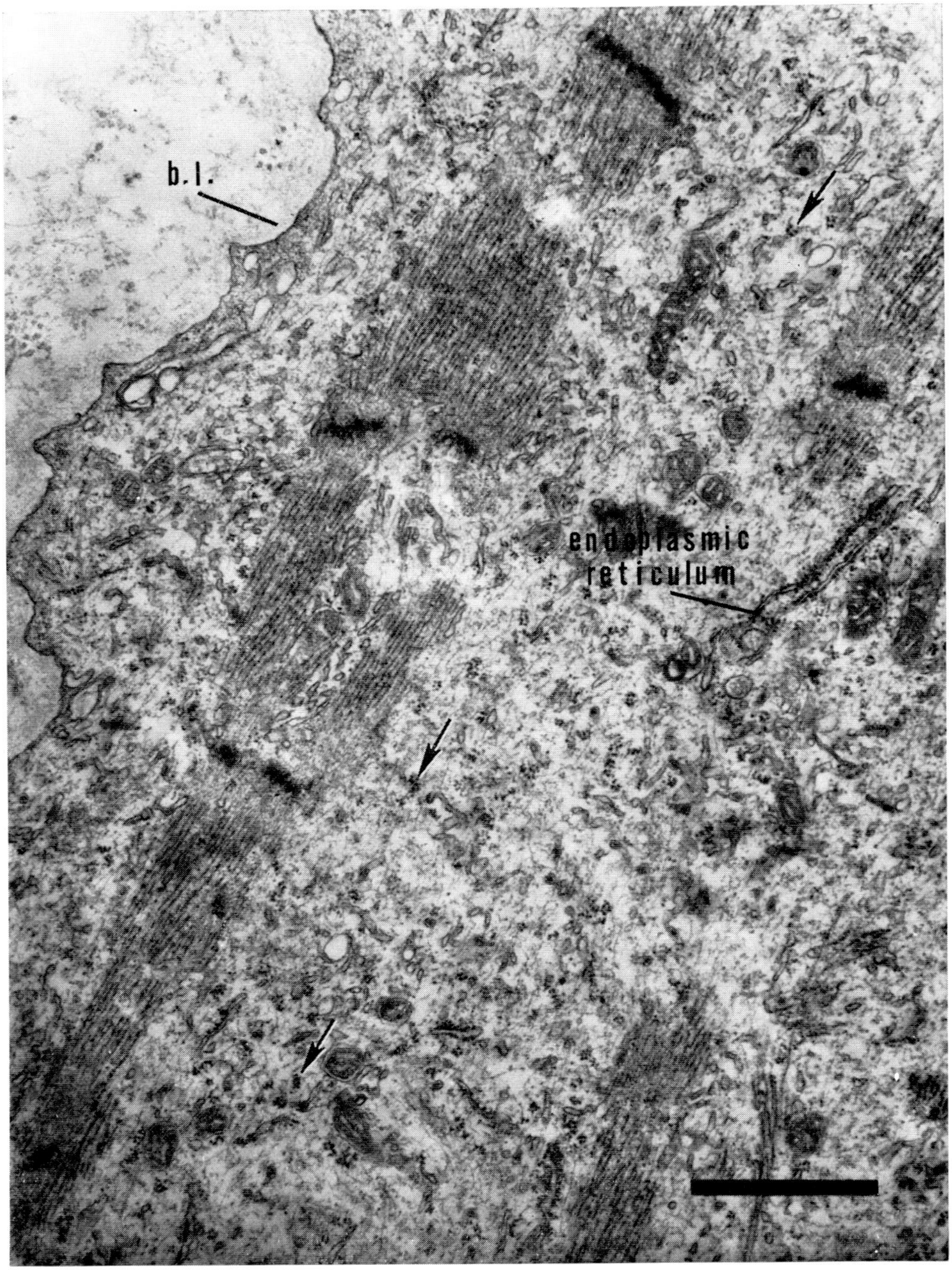

Fig. 3. Electron micrograph of a portion of a cytoplasm of a dedifferentiating muscle fiber in a region similar to those depicted in Fig. 2 (C-E). The cytoplasm is becoming basophilic because of the acquisition of free ribosomes (arrows) and granular endoplasmic reticulum.The reticulum may derive from the numerous smooth-surfaced profiles that characterize the intact muscle. The myofibrils break apart in the A and I band regions. The Z band agglutinates and disappears. Although the myofibrils degenerate, the cell is not degenerating, as can be seen from the fact that the cytoplasm is acquiring new organelles and the nuclei are synthesizing nucleic acids. Part of the basement lamina (b.l.) is seen on the left. The bar equals 1 micron. From an adult newt limb, fixed 17 days after amputation. (*from*: Hay, 1968; copyright by Springer Verlag, Berlin)

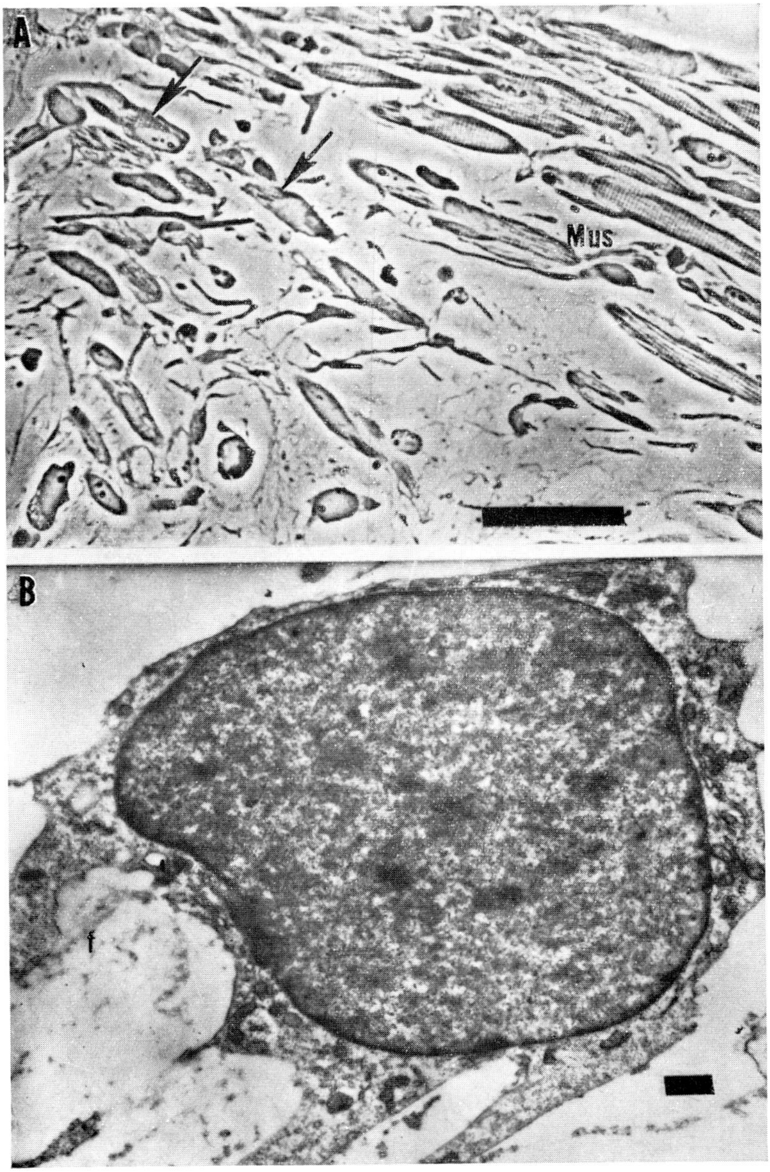

Fig. 4. A. Phase contrast photomicrograph of an unstained plastic section, showing a region of dedifferentiating muscle in a regenerating larval *Amblystoma* limb, fixed 4 days after amputation. The cut end of the limb is to the left and a large bundle of muscle appears in the upper right. Below and to the left of these fibers are two cells (arrows) which have separated from the main mass of muscle (Mus) and yet still contain striations in the cytoplasm. The bar equals 50 microns. B. Low magnification electron micrograph of an early blastema cell from a regenerating larval *Amblystoma* limb, fixed 6 days after amputation. The tip of the limb is toward the upper right. This cell is similar to those labeled m.e. and b.c. in Fig. 2. The nucleus is diffusely granular and the basophilic cytoplasm is rather sparce. The cytoplasm contains endoplasmic reticulum mitochondria and numerous free ribonucleoprotein granules. A few collagen fibrils (f) can be identified in the extracellular ground substance. The bar equals 1 micron. (*from*: Hay, 1959; copyright by Academic Press, New York, N.Y.)

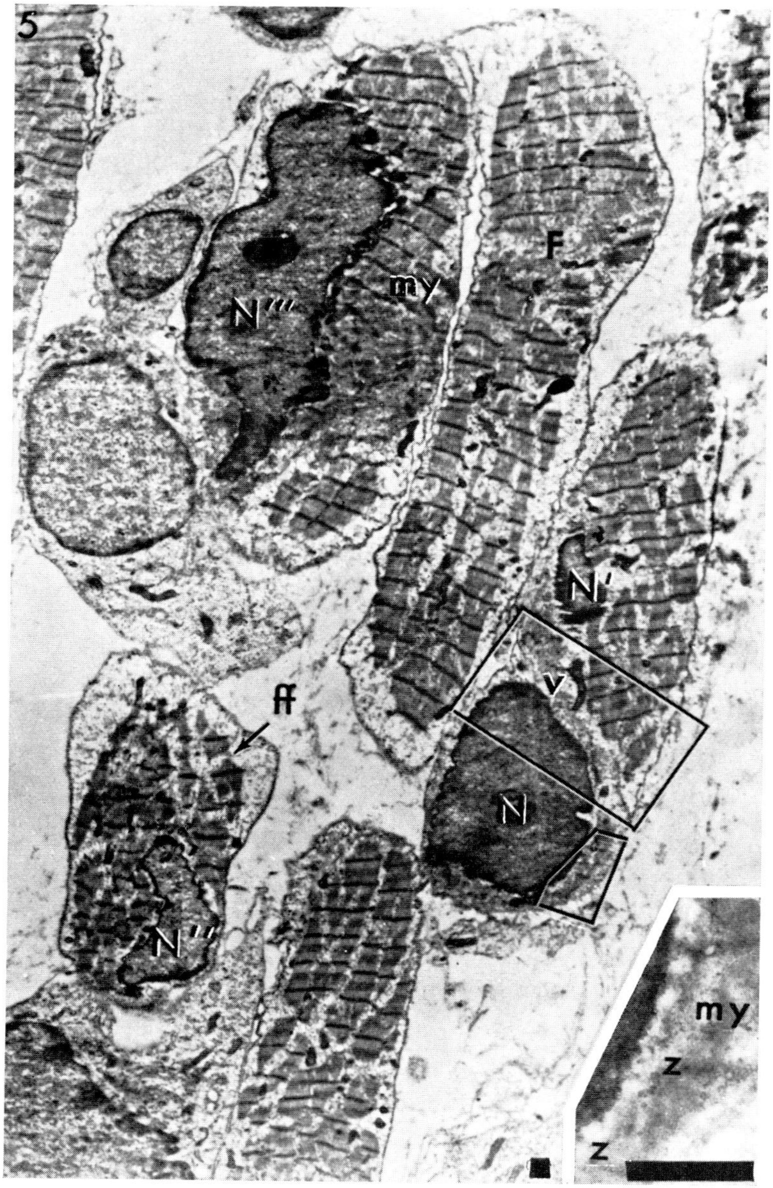

Fig. 5. Another electron micrograph from the earlier work of Hay (1959), showing apparent fragmentation of the muscle fibers into smaller units in a regenerating larval *Amblystoma* limb, fixed in this case 6 days after amputation. The micrograph is oriented so that the distal tip of the limb would be below the picture. A row of vesicles (v) extends across the muscle fiber on the right, demarcating it into two distinct fragments whose nuclei are labeled N and N′. An area similar to that shown in the rectangle is depicted in a more recent electron micrograph in Fig. 6. It was suggested that the small membrane-bound vesicles extending across the fiber are the formative material for the new cell membranes that ultimately demarcate individual muscle units (Fig. 8). The muscle unit whose nucleus is labeled N‴ contains a nucleolus (n). The myofibrils (my) are in the process

obscured because the term blastema often is used only for the cone to paddle-shaped outgrowth visible at 3 weeks (Fig. 7, D).

Both direct regeneration of muscle from muscle, and differentiation of muscle *de novo* occur (Thornton, 1938). In the proximal part of the adult limb (Fig. 7), many muscle fibers do not dedifferentiate completely and muscle-like ribosomes appear (Deck, 1968; Norman and Schmidt, 1967; Hay, 1968; Lentz, 1969). I am using the term 'dedifferentiate' in a relative manner, and I do not by any means wish to imply that I believe all the muscle cells of the limb stump return to the simplified mononucleated cell type we have depicted here. I am quite certain that these partly dedifferentiated muscle fibers in the proximal area will redifferentiate into new muscle without a blastema phase.

However, in the distal area all these mononucleated cells which arise from the muscle that was there previously are individual entities. They proliferate to form an outgrowth which soon begins to redifferentiate into cartilage (Fig. 7, D). Then, new muscles form from aggregates of undifferentiated blastema cells next to the cartilage (Fig. 7, E).

J. C. SLOPER: Does the ordinary muscle fiber in the salamander have a basement membrane around it? If so, you are suggesting that each nucleus with a small bit of peripheral cytoplasm has to become free and the basement membrane must break up to allow its release.

E. D. HAY: Yes, actually that is less of a problem to me than where all that cartilage matrix goes (Fig. 1). The basement membrane appears ravelled (Fig. 8) and then disappears as in hair follicle resorption (Parakkal, 1969).

Perhaps collagenase released by the epithelium, osteoclasts or other cells plays a role (Grillo *et al.*, 1968). Although acid phosphatase is present in the limb stump (Schmidt, 1966), lysosomes are not numerous in the dedifferentiating muscle and free macrophages are not very prominent. Recent work (Miller and Wolfe, 1968) suggests that by far the majority of the acid phosphatase is located in the endothelium or vascular tree.

B. M. CARLSON: You actually said there is some leucocytic response, did you not?

E. D. HAY: In the adult during the first 4 to 5 days after amputation.

A. R. MUIR: Dr. Hay, a moment ago we saw the dissolution of the cartilage matrix; if we relate this to the basement membrane, it seems that the basement membrane is disintegrating during this process.

E. D. HAY: Actually, it is present in the illustrations I showed here. To take micrographs of muscle cells that are in the process of dedifferentiation, one has to go to an area where the tissue is only partly dedifferentiated. In an area where the tissues are completely dedifferentiated, one cannot identify muscle cells or skeletal cells – you cannot tell where the cells came from. So, my pictures of dedifferentiating muscle are selected from the more proximal areas, simply to show a few myofibrils in the cytoplasm. Moving to a transition area between proximal limb stump and blastema, one can find cells derived presumably from muscle which are surrounded only in part by basement membrane (arrows, Fig. 9). More distally, there is no basement lamina around the cells (Fig. 4, B).

To return now to our discussion of the origin of the regeneration blastema, let me review again the series of events that occurs right after amputation. Figure 7 (A) depicts the limb

---

of breaking up. The fibril fragmentation is shown to good advantage in an adjacent cell (ff). This cell, whose nucleus is labelled N″, may be a mononucleated unit. Note that the mononucleated unit whose nucleus is labeled N contains fragmenting myofibrils in its cytoplasm (inset). Thus, this cell must derive from the muscle fiber rather than from a pre-existing 'satellite cell'. The bar equals 1 micron. (*from*: Hay, 1959; copyright by Academic Press, New York, N.Y.)

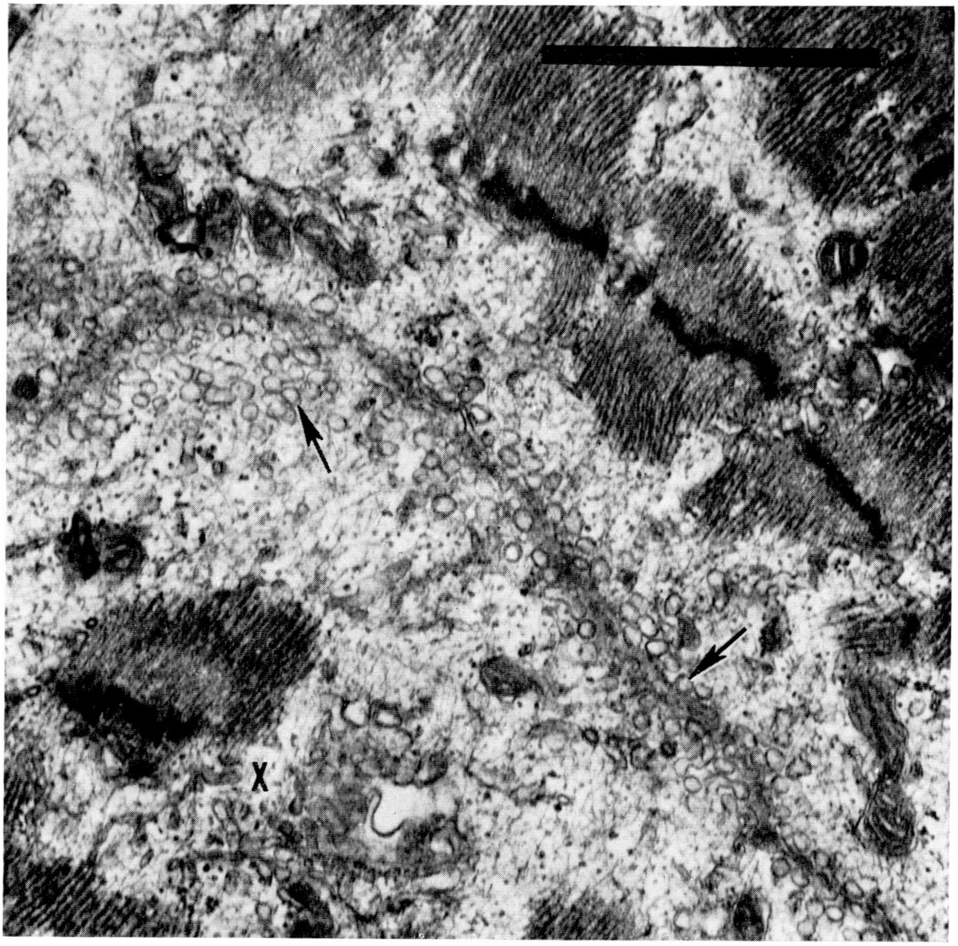

Fig. 6. Higher magnification electron micrograph, showing a region similar to that enclosed in the rectangle in Fig. 5. The areas of low density which can be seen in the low magnification electron micrograph extending across the muscle fibers correspond to zones in which small vesicles (arrows) are concentrated. These vesicles are 300-700 Ångströms in diameter and are like those normally associated with the plasmalemma of muscle cells. Some of them are located in the cytoplasm (arrow on the left), while others (arrow on the right) are associated with a linear density that corresponds either to an obliquely sectioned or to an incompletely formed cell membrane. It was suggested that the vesicles might derive from smooth-surfaced endoplasmic reticulum (X) or from the Golgi complex (Hay, 1959). Interestingly, the image of vesicles fusing with and/or forming membranes across the dedifferentiating muscle fibers is reminiscent of the image of rows of vesicles across coalescing muscle units in the process of differentiation (Hay, 1963). See Fig. 8 also. The bar equals 1 micron.

1 day after amputation. The ends of the radius and ulna go out of the plane of section, but you can see that the epithelium is migrating slowly over the ends of these cut bones, piling up as it does. This is a very low magnification, but in the area distal to the arrow, red blood cells and also a number of polymorphs are present at this time, which will subsequently

disappear. Most of the muscle will break up. Indeed it has already started to break up in the distal area.

By $2\frac{1}{2}$ weeks after amputation (Fig. 7, B), all of that muscle has broken up, the fibers have dissociated, and in their place are numerous regeneration cells, or what we call blastema cells. The bone seems more difficult to get rid of than the cartilage. Some of the bone matrix persists here. Our studies with tritiated thymidine indicate that some of the osteocytes at least are labeled and thus they may contribute to the blastema (Hay and Fischman, 1961). Osteoclasts are present, presumably cleaning up this bony debris. This whole region (distal to the arrow, Fig. 7, C), then, represents a blastema that has arisen backwards at the expense of the old tissues.

What I should like to show you in the next figure is an autoradiograph, at the light microscope level, of an area of muscle like that labeled 'X' in the limb shown in Figure 7 (B). This particular fiber (Fig. 10) was selected after some searching in order to depict 4 labeled nuclei associated with one muscle fiber. It is more common to find that nuclei which incorporate tritiated thymidine are located in a fragment of this sort (arrow, Fig. 10) rather than in a long fiber. The label represents tritiated thymidine which was injected into the newt during the phase of dedifferentiation illustrated in Figure 7 (A). The limb was fixed about 1 hour later, so Figure 10 shows sites of DNA synthesis (Hay and Fischman, 1961).

The issue of satellite cells has arisen since this work was done and therefore recently I have begun to do electron-microscopic autoradiography to determine whether nuclei such as these (Fig. 10) are actually in the muscle fiber, or whether they may be demarcated from it by a cell membrane. In all cases observed so far, the labeled nucleus is separated from the myofibrillar section of the associated fiber either by a row of vesicles or by definitive cell membranes.

Figure 11 shows a nucleus in an autoradiograph in which the silver grains are resolved in the electron microscope. Myofilaments are present in the cytoplasm of the adjacent muscle fragment, which is in the process of dedifferentiating. But, here you can see that the labeled nucleus is separated from that fragment of cytoplasm by 2 definite cell membranes resolved in places as a row of smooth surfaced vesicles which you would expect to see associated with a muscle cell membrane. The muscle cell which is synthesizing DNA has few, if any, myofilaments and scanty cytoplasm. It can only be identified as muscle because of its location within the muscle basement membrane.

I have yet to find any tritiated thymidine labeling over nuclei in the middle of a dedifferentiating fiber (Fig. 12) and what I suspect is that nuclei of this sort are making RNA. This cell is not dying; if you look at the cytoplasm of such cells (Figs. 3 and 12), you find that in addition to the loss of the myofibrils the cytoplasm is acquiring ribosomes and granular endoplasmic reticulum.

It still seems reasonable to conclude, as in our original description (Hay, 1959), that mononucleated cells are cleaving off the fibers (as at N, Fig. 5). The new autoradiographic data make it seem likely, however, that nuclei within the centers of the dedifferentiating fibers (N‴, Fig. 5) are not synthesizing DNA. To make DNA, seemingly, the syncytial state must be dispensed with and cell membranes manufactured which cleave the nucleus and a small amount of cytoplasm off the fiber. I realize this point will come up again in our discussion of satellite cells, and this general discussion is probably not the time to go into it. I can just mention now, however, that we have scanned non-regenerating newt limb muscle without seeing satellite cells (Hay and Doyle, 1970).

H. HOLTZER: The issue here, it seems to me, is whether one portion of this unit, be it mononucleated or multinucleated, can be undergoing regressive changes in one half and regenerative processes in the other half. In short, within the same unit you can have areas of repair as well as areas of destruction. How do you answer this awkward possibility?

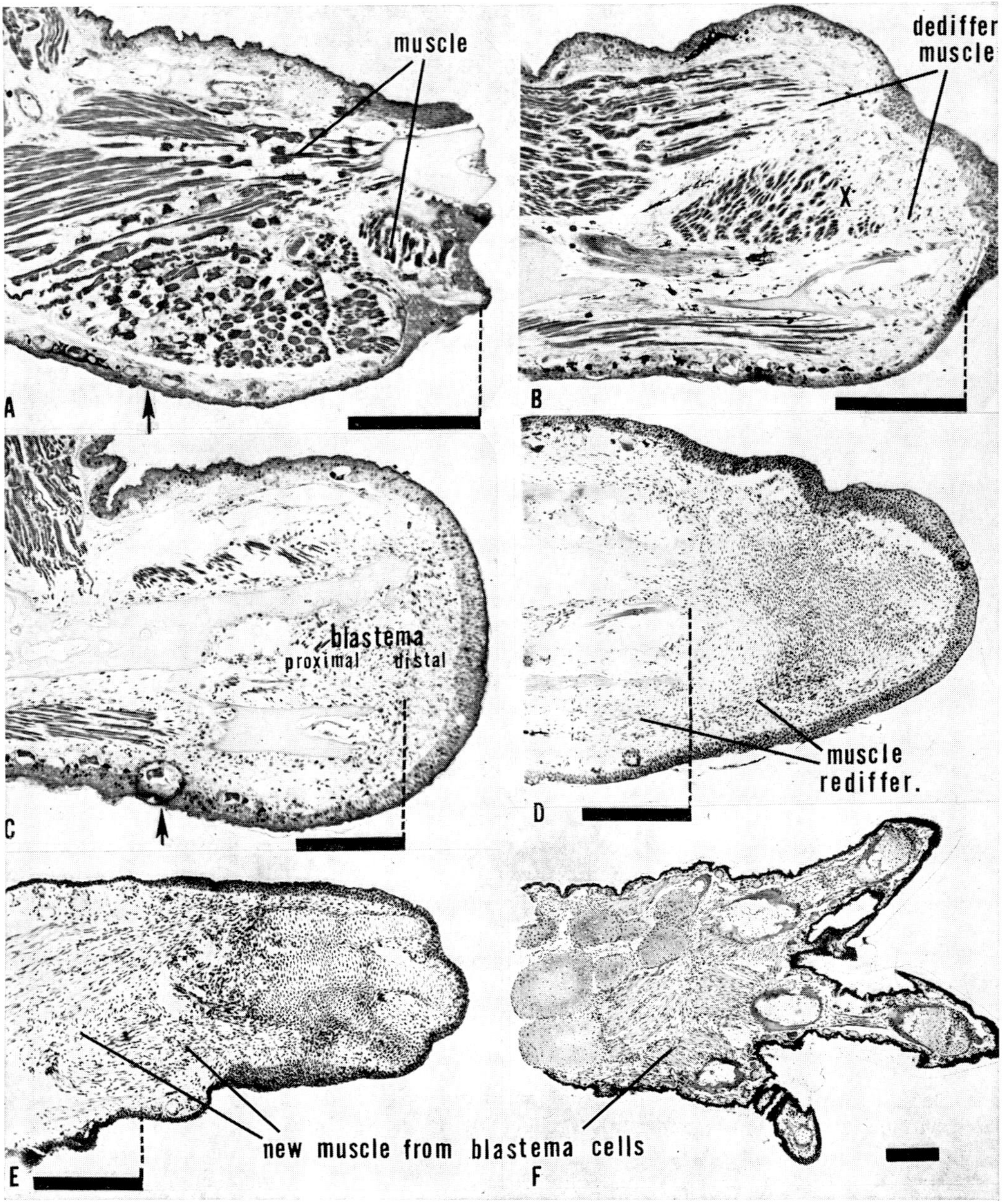

Fig. 7. A series of low magnification light micrographs depicting muscle dedifferentiation, blastema formation and muscle redifferentiation from the limb of the newt (*Triturus viridescens*) following amputation through the radius and ulna. The dotted line indicates the approximate level of amputation. The black bar is equivalent to 0.5 millimeter and it calls attention to the same region of the limb stump in parts A-E. The limbs depicted were fixed at 1 day (A), 18 days (B), 21 days (C), 28 days (D, E) and 8 months (F) after amputation. The plane of the section in A is chosen so as to show to good advantage the large amount of muscle which is present in the forearm at the time of amputation. The process of muscle dedifferentiation will extend proximally to the level indicated by the arrow. By 18 days, the distal 0.5 millimeter of the stump contains blastema cells derived from dedifferentiating muscle (dediffer. muscle, B). (The area labeled X is depicted in Figs. 8-12 in electron

E. D. HAY: I do not find any areas of degeneration; there are no degenerating nuclei. This cytoplasm (Figs. 3 and 12) is not degenerating. The myofibrils are disappearing, but the cytoplasm is becoming intensely basophilic and acquiring granular endoplasmic reticulum as well as free ribosomes.

H. HOLTZER: I thought initially you pointed out that you were going to get dissolution of the myofibrillar material. Surely that is degeneration – is it not?

E. D. HAY: I would limit the term degeneration for purposes of present discussion, to cell death. The term can be used in a variety of ways, so I shall use it any way the group decides, since we have already hit a snag in the nomenclature. In the past, however, when I have used the term 'degeneration' I meant that cells are dying.

J. H. VENABLE: Loss of myofibrils and building up of myofibrils is going on continuously in normal skeletal muscle cells – so I do not see that the word 'degeneration' can apply here solely on the basis of loss of myofibrils.

E. D. HAY: You would not use the term 'degeneration', then, to refer to loss of myofibrils from the cytoplasm?

J. H. VENABLE: No, I would not. I should say that all the phenomena you can see there can be found in normal living skeletal muscle cells.

E. D. HAY: I should not like to use the term degeneration either. 'Resorption', perhaps, might be a better term.

M. REZNIK: Dr. Hay, how can you say that in one muscle cell you know really that there is degeneration, or resorption, of the pre-existing myofilaments and, at the same time, formation of new ribosomes and new myofilaments.

E. D. HAY: I can say that there are new ribosomes present because this muscle contained insignificant ribonucleoprotein granules before the injury to the limbs. The fiber depicted (Fig. 12) was filled with myofibrils before amputation.

J. H. VENABLE: Not that cell.

E. D. HAY: The point that Dr. Venable is making is that I cannot say that this cell is not redifferentiating, is that right?

J. H. VENABLE: No, you made the statement that that cell was filled with myofibrils – you cannot make such a statement on the basis of a thin section because that cell is only there now. You do not know anything about its previous history, nor about what is going to happen next.

---

micrographs.) By 21 days, the level of dedifferentiation extends about 1 millimeter proximal to the arrow (C). The blastema clearly derives from various formed tissues of the old stump (proximal blastema, C). The regeneration cells proliferate and give rise to an outgrowth which can be termed the distal blastema (C). Proximal and distal portions of the blastema contain cells that are indistinguishable in type (Fig. 4, B). In the most proximal part of the blastema, dedifferentiation of muscle is incomplete and these fibers regenerate by reorganization of their internal structure ('budding', Fig. 2, G). In the blastema proper, muscle redifferentiates (muscle rediff., D) from blastema cells in the proximal (black bar) and distal parts of the blastema. Note in the micrograph that the ends of the cut bone still mark the amputation site (dotted line, D). They are covered with differentiating cartilage distally, and the precartilage pattern of the fingers can be detected at this stage of regeneration. The sections at E and F show the large amount of new muscle that forms from blastema cells during the later periods of limb redifferentiation.

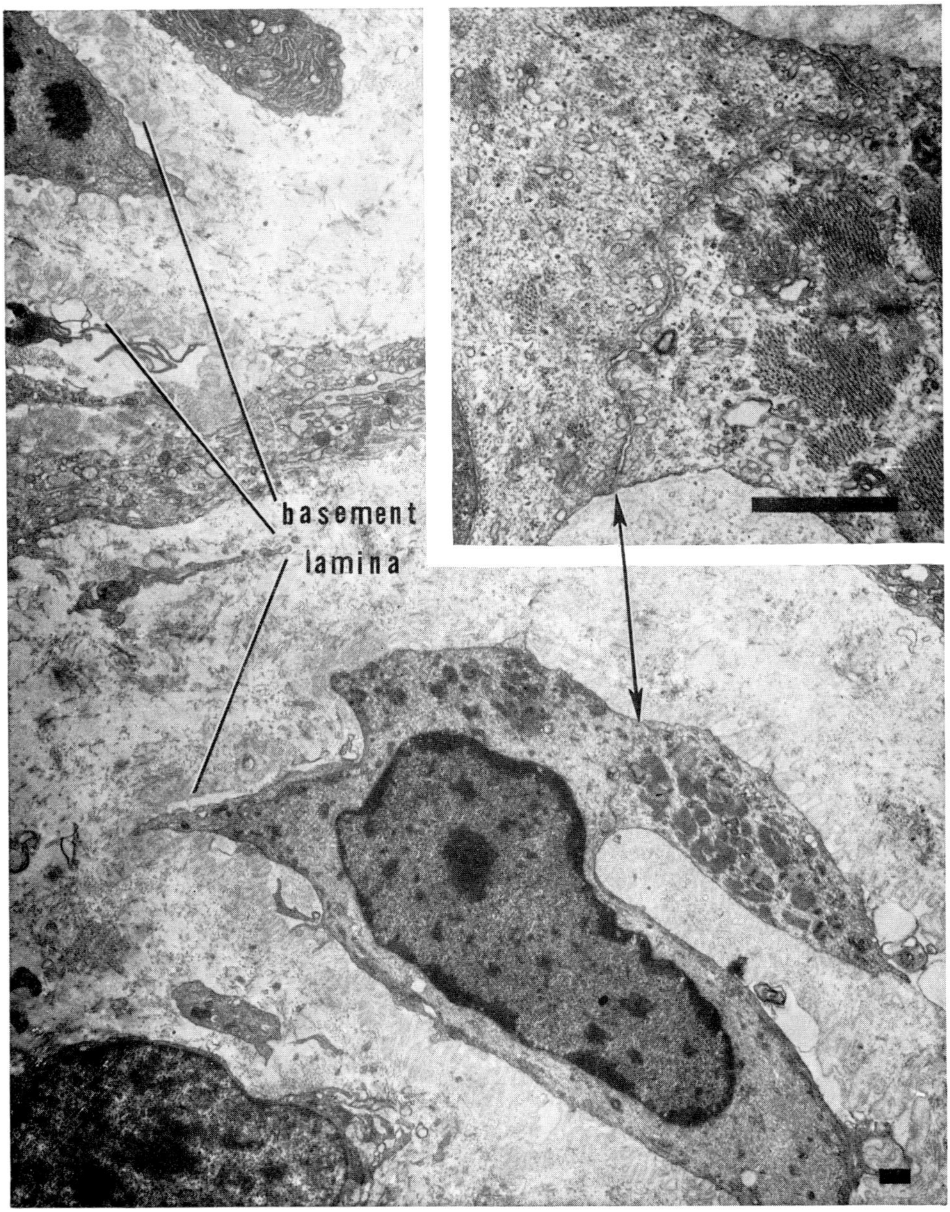

Fig. 8. Electron micrograph of an area of dedifferentiating muscle from a regenerating limb of adult *Triturus viridescens* fixed 17 days after amputation. The basement lamina ('membrane') surrounding intact muscle fibers is a smooth sheath of extracellular material about 1000 Å in width. In areas where the muscle fibers are dissociating, the basement lamina appeared highly irregular in outline, as shown at the top left, middle and lower part of this figure. A pleated basal lamina is also seen during dissolution of hair follicles (Parakkal, 1969). These basement laminas in most areas of the proximal blastema (Fig. 7, C) eventually are resorbed, perhaps due to the action of the collagenase enzymes known to be present in these limbs. The inset shows, at higher magnification, the newly formed cell membranes which demarcate the muscle unit illustrated here into separate fragments. The bar equals 1 micron.

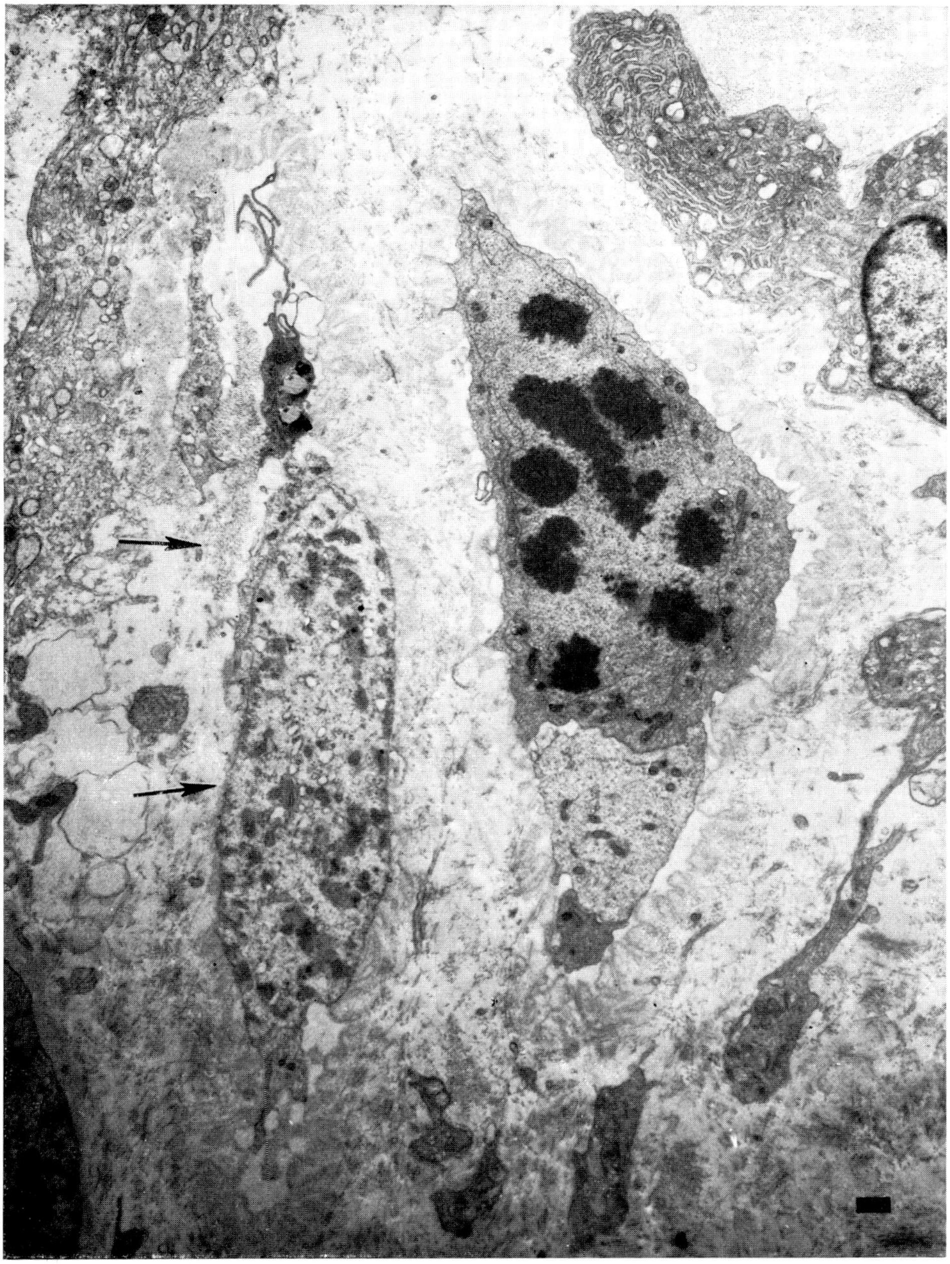

Fig. 9. Another view of the pleated basement lamina during the period of connective tissue dissolution. The area is adjacent to the muscle depicted in Fig. 8. In the region between the arrows, the basement lamina around this particular fragmenting myofiber has completely disappeared. Toward the center of the micrograph, a dividing cell (late prophase) and two adjacent fragments of undifferentiated appearing sarcoplasm can be seen within a pleated basement lamina that is not yet dissolved. Since the satellite cells are rare or absent in this skeletal muscle, it is believed that the three cells depicted within this particular basement lamina arose by dedifferentiation of the muscle fiber itself. The mononucleate cells which arise from muscle seem to take on the appearance of fibroblasts subsequently. Fibroblasts of the connective tissue proper appear in the upper right and upper left hand corners of this micrograph. The bar equals 1 micron.

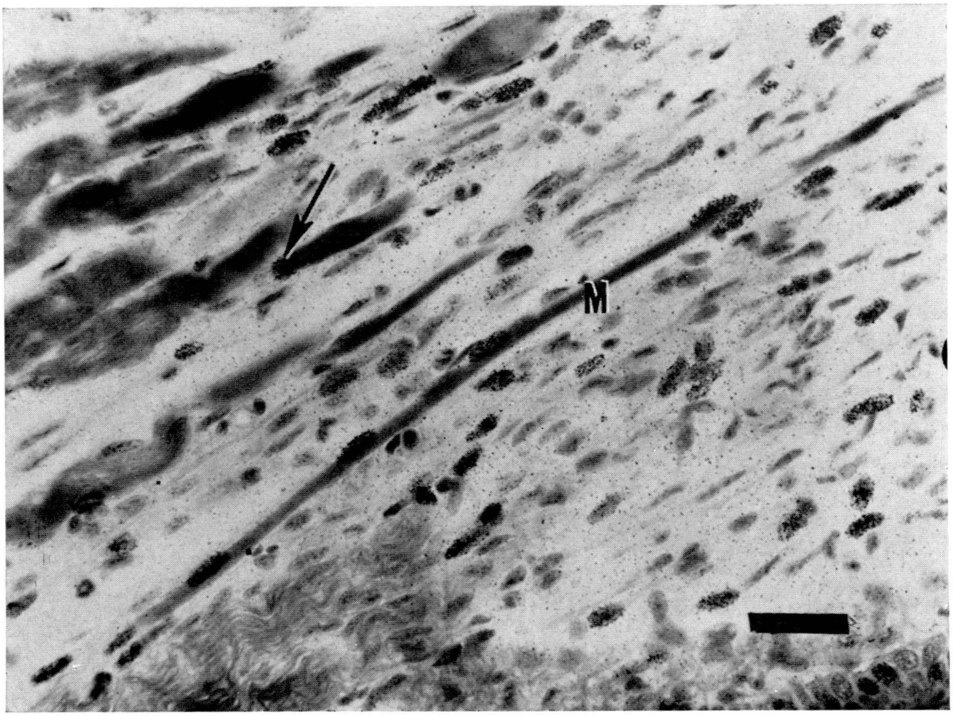

Fig. 10. Light micrograph showing an area of dedifferentiating muscle from a regenerating limb, 12 days after amputation, which was fixed 3 hours after the newt was injected with tritiated thymidine; 4 nuclei within an intact muscle fiber (M) have begun to synthesize DNA. Labeling is commonly associated with fragments of muscle (arrow) or mononucleated cells in the vicinity of the muscle. The bar equals 50 microns. (*from*: Hay and Fischman, 1961; copyright by Academic Press, New York, N.Y.)

E. D. HAY: I can infer a great deal about its history. In looking at these regenerating limbs, one can allow to enter into one's judgement, the time post amputation and the area of the limb one is looking at. This cell (Fig. 12) is from the area marked X in Figure 7 (B) in a limb fixed $2\frac{1}{2}$ weeks after amputation. The dedifferentiated fibers form in the place of all that muscle that was there earlier (Fig. 7, A). One has to conclude, then, that this cell (Fig. 12) was once filled with myofibrils.

J. C. T. CHURCH: When referring to the question of muscle regression, you implied that on the right hand side of that picture (Fig. 7, A) muscle was present, but now (Fig. 7, B) it no longer is. On the presumption that there were basement membranes around the fibers of that muscle, have you any micrographs which would show the basement membranes left behind, where you now have only blastema cells? If the basement membranes are there, would they fill up thereafter with new fibers?

E. D. HAY: There are no basement membranes in the area of the blastema, just as the bony matrix has almost completely disappeared distally, too. It means that most of the old connective tissue has been removed, perhaps by collagenase (Grillo *et al.*, 1968).

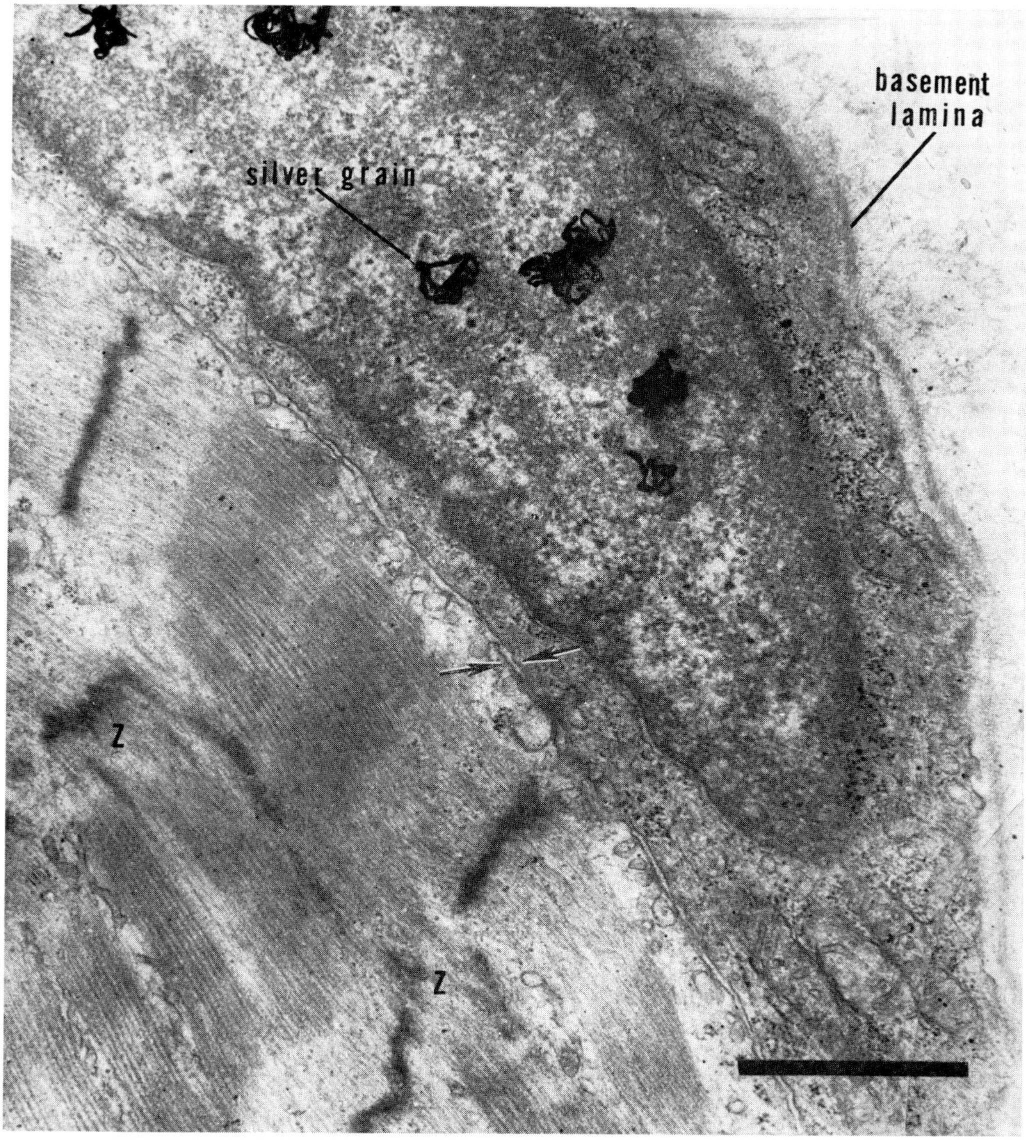

Fig. 11. Electron micrograph-autoradiograph from an area of muscle similar to that depicted in Fig. 10. The animal, an adult *Triturus viridescens*, was injected with tritiated thymidine 1 hour before fixation of the 19-day regenerating limb. All of the silver grains were found to be over nuclei of relatively undifferentiated cells. Where labeled nuclei appear within the myofibers (Fig. 10), they are seen by electron microscopy to be in mononucleate cells demarcated from the fibrillar part of the fiber by plasma membranes (arrows). These labeled cells are located under the basement lamina, however, and therefore clearly are derived from the muscle. It seems likely that such cells cleaved off the fiber proper during the process of dedifferentiation, because 'satellite cells' are rare or absent in this skeletal muscle prior to amputation. The Z band areas of the adjacent muscle cytoplasm are disrupted. The bar equals 1 micron.

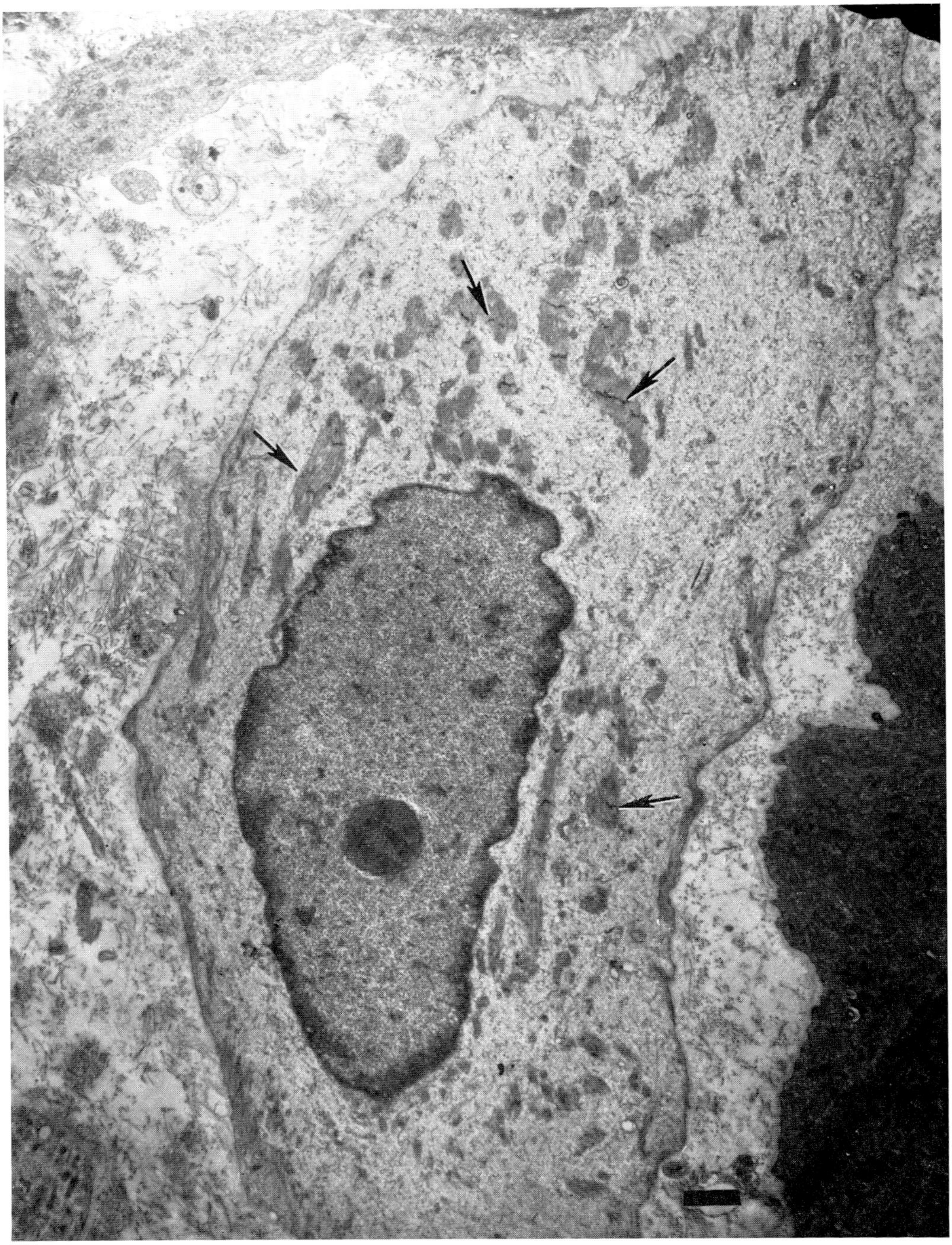

Fig. 12. Electron micrograph of a dedifferentiating muscle unit in an area similar to that depicted in Fig. 11. The nucleus of this partially dedifferentiated fiber was not synthesizing DNA at the time of fixation. Rather, it was probably making RNA and this cell may regenerate directly by cytoplasmic reorganization and 'budding'. The mononucleated cells of the blastema, on the other hand, proliferate as relatively undifferentiated cells and then muscle redifferentiates *de novo* as shown in Fig. 7 (D-F) and Fig. 2 (H). The bar equals 1 micron.

J. C. T. CHURCH: This is interesting because in other systems, at least in the mammalian system I am familiar with – the fruit bat web muscle – the basement membrane is rugged, and survives insult, is not absorbed by macrophages or in inflammatory reaction, and in fact remains behind to be filled out by new fibers, should they come.

E. D. HAY: This is a fundamental difference. Actually, it is also true that wherever the nerves are cut, there may be no Schwann cell sheaths for them to grow back down. Yet the nerves will grow back down into this regenerating bud in the proper way, without having to follow an endoneurial sheath. Thus, the connective tissue guide points that persist and are necessary for mammalian regeneration, which the cells presumably use as a scaffolding for reconstructing the part, are lost in the salamander limb blastema.

J. C. T. CHURCH: Is not the problem here the identification of levels? Can you be sure where your original section level was? Could you indicate where it was?

E. D. HAY: I can follow this bone in other serial sections, but it is difficult to get the longitudinal section exactly. The tip of the bone or amputation level is here, more or less (large arrows, Fig. 7, B and C).

J. C. T. CHURCH: So you take the tip of the bone as the level of the previous transection. The bone, of course, is plastic itself, and might be growing as well.

E. D. HAY: There is no growth of bone. New cartilage is deposited around the transected ends. The bone is not plastic. In any case, when new muscle differentiates *de novo* from blastema cells later (Fig. 7, F), there can be no question of old basement membranes being present as this is clearly in the outgrowing region of the blastema.

D. ALLBROOK: What is the proof of what is happening to that muscle cell? Is it growing or losing myofibrils, or are both processes taking place simultaneously, which is what I rather think?

E. D. HAY: These cells are incorporating nucleic acid precursors and they are dividing, but there is no evidence of redifferentiation of fibers at the earliest stages (Fig. 7, B). When the muscle begins to redifferentiate it looks just like muscle in mammalian systems. You get straps and tube-like structures which are quite different in appearance from those we are looking at here. Redifferentiation of muscle in the outgrowth does not begin in earnest until cartilage has reformed (Fig. 7, E). The myofibrils (Fig. 12) at the stage shown in Figure 7 (B) most likely are regressing, then. However, in the proximal part of the regenerate, it is true that muscle-like ribosomes are produced in partially dedifferentiated fibers (Deck, 1968; Lentz, 1969) so the cells may be 'gearing' up to make new muscle proteins. But to achieve the appearance shown in Figure 12, myofibrils have to be lost at a greater rate than new ones are made. If both processes could occur together, it would have to be in this proximal part of the stump because the distal muscle forms *de novo* (Fig. 2, F).

J. H. VENABLE: One thing bothers me: you point to these profiles (Fig. 5) and you say fragments. To me, this is merely an oblique section through a fiber of which you cannot see either end.

G. PALADE: Two basic points are left open to question in the interpretation of these experiments. The first concerns the origin of the blastema cells; the second, the fate of the muscle fibers at the beginning of regeneration (do they 'fragment' or not?). The approach so far followed may never give a clear-cut answer. Is it possible to graft in a limb, before amputation, either cartilage or muscle with nuclei labeled by radioactivity or otherwise? Cartilage would be better since the graft would bring with it fewer contaminating fibroblasts. Tracing the fate of the labeled nuclei during regeneration may provide answers to the two points mentioned.

E. D. HAY: These questions are very appropriate and, in fact, have been explored extensively in the past few years, particularly by Dr. Steen, who is here in the audience. Rather than going into that in detail now, could we let him present his data as a separate entity? I can say that he has transplanted both labeled muscle and labeled cartilage, usually double labeled by both polyploidy and tritiated thymidine. And he picks up the cells from either cartilage or muscle at later intervals in the blastema.

We actually observed the cartilage cells dedifferentiating to give rise to the blastema in the case of the larval finger (Fig. 1). I think there can be absolutely no doubt as to the origin of the blastema here, for we watched the individual cartilage cells as they were released from the cartilage matrix and so we were able to say that this cell came from that cell. The dramatic thing about watching the living finger is that there can be absolutely no doubt that the blastema arises from the formed tissues by backward regression.

The organization of the limb is more complex, but you can reconstruct a series of sections, as did Thornton (1938) and Butler (1933), in which you actually see stages when the muscle, which occupies the area, is disappearing and is being replaced by blastema cells. Since the nuclei appear metabolically active and can be shown to be incorporating nucleic acid precursors, it would be difficult to argue that the cells are dying, which is the only other alternative. There can be no reasonable doubt as to the origin of the blastema from muscle, cartilage and other stump tissues. I agree that the second question, how the muscle fibers actually fragment, needs more work. We have illustrated the mononucleated cells cleaving off the fibers and have speculated that small vesicles fuse to form the new membranes (Hay, 1959).

G. PALADE: Since it is so difficult to distinguish in tissue sections between number of cell profiles and number of cells, it might be advisable to apply a maceration procedure to the regenerating limb, preferably after removing the epidermis, or the skin. The cells could be set free (*e.g.* by enzymic digestion of the matrix) and then counted and measured. Moreover, one could try to identify cell types by some reliable procedure and thereby decide whether, as postulated, a large number of small muscle cells are produced at the expense of a few large muscle fibers, at the beginning of regeneration.

B. M. CARLSON: How can you get them in single cells?

G. PALADE: Cell separation could be obtained by trypsin (or other protease) treatment. Cell identification could be attempted by using labeled antibodies prepared against specific muscle proteins, and antibody penetration into the cells could be facilitated by fixation.

H. HOLTZER: For over 50 years microscopists have claimed that multinucleated muscle fibers break up into mononucleated cells and that these mononucleated cells contribute to the regeneration blastema. This story was based on observations made with the light microscope and the electron microscope has not as yet added one iota of additional data.

Some years ago (Holtzer, 1959) I tried to follow some of the events in regenerating limbs and tails of salamanders using fluorescein-labeled antibodies against myosin. These experiments were not overly informative. They did reveal that injured fibers underwent degenerative changes in one region while regenerating in another region. They also revealed that a goodly number of nuclei surrounded by bits of cytoplasm lasted for several days and then disappeared; as this cytoplasm bound the antimyosin we assume it came from fragmenting and dying muscle fibers. Certainly in our work there was no evidence of these units reconstituting cell membranes and functioning as viable mononucleated cells.

E. D. HAY: Even in the adult newt, as I showed just a minute ago, the entire distal muscle is completely replaced by blastema cells (Fig. 7, A and B). The whole area, which was previously occupied by muscle, is completely replaced by these growing cells, yet there was no evidence of any degeneration of the muscle nuclei. They were viable as judged by all possible morphological data. The only alternative, other than that they contributed to these cells, is to say that they left the limb. Moreover, the muscle extirpation and transplant experiments by Thornton (1942, 1943), Steen (1968), and others prove that muscle as a tissue gives rise to blastema cells. The point I should like to emphasize is that the morphology is uninterpretable in any other way. Mononucleated cells arise from the muscle (Hay, 1959; see also Hay and Doyle, 1970).

J. C. SLOPER: Have you demonstrated something like this using mitotic inhibitors?

E. D. HAY: I have not done anything with mitotic inhibitors. Colchicine inhibits limb regeneration (Thornton, 1943; Flinker and Sidman, 1953), but it has not been looked at in any great detail.

J. C. SLOPER: Preliminary work that we have done shows it brings the thing to a complete halt.

E. D. HAY: I should think so, since mitosis is necessary for regeneration. The main point of dedifferentiation seems to be to produce cells that can divide rapidly (Hay, 1968; Thornton, 1968).

J. C. SLOPER: That is not so self-evident to me. If, in fact, there was a reasonable contribution of the muscle cells to the structure of the blastema, it would be a very difficult thing to isolate in space. It is very misleading to say: "here is the amputation site", for as you well know, you get degeneration considerably proximal to the amputation point. And you never really know where the heart, as it were, of the blastema is; at least, I could never figure it out at the light microscope level.

Be that as it may, under periodic bouts of colcimide or colchicine, we have never been able to get any that I would consider considerable accumulation of cells in blastema. I merely mention that *a propos* of Dr. Palade's question: "Is there any way of showing that you can get an accumulation of cells simply by dissolution of the pre-existing muscle?"

E. D. HAY: Thornton (1942) and other earlier workers have shown, as I have mentioned, that pre-existing muscle by itself can give rise to a blastema. They removed the other tissue from the limb before amputation.

P. E. CONEN: I am not clear about the difference between outgrowth and blastema. This may be part of our problem.

E. D. HAY: The stages (Fig. 7, A and B) that we have been talking about mainly are all before actual outgrowth, and in the adult limb you can see where the outgrowth is, because it is very gelatinous and very transparent. You can see there is not yet an outgrowth because you follow the limb before you fix it. Beginning 2 to 3 weeks post amputation there will be an outgrowth. Some call this outgrowth, the blastema. The question is, then, whether or not to call the structure that arises backwards to replace the formed tissue, a blastema. I think you have to (Fig. 1). That was the early blastema to which we were referring previously. It is less obvious in the adult because remnants of bone are present (Fig. 7, B).

I should hate to end the discussion without impressing on the audience how remarkable this process is – how the blastema grows out and redifferentiates into a limb; in the case of the newt, this is identical to the one that was amputated and is completely functional in about 6 or 7 months. All of the original muscles have been regenerated (Fig. 7, F). Whether or not they regenerate entirely from cells derived from old muscle, or whether or not, in the blastema, there is a transformation of cells derived from muscle to cartilage, or *vice versa*, we do not know. When I use the term 'dedifferentiation' I am using it in a morphological way and mean the loss of differentiated products, such as myofibrils, accompanied by the acquisition of features typical of growing cells, such as ribosomes, evidence of RNA synthesis, and evidence of cell division.

The question of whether or not cells in the blastema, regardless of their origin, can redifferentiate into muscle or into some other tissue is one which is being pursued actively by a number of investigators, and Dr. Steen is one of them. I know I have not answered many of the questions that were raised, and I had planned to discuss the stages of redifferentiation in more detail. I think it will be more appropriate if we discussed them later informally and went on now to hear about Dr. Steen's experiments because they settle one of the main issues we have been discussing: the origin of the blastema from formed tissues.

A. MAURO: We shall go on with Dr. Carlson's presentation of regeneration in the rat and the frog, the so-called 'higher' vertebrates and in the axolotl; then Dr. Steen will follow. Dr. Bruce Carlson, would you kindly proceed.

THE REGENERATION OF ENTIRE MUSCLES FROM MINCED FRAGMENTS*

BRUCE M. CARLSON

INTRODUCTION

The main purpose of this paper will be to review the major histological features of muscle regeneration in higher vertebrates. Following this, I shall present some recent data on the regeneration of muscles in the axolotl and then compare muscle regeneration in urodele amphibians with that seen in higher species.

Regeneration of the mammalian skeletal muscle fiber has been studied for more than a century, and the histological features of this phenomenon have been well covered in several recent reviews (Betz *et al.*, 1966; Field, 1960; Godman, 1957). As a vehicle for my presentation of the main histological features of muscle fiber regeneration, I am going to describe the regeneration of entire muscles from minced fragments. This experimental system, first described and extensively investigated by Studitsky and his colleagues (Studitsky, 1954, 1959; Studitsky and Ignatieva, 1961; Zhenevskaya, 1962), is in many respects best adapted for the study of muscle regeneration at a higher level of organization than the individual muscle fiber, but at the cellular level the histological events are the same as those which have been studied in other experimental models.

OPERATIVE TECHNIQUE AND GROSS OBSERVATIONS

The technique for producing minced muscle regenerates is relatively simple, and it has been described in detail by Studitsky *et al.* (1956) and Carlson (1968*a*). Briefly, it consists of completely removing a muscle, such as the gastrocnemius, trimming off the tendinous material and mincing the muscle with scissors until it is reduced to fragments about 1 cubic millimeter in size. The minced fragments are replaced into the bed from which the muscle was originally removed, and any overlying muscles and the skin are sutured in place.

Following this procedure, a newly regenerated replica of the original muscle will be produced in a relatively short period of time. After implantation, the mass of implanted fragments is molded roughly into the shape of the muscle by the pressure of the surrounding tissues, and the peripheral area of the minced muscle mass is loosely held together by fibrinous material. Within a week of the operation, the fragments of minced muscle in the peripheral area lose their individual identity, and the presence of a more homogeneous appearing tissue in this area is the first gross indication that muscle regeneration is occurring. During this period tendinous connections are also becoming established at the origin and insertion of the regenerating muscle. These connections are normally made between the remaining cut ends of the original tendons and the proximal and distal ends of the minced

* Supported in part by a grant from the University of Michigan Cancer Research Institute.

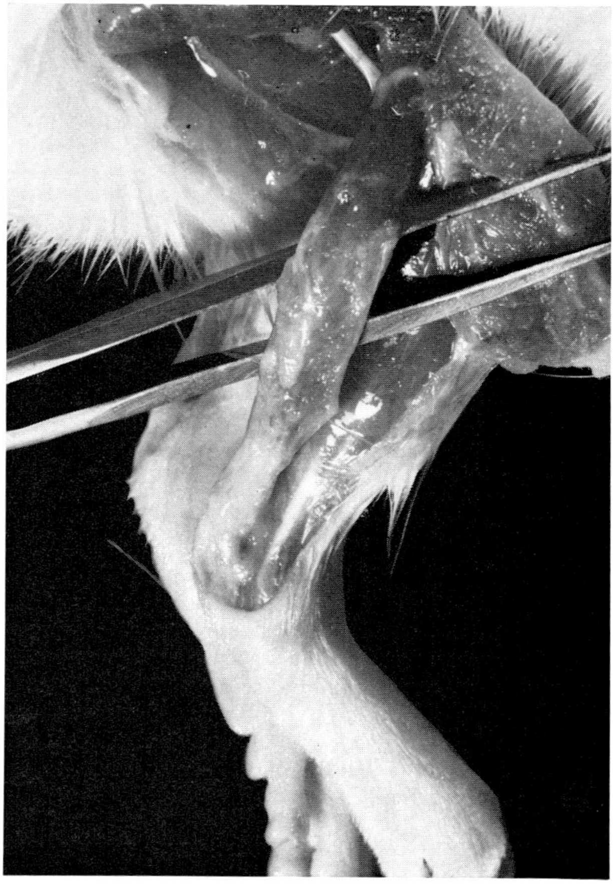

Fig. 13. A 68-day gastrocnemius regenerate in the rat.

muscle mass. Further regeneration is marked by a decrease in the total mass of the regenerate. This is caused by the degeneration of much of the originally implanted muscle fragments and the appearance of newly forming muscle fibers and connective tissue (with a lesser total mass) in their place. The maturing muscle regenerate possesses normal tendinous connections, normal gross relationships with the major nerves and often a fairly normal vascular pattern. In the regenerated gastrocnemius muscle, a band of dense connective tissue runs through the entire regenerate, but most of the proximal half of the regenerate is muscular in the rat. The distal end of the regenerated gastrocnemius muscle is tendinous with little gross evidence of muscle. The total mass of such regenerates is usually considerably less than that of normal control muscles (usually 15 to 25 per cent of the normal weight). A typical regenerate in the rat is illustrated in Figure 13.

HISTOLOGICAL OBSERVATIONS

The histology of muscle fiber regeneration has been studied in detail in the frog and rat (Carlson, 1968*a*). The main features of the regenerative process following mincing are quite similar in these species. In the days immediately following mincing, extensive sarcoplasmic degeneration occurs in the muscle fragments. In the frog stages of nuclear degeneration resembling karyorrhexis are commonly seen during the first 3 or 4 days. Other nuclei appear morphologically intact. In the rat extensive sarcoplasmic degeneration is seen, but nuclear degeneration of the type noted in the frog has not been observed. Vascular injection studies in the rat have shown that the minced muscle mass is completely divorced from a direct blood supply for the first 2 days. The first definite histological indications of regeneration can be seen at 2 to 3 days in the rat and 5 to 6 days in the frog. These occur first at the periphery of the minced muscle mass and then proceed toward the interior so that there is a gradient of regenerative activity ranging from the degenerating, originally implanted muscle fibers in the center to progressively more advanced stages of regeneration towards the periphery. Figure 14 illustrates a low-power view of a 3-day rat regenerate which was perfused with ink (Pelikan C11/1431a) to demonstrate the extent of revascularization. The central area of originally implanted muscle fragments contains no demonstrable vasculature and morphologically appears to be rather inactive (Fig. 15) whereas the peripheral basophilic zone of regeneration is richly vascularized. Of particular interest is a transitional zone between the central region of degenerating muscle fibers and the entirely basophilic zone containing muscle and connective tissue in more advanced stages of regeneration. The main histological features of this zone are best seen in cross section (Fig. 16). As the sarcoplasm of the original muscle fibers begins to fragment (this process is almost always accompanied by an invasion of macrophages), a ring of nuclei begins to appear under the basement membrane of the old muscle fiber. At first they appear to be separate from one another, but soon they become surrounded with increasing amounts of basophilic cytoplasm and finally fuse into basophilic cuffs which completely line the inside of the original basement membrane and surround the remaining fragments of the original sarcoplasm, which has

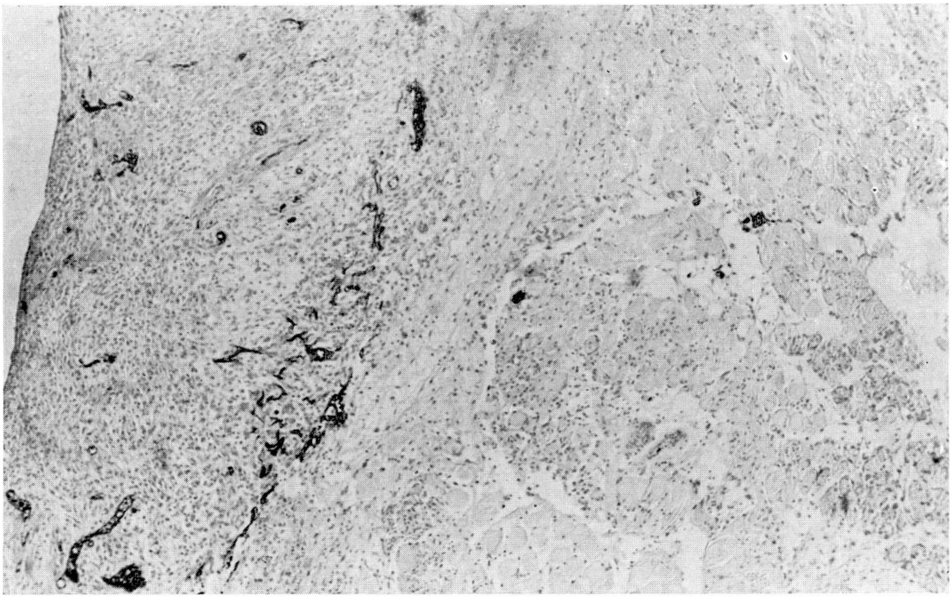

Fig. 14. Section through 3-day minced muscle regenerate in the rat. The regenerate was perfused with ink, and the black material represents blood vessels filled with ink. At the left is the peripheral zone of regeneration. At the right is the central zone of originally implanted muscle fibers. H and E; × 53.

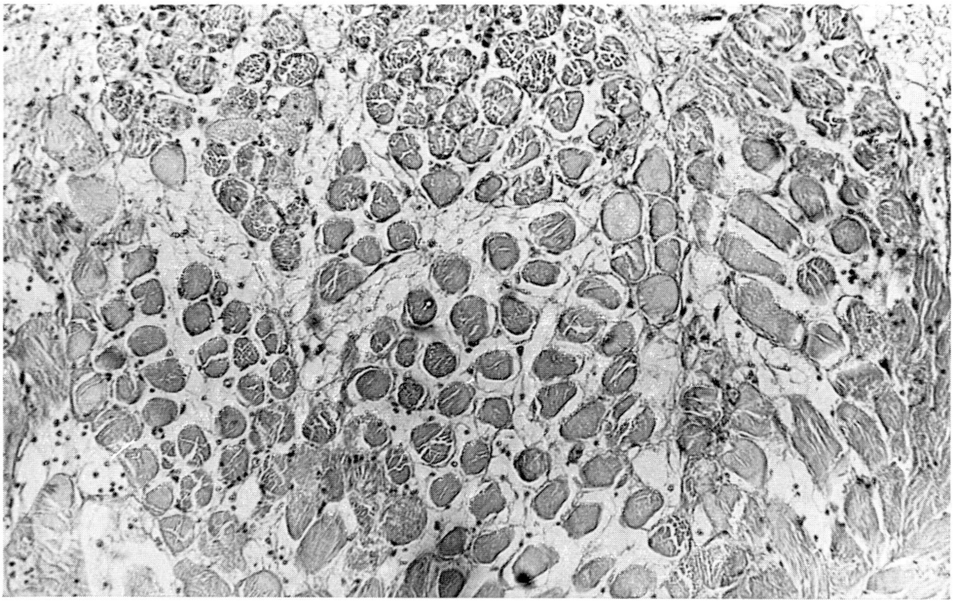

Fig. 15. Central zone of the same regenerate as in Fig. 14. No injected vessels are seen. In the lower part of the photograph are cross sections of originally implanted muscle fibers. Note the virtual absence of nuclei at the periphery of these fibers. The cells at the top are located more peripherally in the regenerate than those at the bottom and exhibit early regenerative changes such as extensive fragmentation of the old sarcoplasm and the appearance of some nuclei under the basement membrane of the degenerating muscle fibers. H and E; × 170.

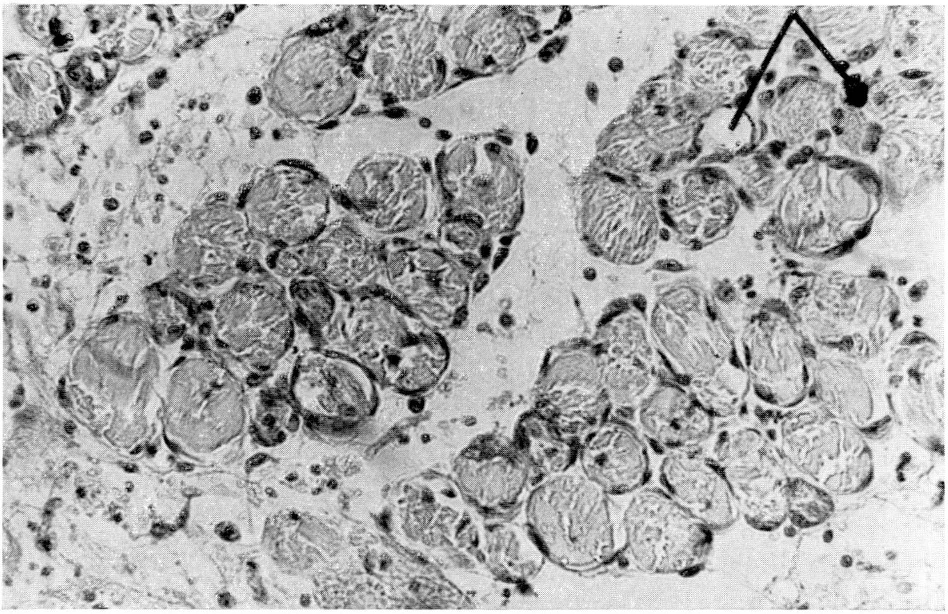

Fig. 16. Transitional zone of 3-day rat regenerate showing pronounced sarcoplasmic degeneration and the formation of early basophilic cuffs. A number of macrophages are associated with the degenerating sarcoplasm. Two injected vessels (arrows) are seen in the most peripheral area of this field. No injected vessels were seen centrally to these. H and E; × 400.

practically disappeared by this stage (see Carlson, 1968*a* for details of this process in the frog). The vascular injection studies have been able to demonstrate only an occasional capillary in the outermost part of the transitional zone. Yet only a couple of hundred microns peripherally, the more advanced regenerating zone is richly supplied with blood vessels. Despite the seeming lack of a direct supply of oxygen and nutrients, the inner part of the transitional zone demonstrates a significant degree of metabolic activity. Autoradiographic studies on minced frog muscle show large numbers of nuclei of very early regenerative stages labeled with tritiated thymidine. Figure 17 illustrates labeled nuclei in the inner part of a regenerate. Most of these nuclei are not surrounded by microscopically visible cytoplasm, and little fragmentation of the cytoplasm of the original muscle fiber has occurred. Nuclei at this same stage of development also exhibit intense incorporation of H^3-uridine.

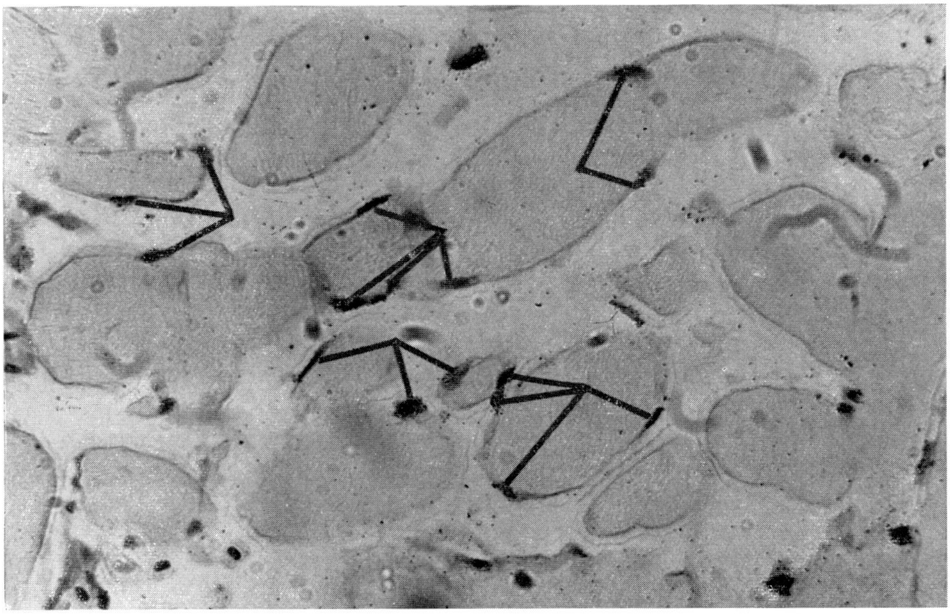

Fig. 17. Autoradiograph of the inner zone of a minced muscle regenerate in *Rana clamitans*. The animal was injected intraperitoneally with 5.0 microcuries of H^3-thymidine per gram body weight on the seventh postoperative day and killed 24 hours later. Most of the nuclei lining the periphery of the old muscle fibers are labeled. H and E; × 400.

As the postoperative interval increases, the thickness of the peripheral regenerating zone becomes greater, and the proportion of originally implanted muscle fibers becomes smaller (Fig. 18). There is a constant relationship between the increasing penetration of blood vessels and the thickness of the regenerating zone. The more advanced regenerating elements (myotubes) are intimately associated with blood vessels (Fig. 19). The relationship between the appearance of basophilic cuffs in the transitional zone and the extent of revascularization remains constant regardless of its relative position within the regenerate.

Further differentiation of the regenerating muscle fibers does not differ from that reported by a number of authors (Allbrook *et al.*, 1966; Betz and Reznik, 1964; Godman, 1957). The early myotubes are characterized by long central chains of nuclei and basophilic cytoplasm which contains large amounts of newly formed RNA (Fig. 20). As the myotubes mature, the

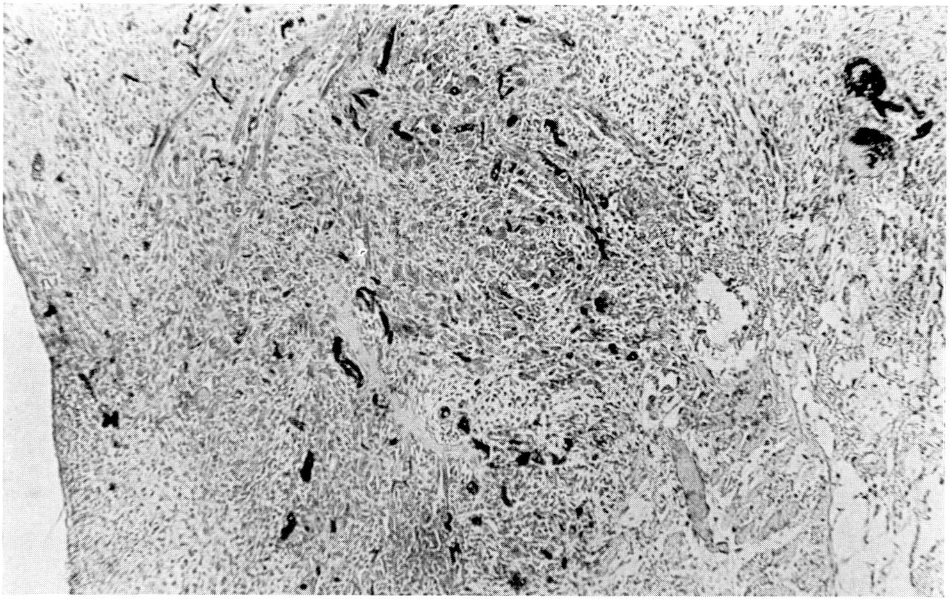

Fig. 18. A 4-day rat regenerate perfused with ink. The peripheral zone (left) is richly supplied with blood vessels. Cross and longitudinally sectioned myotubes are numerous. In the lower right hand corner is the central non-regenerating zone of old muscle. H and E; × 53.

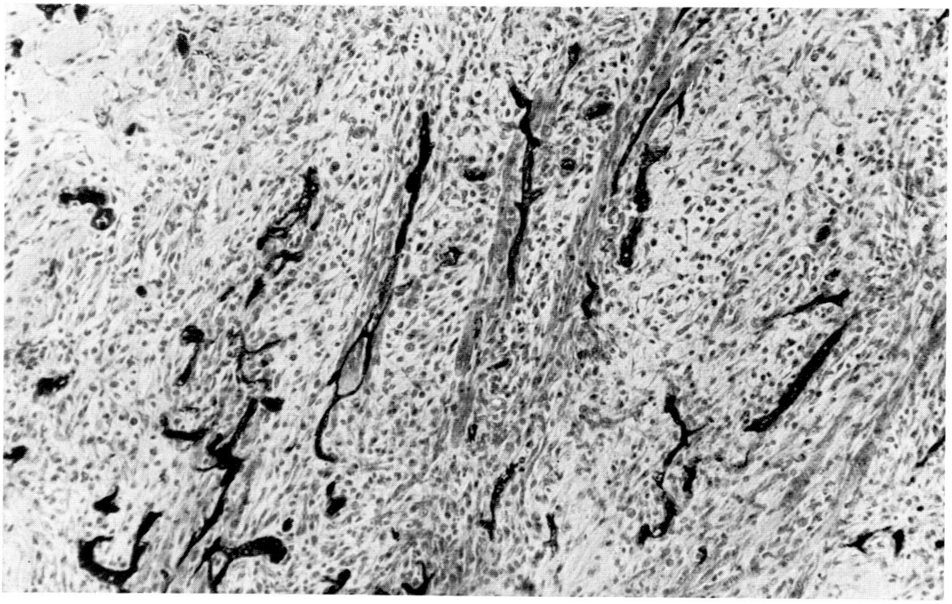

Fig. 19. Peripheral zone of 4-day rat regenerate showing intimate relationship between vasculature and regenerating muscle fibers. H and E; × 170.

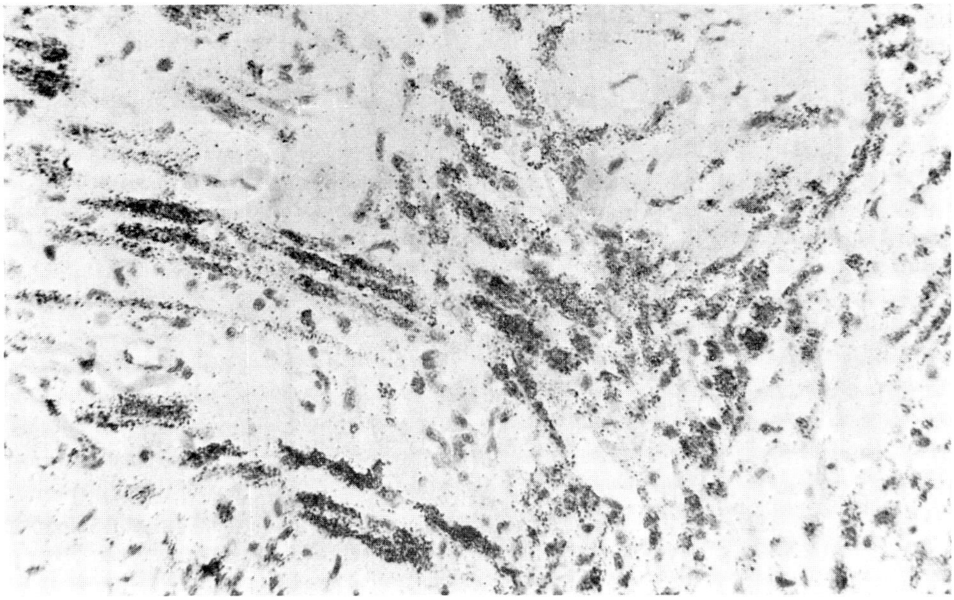

Fig. 20. Autoradiograph of 10-day gastrocnemius regenerate in *Rana clamitans*. The animal was injected intraperitoneally with 5.0 microcuries of H^3-uridine per gram body weight and killed 8 hours later. The long strands of silver grains cover early myotubes. H and E; × 400.

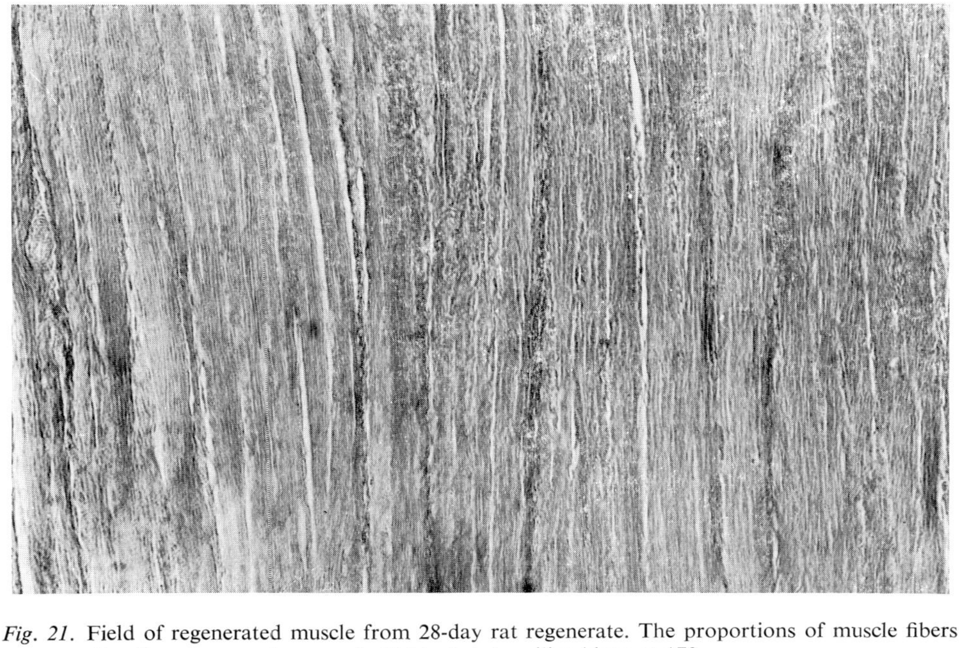

Fig. 21. Field of regenerated muscle from 28-day rat regenerate. The proportions of muscle fibers to connective tissue are nearly normal. Heidenhain's aniline blue; × 170.

intense basophilia of the cytoplasm is supplanted by an ever increasing degree of eosinophilia as the contractile elements of the muscle fiber are being produced. Cross striations become apparent and the nuclei break out of the central chains and migrate toward the periphery of the muscle fiber. Further development of the muscle fibers consists primarily of an increase in diameter and the emergence of contractile ability.

Mature minced muscle regenerates differ histologically from normal muscles in several respects. The total size is considerably less (15 to 25 per cent of control muscles). Although in some instances the proportion of muscle to connective tissue differs little from normal (Fig. 21), in others there is considerably more connective tissue. This is particularly true in the frog, in which dense masses of connective tissue without any muscle fibers are often seen in regenerates. The histological structure of tendons in the regenerated muscles usually differs little from normal tendon except for a very high (95 per cent) incidence of ectopic bone formation in rats.

MUSCLE REGENERATION FROM HOMOTRANSPLANTED FRAGMENTS

Two experimental series have been undertaken in order to determine the ability of homotransplanted muscle fragments to take part in regeneration (Carlson, 1968*b*). Minced muscle (gastrocnemius) was exchanged between sibling Wistar rats and between non-sibling Sprague-Dawley rats. During the first postoperative week, muscle regeneration in the sibling to sibling grafts was intense and could not be distinguished from that occurring in autografts (Figs. 22 and 23). In non-sibling grafts, early regenerative activity was just as intense, but infiltration of small lymphocytes occurred around the regenerating muscle by the end of the first week. Individual muscle fibers regenerated normally, but with time the accumulations

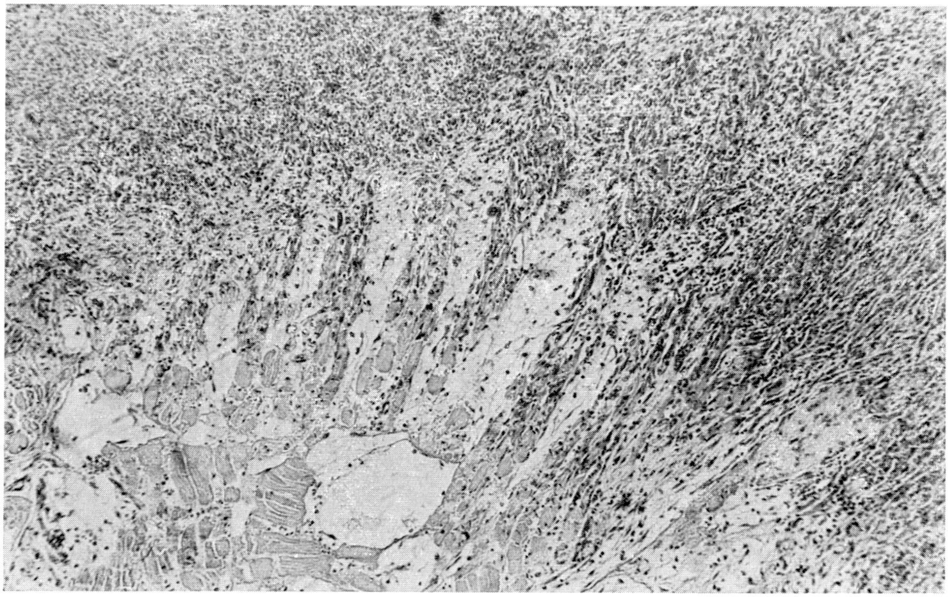

Fig. 22. A 6-day minced muscle regenerate in Wistar rat from sibling donor. The peripheral regenerating zone is on top, and the central zone of old fibers is on the bottom. The transitional zone of early regeneration presents an almost radial appearance. H and E; × 53.

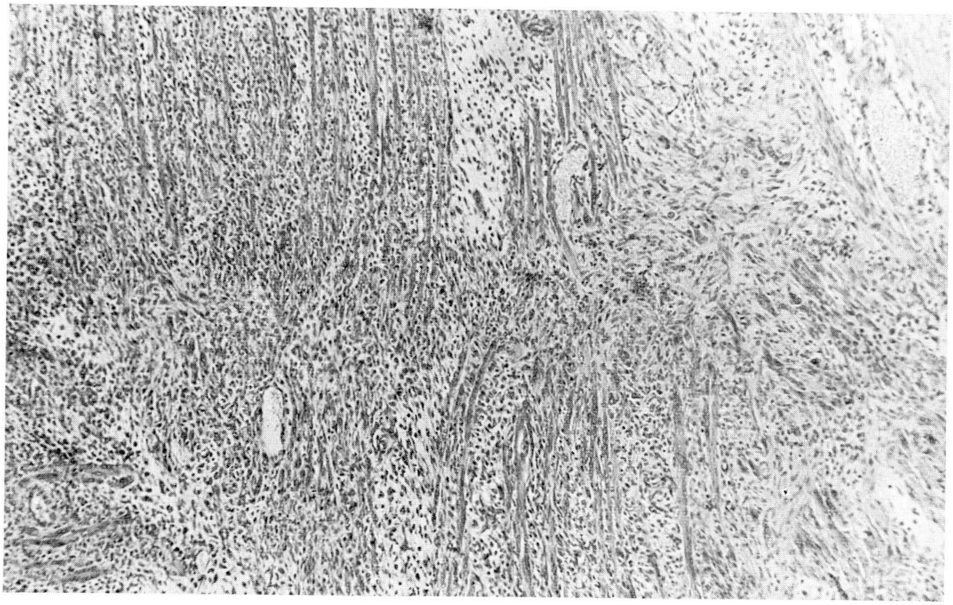

Fig. 23. A 5-day minced muscle regenerate in Wistar rat from sibling donor. Tangential section through peripheral regenerating zone showing numerous myotubes. H and E; × 53.

of small lymphocytes increased significantly, and the total amount of regenerating muscle fibers decreased. Figure 24 illustrates a typical picture of an advanced muscle regenerate arising from non-sibling muscle fragments. The amount of regenerating muscle is considerable despite the intense cellular rejection which is taking place. Figure 25 shows quite mature striated muscle fibers surrounded by small lymphocytes. This, along with the work of Studitsky and Rumyantseva (1964) and Laird and Timmer (1965, 1966), demonstrates that complete regeneration of individual striated muscle fibers can occur even in an immunologically hostile environment.

MUSCLE REGENERATION IN THE AXOLOTL

Of the animals discussed above, the rat has thus far not been demonstrated to be capable of complete limb regeneration. Although normally the adult frog does not regenerate extremities, limb regeneration has been artificially stimulated (Polezhaev, 1946; Rose, 1945; Schotté and Wilbur, 1958; Singer, 1954). Before metamorphosis, however, tadpoles can readily regenerate limbs. As was shown above, the morphology of minced muscle regeneration in these 2 species is essentially the same. Not long ago I became interested in the following problem: In animals normally incapable of limb regeneration ('epimorphic' type of regeneration), minced muscle regenerates according to a characteristic morphological pattern, and such regeneration does not appear to involve or to include complex interactions with the other tissues of the limb. I wondered what would happen if one minced a limb muscle in an animal fully capable of regenerating entire limbs. Would the minced muscle regenerate as a tissue like in the frog and rat or would the great amount of damage inflicted

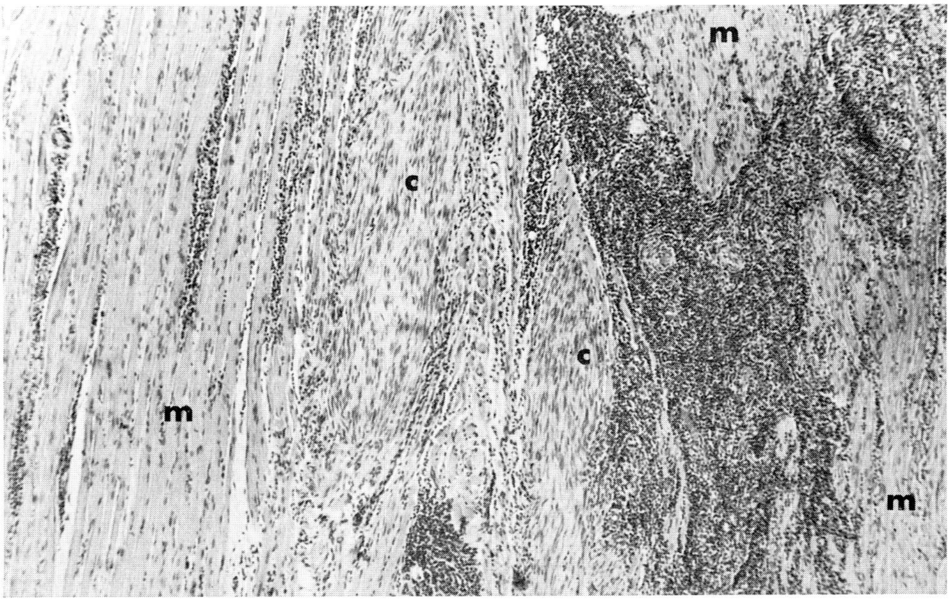

Fig. 24. Low-power view of 32-day regenerate from muscle fragments taken from a non-sibling Sprague-Dawley rat. m = muscle, c = connective tissue. H and E; × 53.

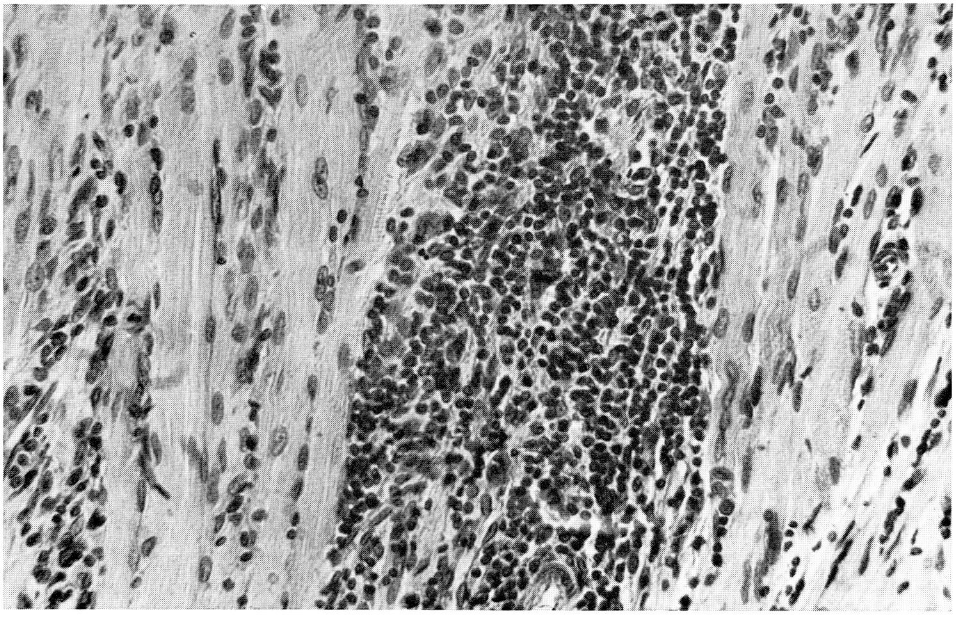

Fig. 25. Higher-power view from same regenerate as shown in Figure 24. × 400.

to the limb be a sufficient stimulus to set off a complete epimorphic regenerative response and cause the development of a supernumerary limb at the site of mincing?

To answer this question, the puboischiotibialis muscle in axolotls was minced in the same manner as was the gastrocnemius of frogs and rats. It was found that following mincing, the course of regeneration in the axolotl is quite similar to that which is seen in the higher forms (Carlson, 1970*b*). By the end of the first week, considerable degeneration of the sarcoplasm of the implanted muscle fragments has occurred, and myoblasts are seen. In the axolotl the phenomenon of cuffing of basophilic cells inside the basement membrane is not so prominent as it is in frogs or rats, but it does take place to a limited extent. By the middle of the second week fusion of myoblasts into multinucleated fibers has occurred (Fig. 26).

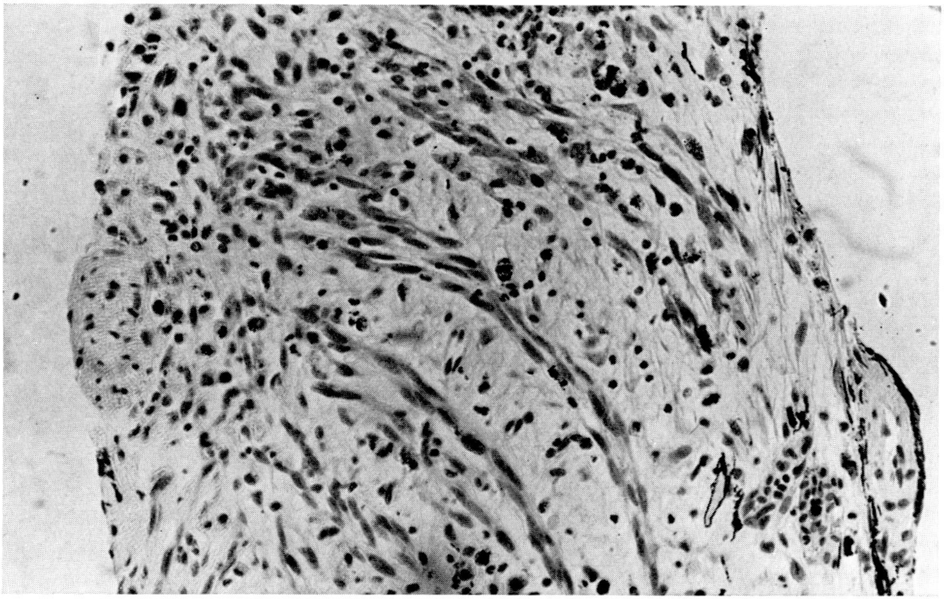

Fig. 26. A 10-day minced muscle regenerate in the axolotl. H and E; × 170.

Central chains of nuclei are not so regularly disposed nor so extensive as those seen in higher species. Within a month, cross-striated muscle fibers with peripherally located nuclei are seen throughout the regenerate (Fig. 27). Maturation of the regenerating muscle proceeds steadily until the individual muscle fibers are indistinguishable from normal (Fig. 28). Although the final regenerate approximates the normal muscle in general form, the total size is usually less and as in other minced muscle regenerates, the amount of connective tissue is greater.

It is thus evident that an entire muscle in an axolotl limb can regenerate from minced fragments without triggering off the process of supernumerary limb formation. Yet the same muscle can also be regenerated after amputation of the limb. There are certain pronounced differences in both the end products and the means by which the end products (*i.e.* the regenerated muscles) are produced. These differences are (1) the presence or absence of a blastema associated with the regenerating muscle, (2) the rates of the regenerative processes and (3) the final morphology of the regenerated muscles.

In the regeneration of minced muscles, a blastema is never seen. Instead, a population of

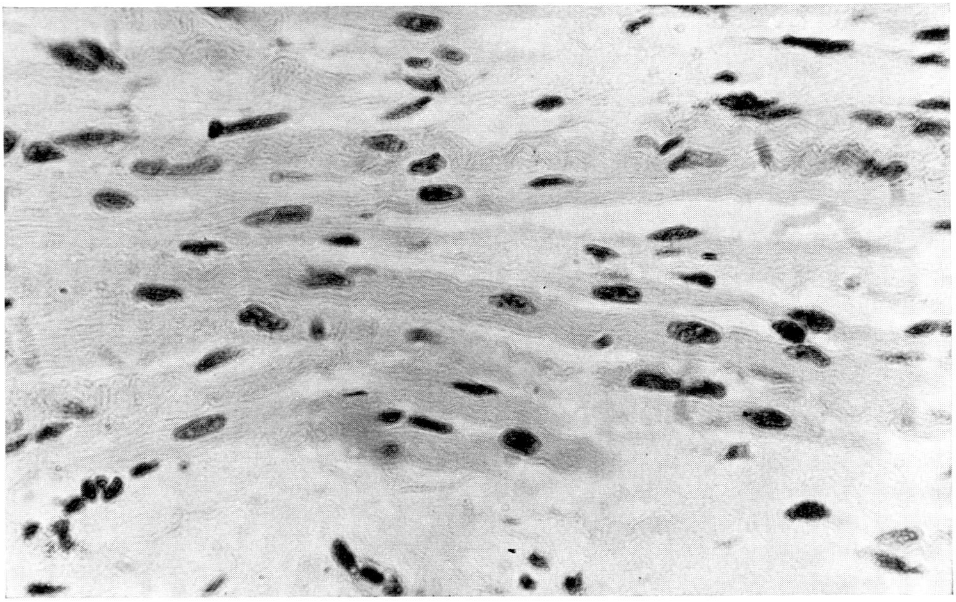

Fig. 27. Striated muscle fiber seen in 32-day axolotl regenerate. H and E; × 400.

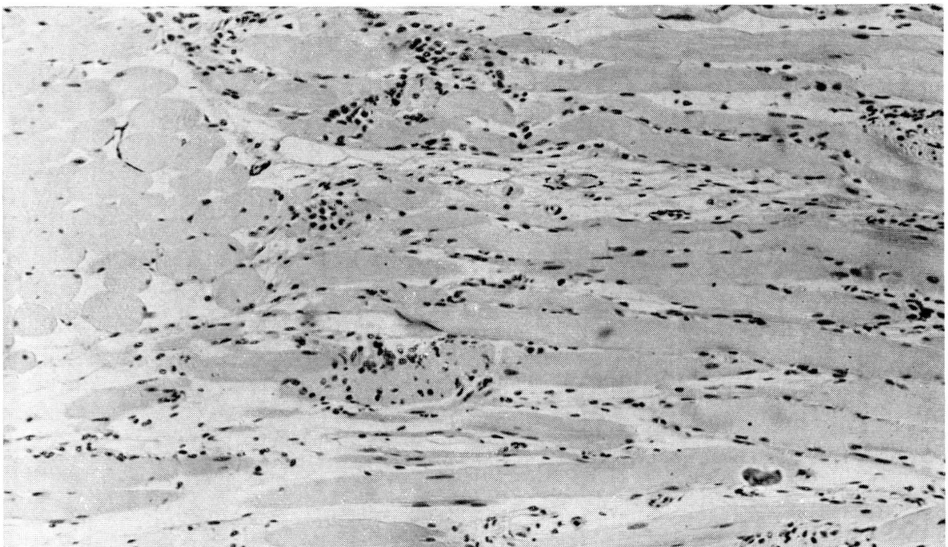

Fig. 28. A 150-day minced muscle regenerate in the axolotl. Cross-sectioned muscle fibers are from a normal muscle which inserts into the regenerate. Longitudinally sectioned fibers are regenerated. Note the greater than normal amount of connective tissue among the regenerated muscle fibers. H and E; × 53.

myoblastic cells (whose origin is still uncertain) is established primarily along the insides of the basement membranes of the degenerated muscle fibers. Within a short time (3 days in the rat, 8 days in the frog and 9 to 10 days in the axolotl) multinucleated myotubes have formed and further differentiation and maturation of the regenerating muscle fibers quickly follows.

In contrast, after limb amputation in the adult axolotl, the most prominent feature is the establishment of a blastema of morphologically undifferentiated cells. At 21° C a blastema is not well established until the end of the first month. Before the establishment of a blastema, the muscle in the limb stump is characterized by pronounced degeneration and dedifferentiation into mononuclear fragments (Hay, 1959, 1962; Lentz, 1969; Thornton, 1938). As a rule multinucleated stages of regenerating muscle fibers are seldom seen during the period of establishment of the blastema, even among the injured muscle fibers of the stump.

Differentiation of muscle in the regenerate itself does not occur until the cartilaginous model of the skeleton has been laid down. Then muscle fibers appear to arise from the fusion of spindle-shaped mononuclear cells which are found alongside the cartilage. Further differentiation of these muscle fibers is morphologically similar to that seen in minced muscle regenerates and cross striations are present midway in the second month. The origin of the muscle precursor cells remains in doubt. Certainly in the early blastema there are no cells which can definitely be identified as descendants or precursors of muscle on the basis of their histological appearance alone. Yet, when the basophilic spindle-shaped muscle precursors can be first identified, there is not always a direct connection between them and the more proximal muscular elements of the limb.

Regardless of the origin of the muscle in limb regenerates, the important thing is that the formation of the regeneration blastema seems to impose a time lag upon the differentiation of muscle fibers in limb regenerates. The location of the muscle precursors and the environmental influences which act upon them during the delay period are questions which remain to be answered. That the delay may be significant, however, is reflected in the morphology of the muscles in limb regenerates when compared with those regenerated from minced fragments. The muscles in regenerated limbs are normally exact duplicates of those which had been present before amputation. In contrast, minced muscle regenerates never exactly duplicate the original muscle. They invariably contain much more connective tissue than normal and the gross morphology, although approximately that of the original muscle, not infrequently deviates somewhat from it. During the earlier stages of muscle formation in limb regenerates, the myogenous elements are always oriented parallel to the longitudinal axis of the muscle. In minced muscle regenerates the myoblasts and myotubes are scattered randomly. It is not until the appearance of myofibrils and cross striations that the regenerating fibers become uniformly oriented parallel to one another.

Although the data presented above may reflect a difference at a higher level of organization than the individual muscle fiber, the difference does seem to be a significant one. Thus on the basis of the experiments conducted above, there seem to me to be 2 means by which entire muscles can be regenerated, at least at the level at which I am working. There is the regeneration of muscle as a tissue (following simple wounds or mincing), which has been described in mammals, birds, lizards, anuran and urodele amphibians, and there is the regeneration of limb muscles as part of an epimorphic process of regeneration. Further work is presently being conducted in my laboratory to determine other similarities or differences between the 2 proposed types of regeneration.

HISTOLOGICAL OBSERVATIONS ON THE REGENERATION OF MAMMALIAN AND AMPHIBIAN MUSCLE

BRUCE M. CARLSON

As I see it, the purpose of my presentation is threefold: (1) to present some examples of muscle regeneration as a process of tissue repair, (2) to compare this type of regeneration, which is commonly seen in mammals, frogs and other vertebrates, with the type of muscle regeneration seen in the regenerating limb of the salamander, and (3) to raise some questions suggested by this work.

The bulk of the material which I would like to present will be some light-microscopic preparations of regenerating muscle in rats, frogs, and axolotls. Most of my research on muscle has been conducted at the tissue and organ level, and thus it has not been directly concerned with the origin of myoblastic cells. It should, however, expose a number of areas in which our knowledge of mechanisms regulating the regenerative process is deficient. During the course of this presentation these problems should become quite obvious.

I shall present mainly results obtained during recent investigations of the phenomenon of muscle regeneration from minced muscle fragments. This procedure was originally devised by Studitsky in Moscow during the early 1950's (Studitsky, 1952*a*, *b*). Although it has been extensively investigated in his laboratory, few other people have used this experimental model of muscle regeneration (Gallucci *et al.*, 1966; Carlson, 1968*a*; Allbrook, personal communication). Briefly, one can produce a minced muscle regenerate by removing an entire muscle from the limb of an animal, mincing it with scissors until the fragments are about 1 cubic millimeter and then replacing the minced fragments back into the limb.

Following this procedure, a new model of the muscle will regenerate within a relatively brief period. At a gross level, the regenerate is quite similar to the muscle which was removed. It has normal tendon connections, and its gross relationships to nerves and the vascular supply is surprisingly close to normal. The total mass of a muscle regenerate is usually considerably less than normal (15 to 25 per cent of normal), and it contains a greater proportion of connective tissue. These latter constitute the main limitations on the experimental model. I shall describe first the regeneration of the gastrocnemius muscle in the frog (*Rana pipiens*), then in the rat and finally the regeneration of the puboischiotibialis (p.i.t.) muscle in the axolotl (*Amblystoma mexicanum*).

Figure 29 illustrates the typical histological picture seen 2 or 3 days after implanting the minced gastrocnemius muscle into the leg of a frog. All of the muscle fibers in the figure are those which were originally minced and implanted. During the early postimplantation period (3 to 4 days), many of the nuclei of the muscle fragments go through a characteristic set of degenerative changes (Fig. 30). The nuclear chromatin appears to condense into a number of dense globular fragments by a process similar to that described as karyorrhexis. At this early period most of the cells surrounding the muscle fibers themselves are erythrocytes (nucleated in the frog), and leukocytes – primarily of the mononuclear variety.

D. ALLBROOK: Could you tell us if there are any membranes around those things?

B. M. CARLSON: Yes. The membranes of the old muscle fibers persist for several days.

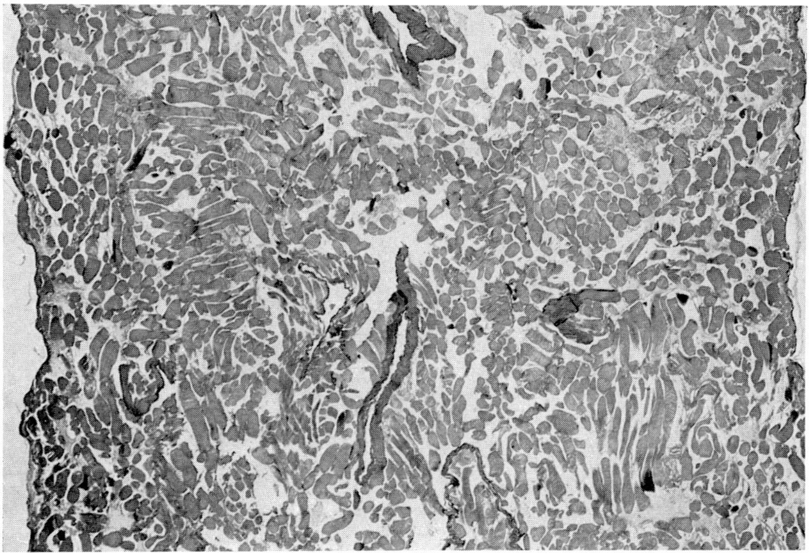

Fig. 29. A 60-hour frog regenerate. All tissue present represents originally implanted muscle and connective tissue fragments. H and E; × 20. (*from*: Carlson, 1968*a*)

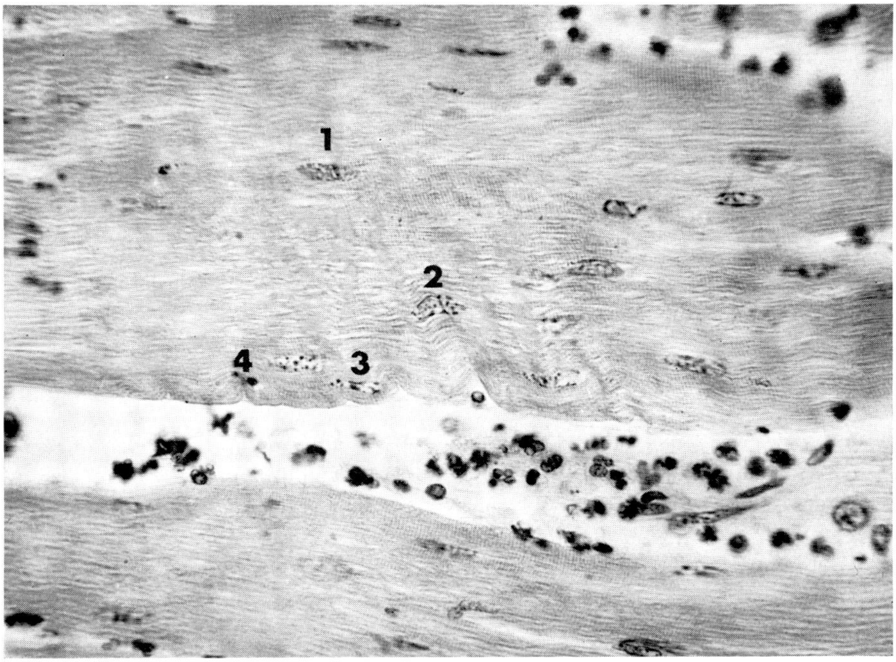

Fig. 30. A 3-day frog regenerate. Successive stages of nuclear degeneration in an implanted muscle fiber (numbers 1 to 4). H and E; × 440. (*from*: Carlson, 1968*a*)

D. ALLBROOK: I beg your pardon, I meant the nuclei.

B. M. CARLSON: No. In current studies on the regeneration of muscle in diffusion chambers, similar degenerating nuclei have been seen. Figure 31 is an electron micrograph taken by one of my students, Mr. Gene Trupin. The chromatin is condensed into a very dense mass, and it is no longer surrounded by a nuclear membrane.

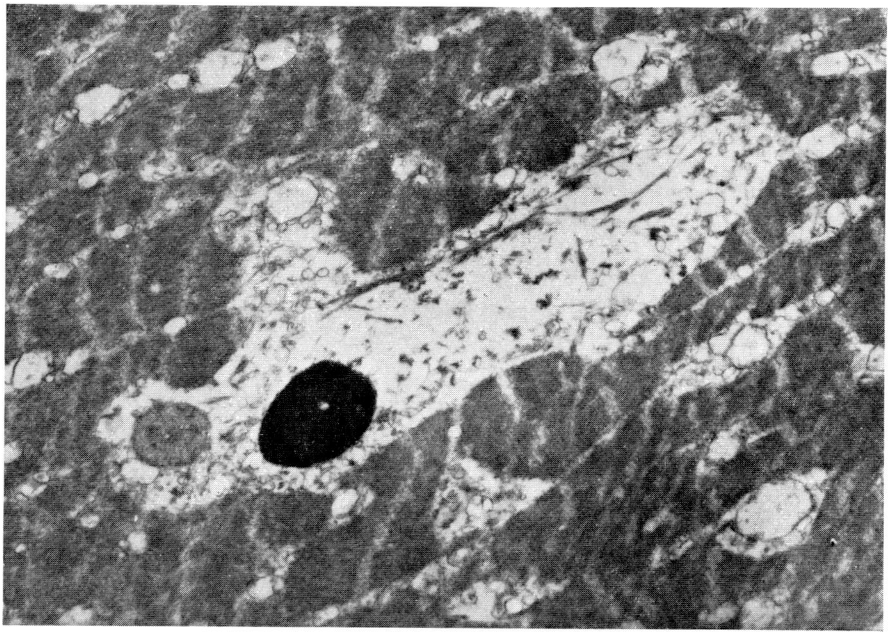

Fig. 31. Advanced case of nuclear degeneration in the frog. This stage is similar to number 4 in Fig. 30. The chromatin has condensed into extremely dense masses, and the nuclear membrane has disappeared. This sample came from a piece of muscle from *Rana catesbeiana* which had been placed into a millipore diffusion chamber and implanted into the peritoneal cavity for 8 days. Histologically the pattern of nuclear degeneration at this time was indistinguishable from that described in Fig. 30. The muscle was fixed in 3 per cent glutaraldehyde in 0.144 N cacodylate buffer, post-fixed in 1 per cent osmium tetroxide, stained en bloc with uranyl acetate, and embedded in Epon. Electron micrograph taken by Gene Trupin. × 19,500.

Starting about 6 or 7 days after implantation of the minced fragments in the frog, we begin to see replacement of the old muscle fibers by newly regenerating tissue. Figure 32 shows a longitudinal section through the proximal part of a 7-day regenerate. Characteristically, the center of a regenerate at this stage contains old muscle fibers which are undergoing certain degenerative changes. On the other hand, the periphery is characterized by the absence of old muscle and by the presence of a great amount of cellular activity. Histologically, the central region of tissue sections is very eosinophilic (even to the naked eye) because of the preponderance of old muscle fibers. The peripheral regenerating area is quite basophilic. It contains both fibroblasts, which are producing connective tissue fibers, and the myogenic elements. Autoradiographs of regenerates at this stage show great numbers of cells incorporating thymidine and uridine.

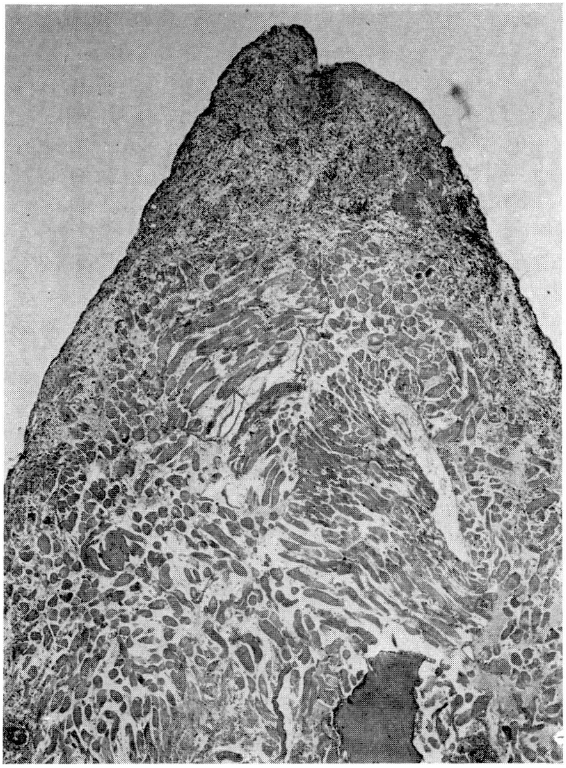

Fig. 32. A 7-day frog regenerate. Low-power view illustrating early regenerative changes occurring at the periphery. In the center are the originally implanted minced muscle fragments. The large homogeneous mass at the bottom of the regenerate is a piece of old dense connective tissue. H and E; × 20. (*from*: Carlson, 1968*a*)

I shall now concentrate upon the regeneration of muscle fibers themselves within this system. The origin of the myogenic cell has not been unequivocally determined, but in my laboratory Trupin (unpublished observations) has studied the early stages of minced muscle regeneration in the frog with the electron microscope. During the first week, 2 major morphological processes occur under the basement membrane of the muscle fiber. One is the degeneration of the sarcoplasm and its removal by phagocytic cells (primarily macrophages). The other is the appearance under the basement membrane of spindle-shaped mononucleated cells with structural characteristics similar to those seen in satellite cells. Trupin noted that the morphology of these cells differs considerably from the invading macrophages as well as from cells normally situated outside the muscle fiber, such as fibroblasts and endothelial cells. Because of the low incidence of satellite cells in normal frog muscle (1 or 2 satellite cells in several hundred nuclei) and the occasional finding of cytoplasmic bridges between these mononuclear cells and the degenerating muscle fiber, Trupin believes that these cells are derived from the muscle fibers themselves. Trupin has also noted a certain degree of phagocytic activity in these cells. At 2 to 3 days after implantation, small areas of these cells partially envelop fragments of degenerating muscle, and by 4 to 5 days, small masses of fibrillar material can be seen within membrane-bounded vesicles in these cells. Additional experiments have shown that these cells take up small clumps of colloidal carbon (Pelikan ink). The phagocytic ability of these cells, however, does not even approach that of the invading macrophages. Early in the second week, these cells begin to fuse under the basement membrane of the old muscle fiber, and shortly thereafter early myofilaments can be seen in their cytoplasm. Light-microscopic autoradiographs at this early stage show that a high percentage of the nuclei at the periphery of the old muscle fibers incorporate H^3-thymidine (Fig. 17), but the identity of these nuclei has not yet been determined at the electron microscopic level.

Figure 33 illustrates the histological association between the newly formed syncytial elements of the regenerating muscle and the old muscle fiber. In frogs this picture is most prominent from 7 to 9 days after the initial operation, and it is usually seen in the transitional

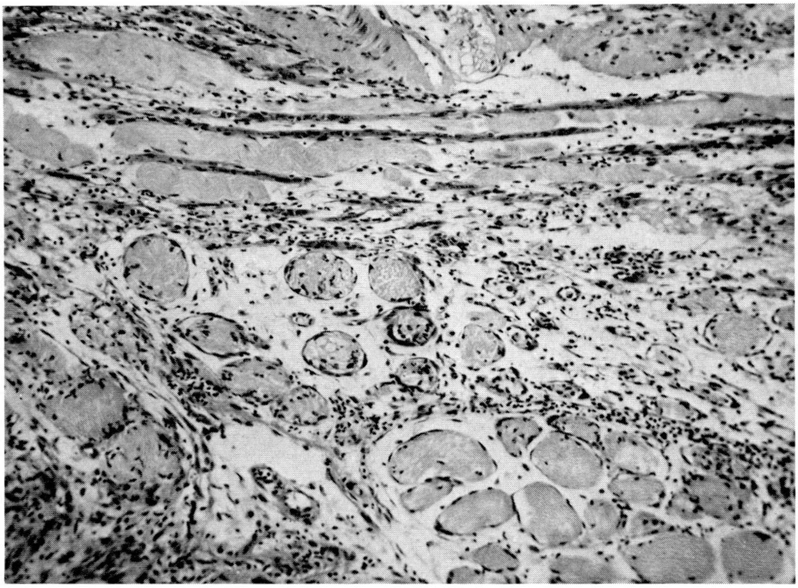

Fig. 33. An 11-day frog regenerate. Low-power view through part of the transitional zone showing the appearance of newly regenerating elements around the periphery of the degenerating fibers of the originally implanted muscle. H and E; × 100. (*from*: Carlson, 1968*a*)

zone between the originally implanted muscle fibers in the center and the peripheral basophilic regenerating zone. In Figure 33 the phenomenon of cuffing of the basophilic myogenic cells around the old muscle fibers can be seen in both cross and longitudinal sections. Figure 34 illustrates at higher power a portion of Figure 33 and shows longitudinal sections of the multinucleated basophilic straps. These structures have been given many names – myotubes, sarcoblasts, myosymplasts, etc. Most of the cells outside the muscle fibers in this figure are fibroblasts, macrophages and mononuclear leukocytes. Within the myotubes (arrows, Fig. 34) is a rather unusual type of cell, seen so far only in frog tissue. In this type of cell, the nucleus is surrounded by a thin rim of cytoplasm. Around these cells are clear 'halos'. The halos appear to be completely surrounded by the sarcoplasm of the myotube. The identity and function of these cells is not known, but they look like lymphocytes. It is possible that these cells were either trapped within the developing muscle fiber or that they wandered into the myotube in a fashion similar to the phenomenon of emperipolesis which has been described for lymphocytes (Humble *et al.*, 1956; Moore and Hlinka, 1969).

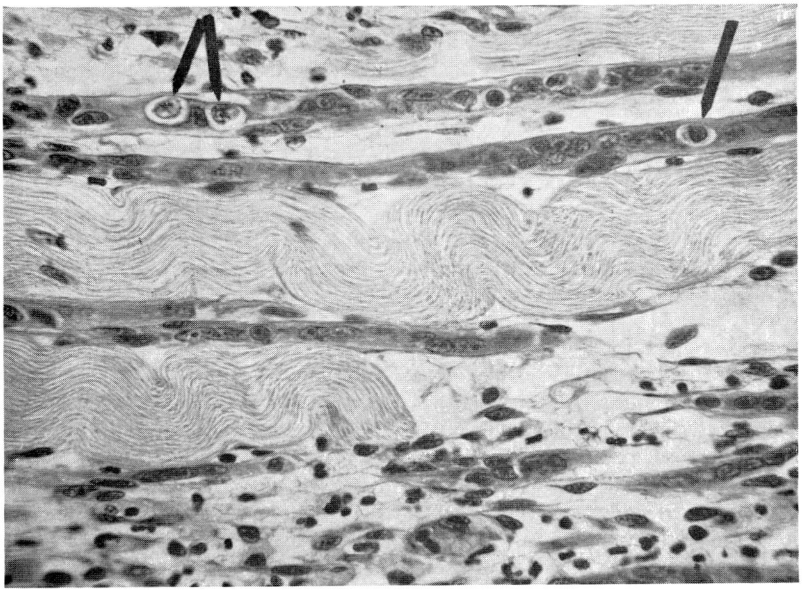

Fig. 34. An 11-day frog regenerate. Higher-power view through Fig. 33 showing the relationship of the newly forming basophilic bands of regenerating muscle to the old muscle fibers. 'Halo' cells are indicated by arrows. H and E; × 400. (*from*: Carlson, 1968*a*)

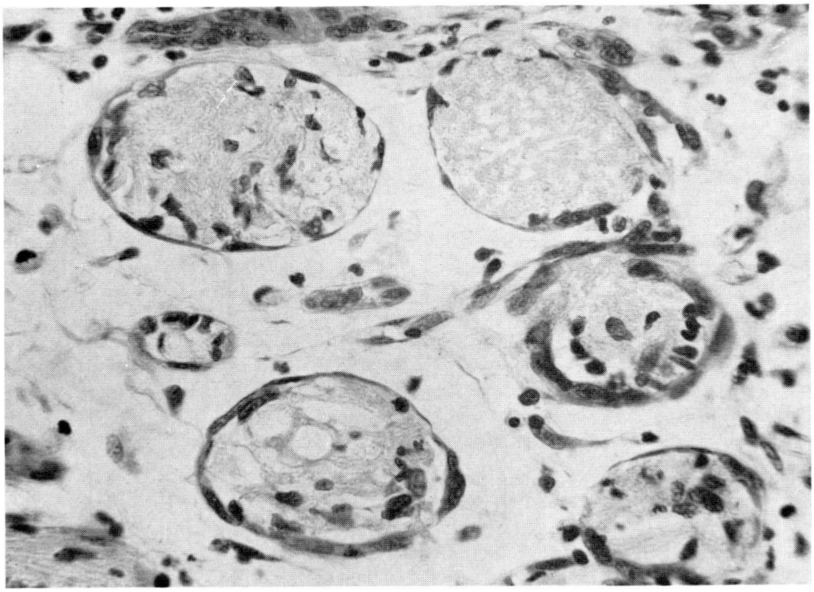

Fig. 35. An 11-day regenerate. Higher-power view through Fig. 33 showing in cross-section the formation of syncytial basophilic cuffs around the degenerating sarcoplasm of the originally implanted muscle. H and E; × 400. (*from*: Carlson, 1968*a*)

In Figure 35, the phenomenon of basophilic cuffing is seen in cross section. The degree of development of the basophilic cuffs is roughly correlated with the extent of degeneration of the sarcoplasm of the original muscle fiber and its removal by macrophages. After the old sarcoplasm is completely removed, the regenerating myotube usually becomes a solid structure which is still located within the basement membrane (Fig. 36).

Further development of the muscle fiber occurs in much the same fashion as that described in mammalian systems, but the rate of development is considerably slower. Figure 37 shows a myotube at 11 days. Its cytoplasm is intensely basophilic, and there is no evidence of myofibrils. By 15 days (Fig. 38) the nuclei are aligned long central chains, and considerable cytoplasmic eosinophilia is present. Myofibrils are visible peripherally, but they are still not seen in the more basophilic interior of the fiber.

By 3 weeks (Fig. 39), cross striations are seen in most of the fibers, and the nuclei begin moving toward the periphery. During the third week, many of the regenerating muscle fibers become vacuolated and the nuclei become pycnotic (Fig. 40). The significance of this is not yet known, but a possible relationship between this and inadequate innervation is presently under investigation. By 50 to 60 days after the initial operation, differentiation of the regenerated muscle fibers is essentially complete (Fig. 41). In most cases the muscle fibers are oriented parallel to one another, but occasionally the orientation is irregular. Figure 42 shows a typical gross regenerate in a frog. In terms of overall size, frogs regenerate more poorly than any animal I have studied so far, but the leisurely pace of regeneration and the simplicity of the system makes it an ideal experimental animal for studying certain aspects of the regenerative process.

A. SANDOW: Before you leave that, do these regenerated muscles contract?

B. M. CARLSON: Most of the frog regenerates I have tested have not. In 2 regenerates definite

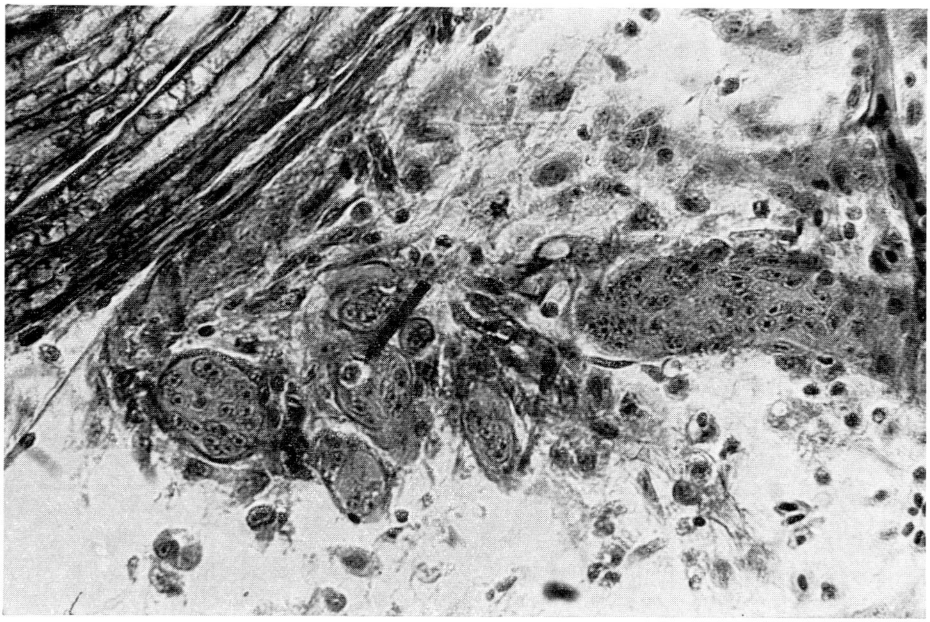

Fig. 36. A 9-day frog regenerate. Cross-section of early myotubes. Each is surrounded by a basement membrane. Halo cell (arrow) is seen within one of the myotubes. The diagonally running elements are portions of tibial nerve. Heidenhain's aniline blue; × 400.

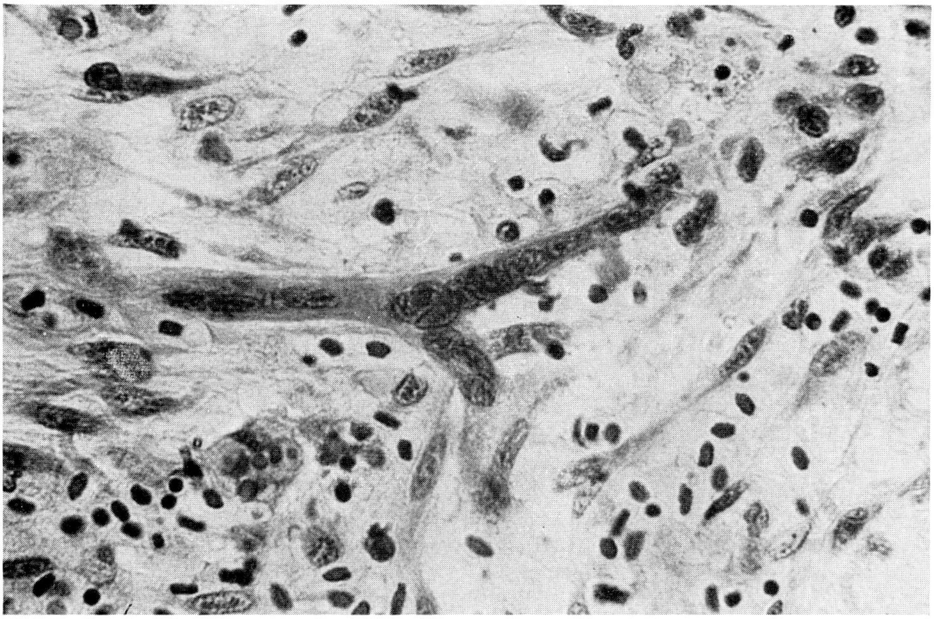

Fig. 37. An 11-day frog regenerate. Early myotubes with central nuclear chains. Myofibrils and cross-striations are not yet distinguishable. Most of the surrounding mononuclear cells are fibroblasts and blood cells. H and E; × 700.

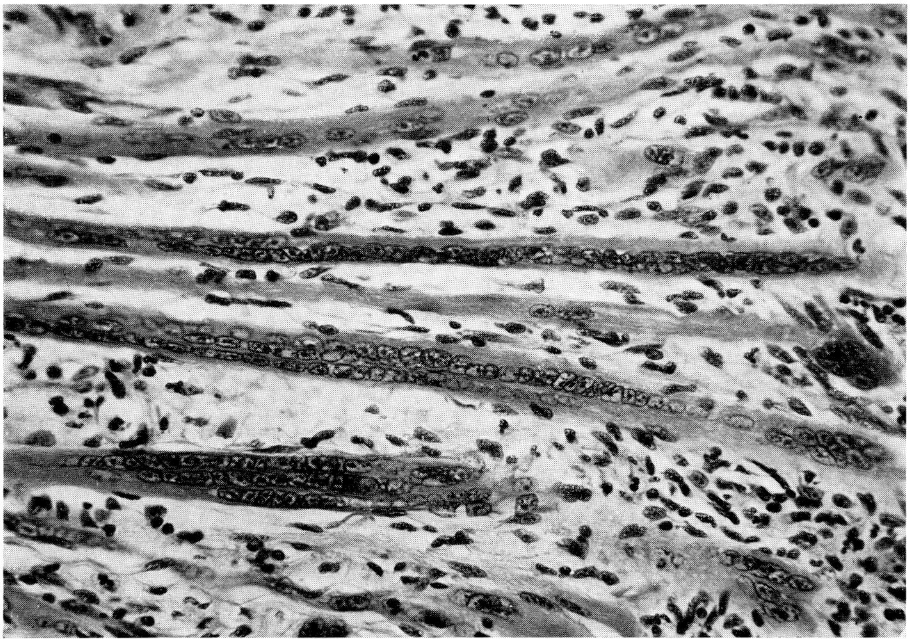

Fig. 38. A 15-day frog regenerate. Myotubes with long nuclear chains. Myofibrils are evident, and the cytoplasmic staining is becoming more eosinophilic. H and E; × 340. (*from*: Carlson, 1968*a*)

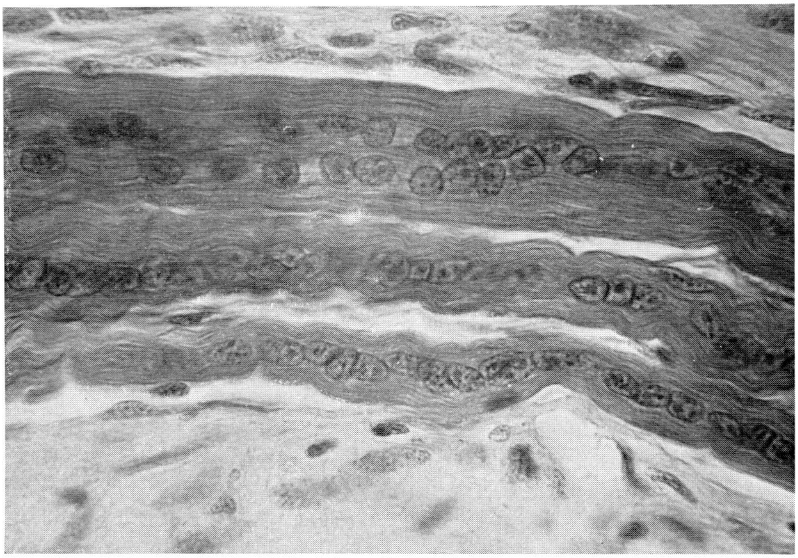

Fig. 39. A 26-day frog regenerate. Stages in the maturation of regenerated muscle fibers showing the appearance of cross-striations and the peripheral migration of nuclei from the central chains. H and E; × 800. (*from*: Carlson, 1968*a*)

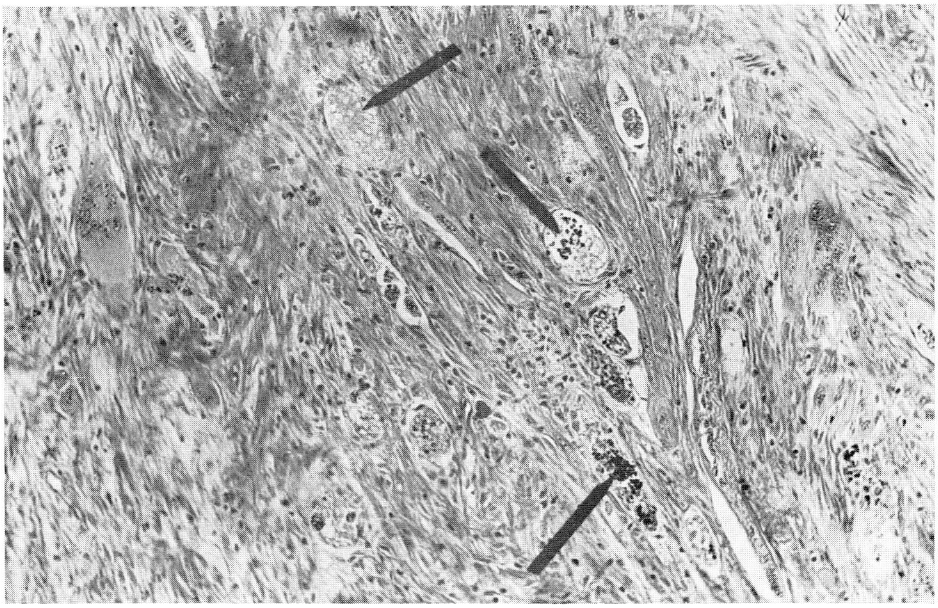

Fig. 40. A 15-day frog regenerate. In this field are numerous degenerating myotubes (arrows) in which vacuolation of the cytoplasm and nuclear pycnosis are occurring. Other normal myotubes are still present. Heidenhain's aniline blue; × 170.

Fig. 41. A 54-day frog regenerate. High-power view of regenerated muscle fibers. Normally fiber orientation is strictly parallel, in contrast to the apparent disorganization in this photomicrograph. Heidenhain's aniline blue; × 400. (*from*: Carlson, 1968*a*)

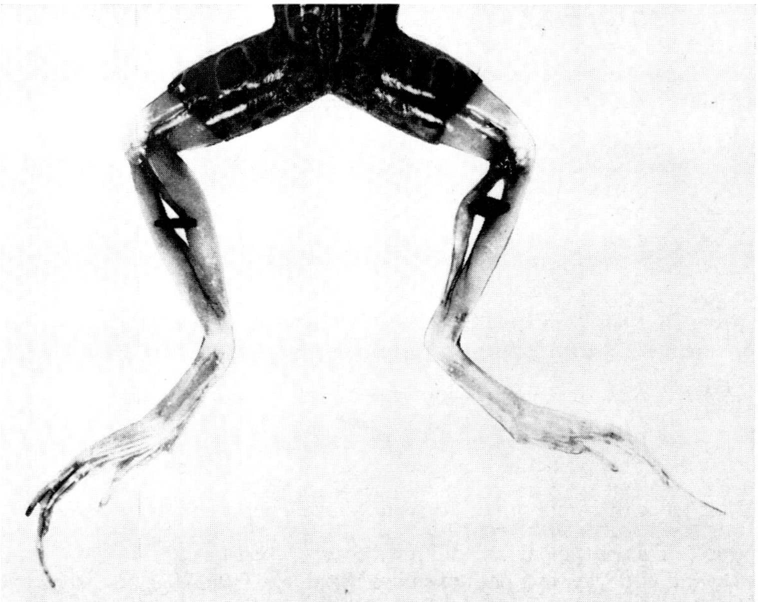

Fig. 42. An 18-day frog. The normal gastrocnemius muscle is on the left; the regenerate is on the right. (*from*: Carlson, 1968*a*)

contractions have been observed following stimulation of the nerve, and 1 regenerate, which had been removed from the animal before I saw it, spontaneously contracted for close to 1 hour. My general impression, however, is that frog regenerates usually possess a poor degree of function, at best. This is due to the large proportion of connective tissue in the regenerates.

D. ALLBROOK: Dr. Carlson, could you tell us if there are any nerves in that muscle?

B. M. CARLSON: I, myself, have not done any specific nerve staining. This has been extensively investigated in Studitsky's laboratory (Studitsky *et al.*, 1963). They have demonstrated the return of motor end plates in rat muscle regenerates.

P. E. CONEN: You said the regenerate was 15 to 25 per cent of the original muscle mass and this is 15 to 20 per cent muscle. Is most of the remainder scar tissue?

B. M. CARLSON: Yes, particularly in the frog. Out of 300 to 400 frogs I have seen 1 or 2 in which the entire regenerate was nothing more than a band of connective tissue. In others at the opposite end of the spectrum actual muscle may make up from 50 to 75 per cent of the mass of the regenerate. In most of the frogs about 30 to 40 per cent of the regenerated tissue is muscle.

D. ALLBROOK: Do you fix your tissue with its 2 attachments intact, or is it dissected free prior to being fixed?

B. M. CARLSON: In most cases the ends of the regenerates are secured to cardboard during fixation, but some are not.

R. BISCHOFF: I wonder if you could make some attempt when you have this muscle minced up and before you put it back, by using enzyme treatments such as collagenase, to remove some of the single cells and then put them back together with the muscle fiber fragments, or just put the muscle fiber fragments back alone.

B. M. CARLSON: That has already been suggested to me, and I think it is a good idea. Perhaps an even better idea would be to use muscle which has been grown in culture from single cells if a sufficient mass of material could be obtained.

P. E. CONEN: Dr. Allbrook, could you comment on that?

D. ALLBROOK: Yes. In our anatomical laboratories in Perth, Western Australia, we have been doing something similar to this. In December last year R. J. Russell spun down trypsinized neonatal and adult rat skeletal muscle. He obtained debris-free basement membrane, external lamina or glycocalyx with adherent cells (Fig. 43). His light- and electron-microscope pictures are not incompatible with satellite cells. We are currently attempting to grow these cell isolates *in vitro*, in diffusion chambers, and *in vivo*. Results more definite than our present hopeful signs should be available in 1970/71.

E. ZAIMIS: When you mentioned continuous contractions, did you mean the muscle fibrillates?

B. M. CARLSON: In this particular case, yes.

D. ALLBROOK: Does that imply that it does not have a nerve supply?

B. M. CARLSON: This particular one did not because when I saw it, it had already been cut free of the leg and was lying in a bath of saline. Studitsky (1963) has published kymographic tracings of contractions of regenerates after stimulating the nerve.

D. ALLBROOK: I have looked for muscle nerve in my experiments with these implant regenerate regenerates in mammals. I find them very difficult to find, though I have no doubt

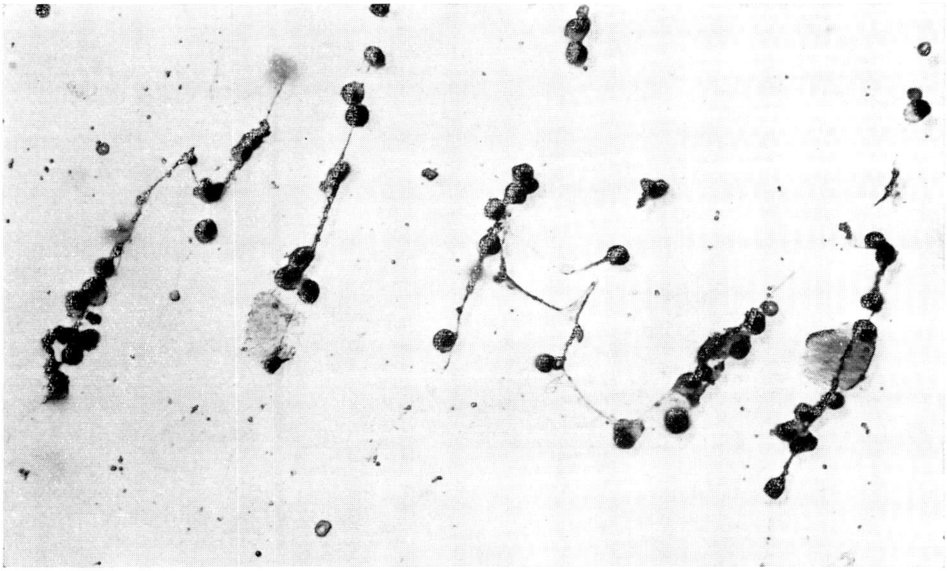

Fig. 43. Trypsinized 1-day-old rat muscle showing glycocalyx fragment and attached cells. Electron microscopy showed the cytoplasm to be free of rough endoplasmic reticulum and consistent with that of a satellite cell.

that some nerve endings are there. My reason for believing this is that in 1951 I studied isotopic autografts of skeletal muscle. Some grafts I implanted into a denervated muscle belly. Regeneration took place within the graft in a similar way as in the experiments reported by Le Gros Clark (1946), but the fibers did not mature (grow in girth). However, when a muscle nerve was implanted either into the graft or near it, the regenerated muscle formed demonstrable myoneural contacts. Furthermore, electromyographic studies demonstrated genuine interference patterns and end-plate potentials on stimulation of the implanted muscle nerve.

B. M. CARLSON: Photomicrographs of nerve terminations in regenerating muscle are found in monographs by Studitsky (1959) and Striganova (1961).

I have a few autoradiographs of minced muscle regeneration in the frog which may be of interest. In studying the frog, I was struck by the fact that the basophilic cuffs form around 'degenerating' muscle cells. I was interested to see what types of activity one sees in the nuclei associated with the degenerating muscle fibers and in the basophilic cuffs. Figure 17 shows a frog regenerate 7 days after the operation. It had received H^3-thymidine 24 hours previously. There is a rather great amount of uptake of isotope by nuclei associated with the old muscle fibers as well as in nuclei of some free cells, but the labeling is most significant in nuclei located at the periphery of the minced muscle fibers.

S. A. SHAFIQ: Could you give some technical details of thymidine administration?

B. M. CARLSON: This animal was given 5 microcuries per gram intraperitoneally at 9 a.m. and was killed 24 hours later.

S. A. SHAFIQ: How long after implantation of the minced muscle was the thymidine given?

B. M. CARLSON: It was injected 7 days postoperatively and the animal was killed at 8 days.

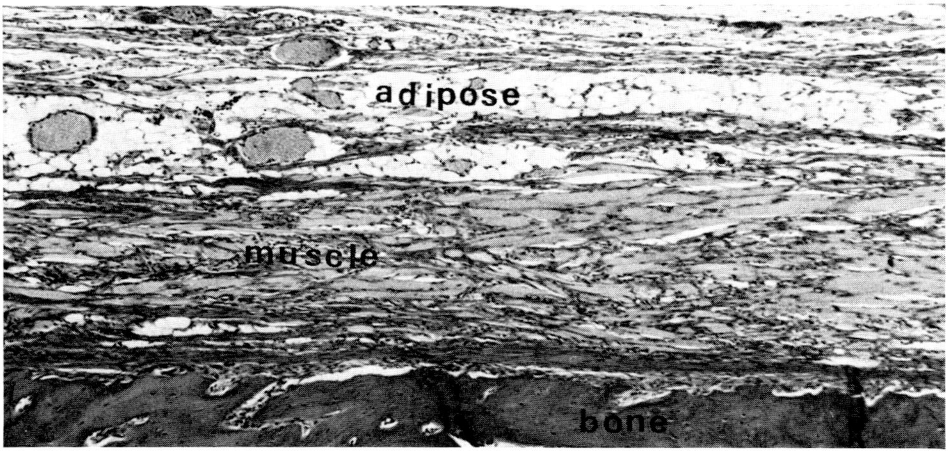

Fig. 44. Rat anterior tibial muscle sectioned into 250 to 50 micron fragments and implanted into isotopic site 5 weeks previously. Fat, veins, and new muscle are in the site of the extirpated muscle.

R. Bischoff: Have you ever labeled earlier to see if there is any uptake of this by degenerating nuclei?

B. M. Carlson: I have, but the slides have not yet been finished.

M. Nameroff: Can a muscle, which has once regenerated under these conditions, regenerate again?

B. M. Carlson: I have not tried this myself. One of my students is working on this now.

M. Reznik: To answer the question of Dr. Nameroff: We have tried to study the regeneration of previously regenerated muscle *in vivo*, but we found it very difficult to analyse. It seems, however, that previously regenerated muscle, for 2 or 3 months at least, does not regenerate again.

B. M. Carlson: I would be very interested to learn if there are satellite cells in regenerated muscle.

D. Allbrook: Yes. I have done some similar experiments, Dr. Carlson. I used rat tibialis anterior, and reimplanted isotopically fragments of regenerated muscle fibers. I also found considerable regeneration (Fig. 44). Electron microscopy clearly shows satellite cells and ordinary myonuclei in the regenerated fibers (Fig. 45).

J. C. T. Church: In the bat web muscles which we have been crushing, there is total destruction of the fiber sarcoplasm without the loss of basement membranes or connective tissue, and one has been able to watch the process of complete restitution of all the elements, together with satellite cells, so that by 4 weeks it looks normal again. The bat web, of which I have pictures (Figs. 78-80), allows accurate localization of lesions, so that one can inflict double injuries, or more. After double injury, we find the same pattern of regeneration, and the same restitution of all elements, including satellite cells. I have also been able to do some quantitation to show this.

B. M. Carlson: Very good. I think you have about the nicest system I have seen in which this can be done accurately.

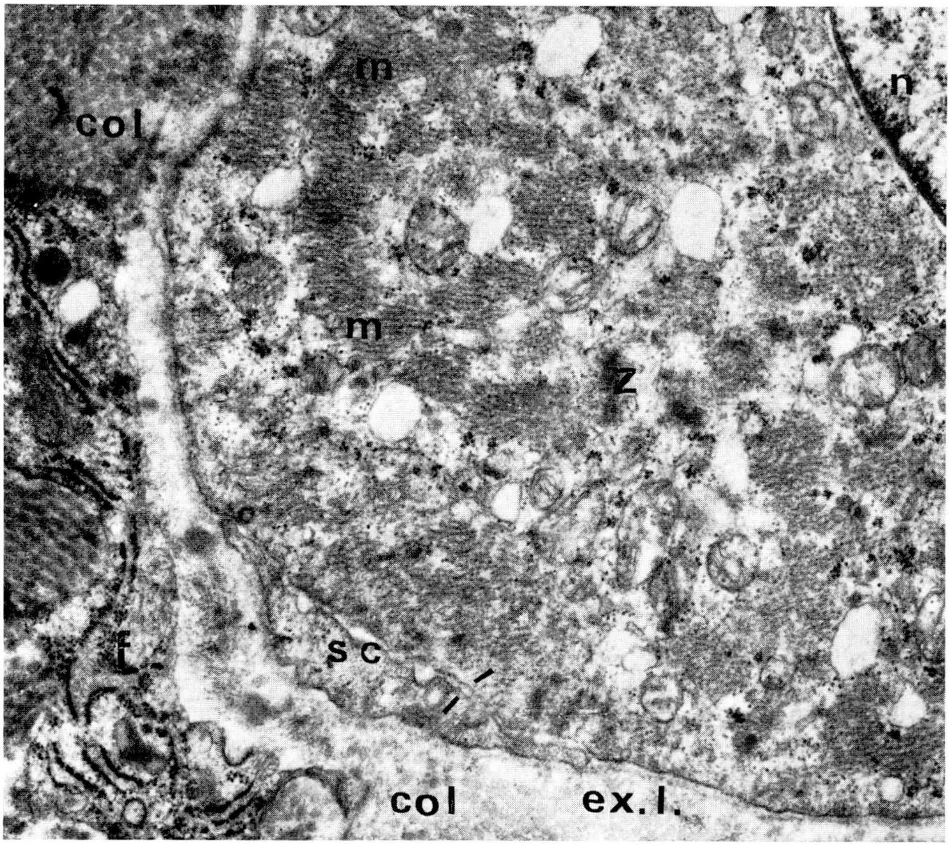

Fig. 45. Transverse section of regenerated muscle fiber 39 days after implantation of 120 micron fragments. Cytoplasm of a satellite cell could be the reason for the appearance marked sc. A fibroblast is on the left (f).

To return to frogs, Figure 46 is a frog regenerate taken 10 days after the initial operation. The frog was injected with 5 microcuries of H^3-uridine per gram and the muscle was fixed 8 hours later. One can see a great amount of uptake around the nuclei at the periphery of the degenerating muscle fibers.

Older regenerates do not show a particularly great uptake of tritiated thymidine except in fibroblasts. No silver grains are found over nuclei within multinucleated stages of regenerating muscle fibers. Figure 47, however, shows silver grains over a high percentage of nuclei located within myotubes when the frog was given H^3-thymidine at 8 days and allowed to regenerate for an additional 6 days.

D. ALLBROOK: So what you are showing, if I can interpret here, is that you have a tremendous number of labeled nuclei at around 7 days, but very few at 19 days.

B. M. CARLSON: Yes. I have seen only 1 or 2 labeled nuclei which I interpret to be muscle nuclei at 20 days. Since there was a 24-hour interval between administration of isotope and fixation, this would allow time for incorporation of a labeled mononucleated cell into the

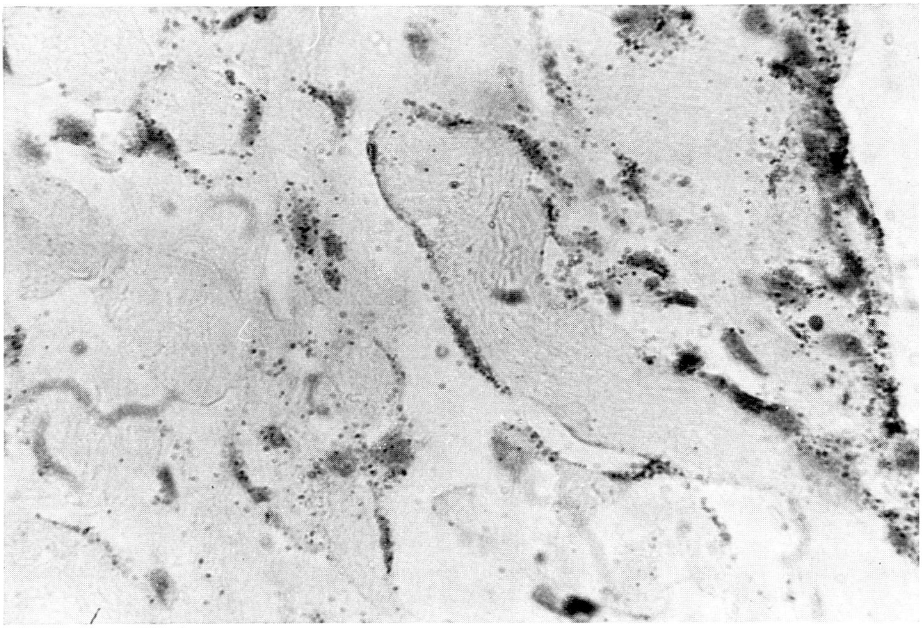

Fig. 46. A 10-day frog regenerate. This animal was injected intraperitoneally with 5 microcuries of H^3-uridine per gram. The muscle was removed 8 hours later. Note the high incorporation of isotope in cells situated at the periphery of the degenerating muscle fiber. H and E; × 700.

muscle fiber. On the other hand, the labeled cells may have actually been satellite cells. This distinction cannot be made with light microscopy.

D. ALLBROOK: I take it that the label is not in muscle, but in fibroblasts.

B. M. CARLSON: The vast majority of labeled nuclei at 20 days belong to fibroblasts, but in minced muscle regenerates in frogs, occasionally early stages of regeneration are seen quite late after the mincing operation. Thus some labeling of muscle precursors would not be unexpected.

I would next like to discuss the morphology of minced muscle regeneration in the rat. In my work I have repeated Studitsky's (Studitsky *et al.*, 1956) procedure of removing the entire jumping complex of muscles (gastrocnemius, soleus and plantaris), mincing them and replacing about 2/3 of the mass into the bed of the removed muscles. The muscle fragments are kept in place by rejoining the previously transected biceps femoris over the proximal half and then suturing the skin incision over that.

The course of regeneration in rats is basically the same as in frogs except that it occurs about twice as quickly and the amount of muscle regenerated is about twice as much.

Even more than in the frog, minced muscle in the rat demonstrates a zonal pattern of regenerative activity. Recent perfusion studies have indicated a close geographic relationship between the ingrowth of the vasculature and the initiation of the regenerative process. Examination of histological sections through a 3-day minced muscle regenerate in a rat (Fig. 14) reveals a tremendous amount of disorganization. Small muscle fragments are scattered in randomly oriented fashion throughout the minced mass of the implant, and at 3 days the majority of them are still completely divorced from a blood supply. A certain amount of regeneration and a corresponding vascular supply is seen in some peripheral

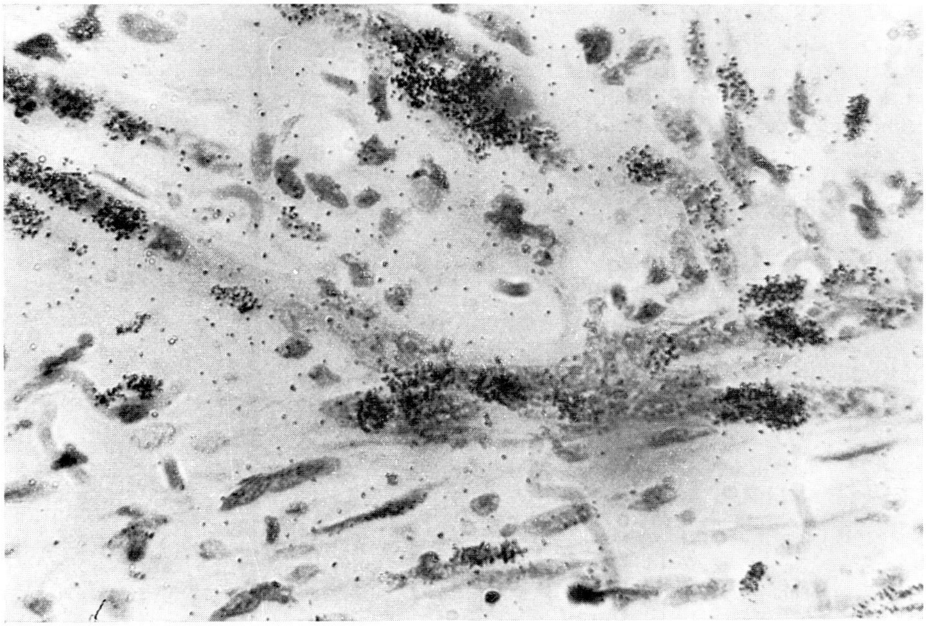

Fig. 47. A 14-day regenerate. This animal was injected intraperitoneally with 5 microcuries of H^3-thymidine at day 8 and killed 6 days later (at 14 days). Many nuclei within myotubes contain large amounts of isotope. Because labeling of nuclei within myotubes at this stage is not seen in short-term labeling experiments, it is assumed that labeled myogenic cells become incorporated into the myotubes. H and E; $\times$ 700.

areas. Closer examination of a regenerate at this stage reveals a very inactive central zone (Fig. 15), in which the predominant feature is the originally implanted muscle fibers with a few inflammatory cells scattered among them. The most striking observation about this zone is the paucity of demonstrable nuclei associated with the old muscle. It is not uncommon to see fields of from 10 to 15 cross-sectioned muscle fibers with no nuclei visible at their periphery. Yet, farther toward the periphery of the regenerate as a whole, in what I call the transitional zone (Fig. 16), it is not uncommon to see fields of cross-sectioned muscle fibers, each of which is surrounded by a syncytial cuff of cells containing many nuclei. The prominent overall features of the transitional zone are (1) the first appearance of cells which can definitely be considered to be early stages of regenerating muscle, (2) the breaking up and disappearance of the sarcoplasm on the originally implanted muscle fibers (normally associated with invasion of the degenerating fibers by macrophages) and (3) the ingrowth of vasculature in the outer (but not the inner) portions of this zone. Even in a 3-day regenerate the peripheral zone contains multinucleated myotubes. These are normally intimately associated with blood vessels (Fig. 19), no matter what the age of the regenerate. Other characteristics of the peripheral zone are a complete absence of the originally implanted muscle fibers and considerable fibroblastic activity leading to the formation of connective tissue.

The ability of minced muscle fragments to survive for several days in the absence of a direct blood supply and the relationship between vascular ingrowth and the initiation of histologically demonstrable regeneration are merely superficial indications of an extremely important, but virtually unexplored field, namely the metabolic adaptations enabling severely traumatized muscle to survive and later repair itself. Several enigmatic features seen in these early regenerates require explanation. One concerns the breaking up of the sarcoplasm of the old muscle fibers. When they are far removed from a blood supply, hyaline and vacuolar degeneration of the sarcoplasm is seen, but not until blood vessels have penetrated within a couple of hundred microns of a given muscle fiber does it begin to fragment. This is almost always associated with the presence of macrophages within the sarcoplasm. Are macrophages required for the removal of the degenerating sarcoplasm? Allbrook *et al.* (1966) favor this viewpoint. Closely associated with the breakdown of the sarcoplasm of the old muscle fibers is the appearance of the syncytial cuffs of basophilic cells inside the basement membrane of the degenerating muscle fiber. This phenomenon brings up 2 major questions. First, where do the nuclei which form the cuffs come from? It should be recalled that nuclei are not numerous around the old muscle fibers in the central region of the regenerate. The second question is: what stimulates the formation of the basophilic cuffs? Is it the disappearance of the sarcoplasm of the original muscle fibers (basophilic cuffs are almost never seen around intact fibers) or is there some relationship to the ingrowing vasculature (a possible relationship to increased supply of oxygen and metabolites or even to cells in the blood)? It seems likely that both may be involved.

P. E. CONEN: The inference that you have a vascular supply is because the metabolism of these cells continues.

B. M. CARLSON: A considerable degree of metabolic activity appears to be present even in the absence of a demonstrable blood supply according to autoradiographic and histochemical work being done in my laboratory, but the actual histological evidence of a progressive phase in regeneration is only seen when there are blood vessels within a few hundred microns of the regenerating tissues.

P. E. CONEN: I wonder what would happen if you did a homograft of the entire muscle.

B. M. CARLSON: I have not transplanted any entire muscles from one animal to another. In the case of homografts of minced muscle, the central avascular region is never invaded by lymphocytes whereas after the first week, rather extensive fields of small lymphocytes are seen in the vicinity of the regenerating muscle fibers. Thus there is a definite difference in the reaction of the host to homografted muscle as compared with autografted muscle.

P. E. CONEN: The reason I ask this question is that many people now believe that cartilage is not normally "known" to the body cell, but if cartilage is scraped free into the knee joint, then cartilage antigens are accessible. Antibodies to them may then be produced, there are increased numbers of pyronine-positive cells in the enlarged popliteal lymph nodes, and the animals get arthritis.

I wonder whether your changes are due to purely vascular problems if there were any added immunologic process.

B. M. CARLSON: I am inclined to suspect that immunologic processes play a small role in autografted material, despite their apparent role in the case of damaged cartilage.

J. C. SLOPER: Are you using a genetically homogeneous strain for these types of grafts?

B. M. CARLSON: Most of my autografting experiments have been done upon Wistar rats from Albino Farms, Red Bank, New Jersey. I have performed homografts between littermates of this strain and between non-littermate Sprague-Dawley rats.

J. C. SLOPER: Do you see any evidence of graft rejection as you might expect using other systems?

B. M. CARLSON: In autografts or in homografts?

J. C. SLOPER: Both.

B. M. CARLSON: Very little in autografts. In some early autograft regenerates I see occasional mononuclear cells scattered among the early myotubes, but these are larger than the typical small lymphocytes taking part in graft rejection processes. In homografts, regeneration seems to proceed quite normally for about the first 5 or 6 days, with little evidence of rejection. Then after this, a rather substantial influx of lymphocytes is seen in association with the newly regenerating muscle fibers. This process is described more fully for the rat in Carlson (1970*a*). Work in progress in my laboratory indicates that in the frog, regenerates arising from homografted minced muscle are similarly attacked by small lymphocytes (Hall, unpublished data).

J. C. SLOPER: Do I understand that you do not see any difference in these (basophilic) cuffs in homografts and autografts?

B. M. CARLSON: No. In rats I can detect no difference between them until about 7 days.

L. P. ROWLAND: If you put the mince back somewhere else than in the muscle bed, what happens?

B. M. CARLSON: I have tried placing minced pectoral muscles into the tongues of rats and have gotten a small amount of regeneration. In the dog, Ignatieva (1960) has implanted minced muscle taken from the forelimb into subcutaneous pockets on the thorax and back. Although rather extensive regeneration was described for the first 2 weeks, the regenerating muscle was eventually replaced by connective tissue. No innervation of the regenerating muscle was observed.

D. ALLBROOK: Can you tell what happens if you take out the popliteal gland before homografting; do you get continued growth or does the host just go on rejecting the implant?

B. M. CARLSON: I have not done this.

D. ALLBROOK: This seems to me a critical experiment to do. I do not know what Dr. Holtzer thinks about that.

B. M. CARLSON: To date I have only implanted minced muscle from one animal into another. I have not yet done any additional work in this area.

D. ALLBROOK: I think this is just fascinating, this facet.

B. M. CARLSON: I have also tried transplanting muscles from frogs to rats, rats to frogs, frogs to axolotls, etc., and have not yet seen any regenerative responses. This work is very preliminary, however, and I can not make any final comment on it.

D. ALLBROOK: How fine is the mincing?

B. M. CARLSON: I have done most of my work with minced fragments about 1 millimeter in diameter. I have not yet tried homogenates. In recent work I have implanted entire muscles (autografts) or pieces about $5 \times 5 \times 10$ millimeters. The vast majority of this latter material develops an acute inflammatory response with great numbers of polymorphonuclear leukocytes around the muscle fibers. In isolated peripheral areas small fields of regenerating muscle are seen.

D. ALLBROOK: Has Studitsky gone any further than this?

B. M. CARLSON: No, he has not, as far as I am aware.

D. ALLBROOK: Have you tried control experiments, *e.g.* putting gelatine in place of muscle implant?

B. M. CARLSON: I have tried a number of controls. In the frog, if you remove a muscle and replace nothing, no structure will form. If you replace the muscle with Gelfoam, for instance, you apparently provide a substrate for the migration of fibroblasts, and frequently after this treatment a thin band of connective tissue will be seen between the Achilles tendon and the proximal attachments. After removal of the jumping complex of muscles in the rat without any replacement, no muscle is seen, but the Achilles tendon regenerates. This tendon does not attach to the proximal muscle attachments, but rather extends 5 or 6 millimeters proximally and attaches to the underlying tissues.

D. ALLBROOK: Do you see any muscle at all?

B. M. CARLSON: Absolutely none. I have never seen muscle in a control.

D. ALLBROOK: I did something similar in 1951, and put Gelfoam into the space left after pieces of muscle had been excised. There was no regeneration of fibers into it. The Gelfoam was contracted down as scar tissue. There were a few small islands of muscle giant cells and a few isolated fibers (Fig. 48).

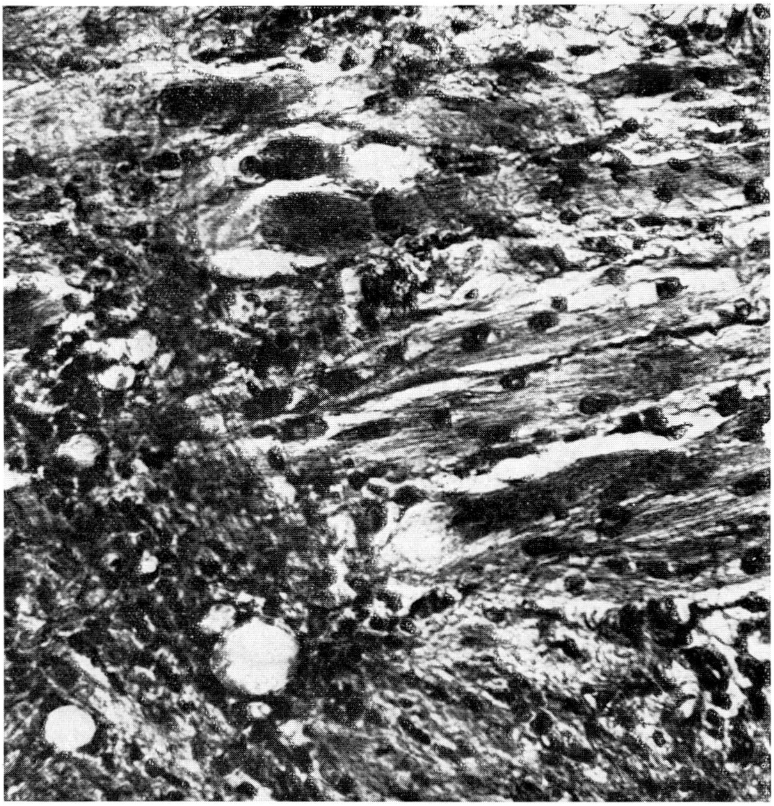

Fig. 48. Edge of 'Gelfoam' implant with muscle fibers ending at muscle/implant interface, 3 months after implantation of 'Gelfoam'.

B. M. CARLSON: Giant cells will often appear, especially in homografts, but you do not often see these in autografts.

Now to return to my description of regeneration in the rat, from 4 to 10 days there is a

Fig. 49. A 17-day rat regenerate. Originally implanted muscle fibers are completely replaced by newly regenerated muscle and connective tissue. H and E; × 20. (*from*: Carlson, 1968*a*)

progressive increase in thickness of the peripheral zone of the regenerate and a decrease in size of the eosinophilic central area as the old muscle fibers are replaced by regenerating tissue. By 10 days one seldom sees any differentiated elements of old muscle left. The regenerate consists entirely of connective tissue and new muscle fibers. Figure 49 is a low-power view through a 17-day regenerate showing complete replacement of the originally implanted muscle fibers by new tissue. Figure 50 illustrates a regenerate taken 28 days after

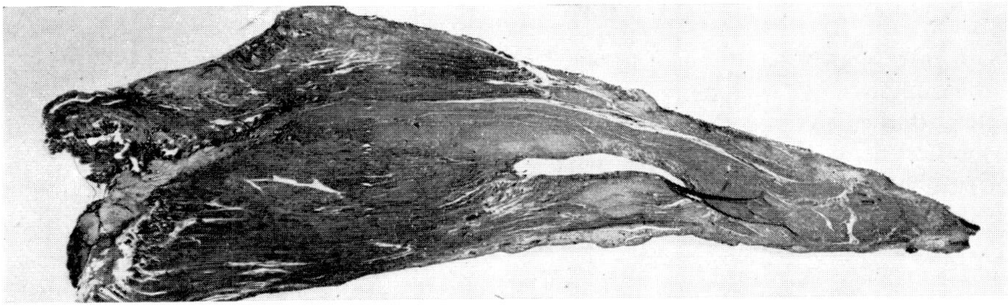

Fig. 50. A 28-day rat regenerate. The right-hand portion of the regenerate is entirely connective tissue (Achilles tendon), whereas most of the larger half (proximal part) is composed of muscle. Heidenhain's aniline blue; × 7.5. (*from*: Carlson, 1968*a*)

the operation. It contains a high percentage of muscle proximally, and the distal part is entirely tendon. A higher-power view of this muscle is illustrated in Figure 25. In most rats about 50 per cent of the mass of a regenerate is muscle. Of course the Achilles tendon is completely connective tissue. In about 95 per cent of regenerates over 30 days old, nodules of bone and cartilage are seen in the regenerated tendon (Fig. 51).

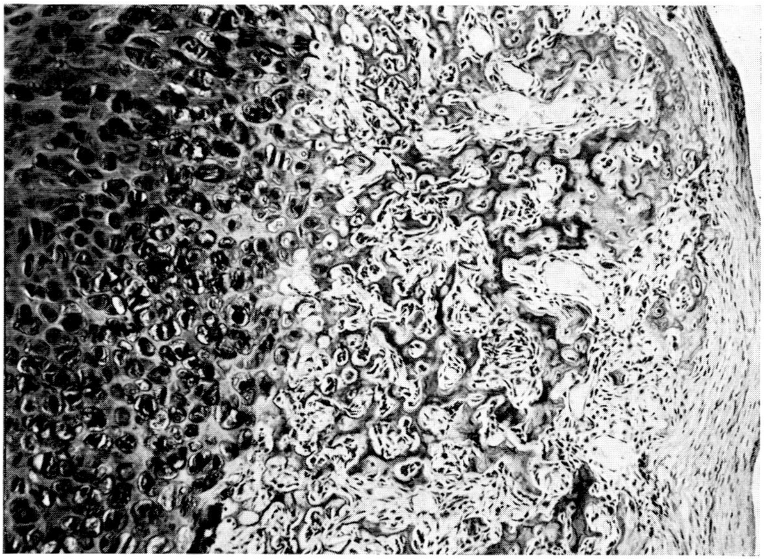

Fig. 51. A 25-day rat regenerate. A portion of a nodule containing hyaline cartilage and cancellous bone found in the Achilles tendon of the regenerate. H and E; × 250 (approximately).

D. ALLBROOK: It is a very much better 30-day regenerate than I have seen in my anterior tibial muscle material.

B. M. CARLSON: I would say that about 25 per cent of the cases will look like this; 50 per cent will not be so good and the last 25 per cent will contain a considerable amount of connective tissue with scattered bands of muscle within the total regenerate.

D. ALLBROOK: You might have some nerve in there. Would it be possible to leave the nerve *in situ* and intact?

B. M. CARLSON: My handling of nerves is done as follows: There are 3 nerves associated with the muscle – the tibial nerve which sends branches into the muscles, the peroneal nerve passing over the proximal end of the muscle and the sural nerve which runs along the surface of the gastrocnemius muscle. When I remove the muscle, I am quite careful to free as many small branches of the tibial nerve as I can going into the muscle. These then lie free in the cavity left by the removal of the muscle. Presumably the regenerating muscle becomes innervated through these branches.

D. ALLBROOK: The nerve to the medial head of the gastrocnemius in a rat is a long fine entirely motor nerve, and you could do a beautiful controlled study on the influence of this particular nerve on that particular regenerating muscle.

B. M. CARLSON: You really could. I would refer you to a paper by Zhenevskaya (1960), who demonstrated that it is the motor supply to muscle which is apparently the important factor in maintaining muscle regeneration. This is in distinct contrast to the non-specificity of the nerve fiber requirements in the early regenerating amphibian limb (Singer, 1952). I strongly suspect, however, that the requirement for nerve fibers in the early regenerating limb is a different type of phenomenon from that required for the regeneration of minced muscle or the differentiation of muscle in the later stages of the regenerating limb.

P. E. CONEN: Can this be done on the medial and lateral heads of the gastrocnemius, which I believe are of different fiber types?

B. M. CARLSON: I have done nothing along this line. Dr. Allbrook, have you done anything on fiber typing in this respect?

D. ALLBROOK: No.

B. M. CARLSON: Could you explain, Dr. Conen?

P. E. CONEN: If the regenerate experiments were done on 2 different muscles, one of fiber type I (red, mitochondrial-rich) and the other of fiber type II (white, mitochondrial-poor), you could determine which fiber type regenerated best, or at least survived. Such experiments would also allow exchange of minced muscle such that type I minced muscle would be placed in the muscle bed and nerve supply of an excised type II muscle and *vice versa*.

M. ALOISI: I would like to know your opinion about the factors which control the orientation of the myoblasts and the new fibers. I performed the same kind of experiments and I became interested in the way in which the regenerating muscle acquired the correct orientation. If you disturb the maturation of collagen tissue with lathyrogenic substances, the regenerated muscle fibers are no longer oriented properly.

B. M. CARLSON: In the early stages of regeneration, the muscle fibers are almost randomly oriented. The direction of orientation of early myotubes appears to correspond with the orientation of the originally implanted muscle fibers. But in time (usually during the second and third week) the muscle fibers become oriented parallel to the long axis of the limb. It is interesting that there is a close correlation between the 'straightening out' of the regenerating muscle fibers and their acquisition of cross striations. I suspect that this realignment of the interacting muscle fibers is mediated by the mechanical tensions exerted upon the regenerate by the functioning limb. Current work in my laboratory indicates that immobilization of a rat's leg by a plaster cast delays the realignment of the muscle fibers.

F. BOWDEN: What is the functional state of this material? How does it compare with the frog?

B. M. CARLSON: I have not tested this at all. Studitsky (1963) has published kymographic tracings of the contraction of mammalian minced muscle regenerates. From my experiments I would say that, practically speaking, such regenerated muscles in the rat are not functional because the regenerated muscle is rather tightly bound down to the underlying tissues by connective tissue adhesions.

D. ALLBROOK: I have once seen localized contraction on stimulation of the peroneal nerve in my rat anterior tibial muscle implant experiment.

A. SANDOW: Has anybody looked at these with the electron microscope?

D. ALLBROOK: Yes, I have. The electron-microscopic studies done by me show well-developed muscle fibers 39 days after implantation of fiber fragments 120 microns in length. The regenerated fibers show myofibrils composed of thick, thin, and intermediate filaments, and a well-organized sarcotubular system. The fibers are like those of a newborn rat without tight packing of myofibrils (Fig. 52).

A. SANDOW: Is this sarcoplasm particularly unusual? May I ask another question? You mentioned the difference in speed in regeneration in relation to the rat and the frog. Is this a species or a temperature difference?

B. M. CARLSON: I have not yet been able to determine this. So far, attempts to obtain muscle regeneration in transplants between homothermic and poikilothermic animals have failed.

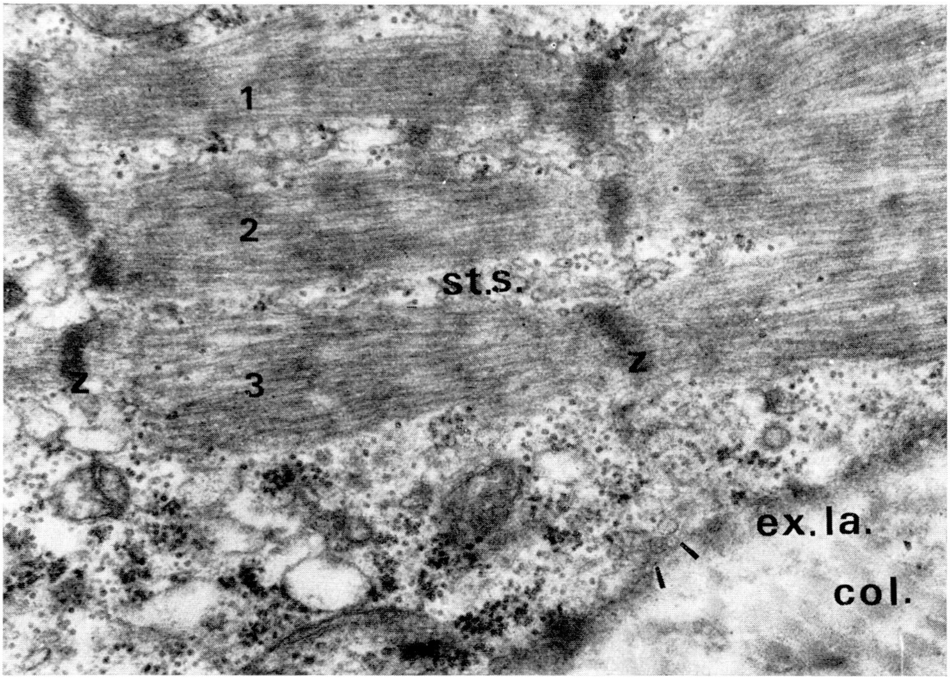

Fig. 52. Fragments of 120 microns 39 days after implantation; 3 myofibrils along 2 sarcomere segments. Note the sarcotubular system related to Z-material and its relation to the plasmalemma and external lamina and collagen fibrils.

M. Aloisi: I would emphasize that there are other, more important differences, species differences. If, for instance, you follow diffusion chamber experiments, you can appreciate such species differences. Rat muscles grow quite well in rat hosts and to a lesser degree in guinea pig hosts. On the contrary guinea pig muscles do not grow as well either in an autologous peritoneal cavity or in rat hosts. So, here it is not a question of temperature, but of a species difference in the capacity of muscle cells to grow and regenerate.

B. M. Carlson: Since guinea pigs were mentioned, Studitsky (1963) finds that minced guinea pig muscle will not regenerate unless thyroidin is placed in their food.

To finish my presentation on regeneration in the rat, Figure 13 illustrates a typical regenerate. The tendinous connections and gross relationships with nerves and blood vessels approximate the normal condition. I must again stress that the regenerate has been dissected free of the surrounding tissues. An undissected regenerate is normally tightly bound to the underlying tissues.

D. Allbrook: In our mammalian studies, the regenerated muscle is usually non-functional because it is completely bound down by fibrous tissue. Unless one does a nice artificial dissection, like you have done, the regenerate is rather undramatic.

B. M. Carlson: That is very true. But it is interesting that in the frog a regenerated gastrocnemius muscle is not bound down as it is in the rat. It therefore makes a much cleaner system for certain types of analysis because one can be assured of absolutely no contamination of the regenerate by neighboring muscles. The main problem with the frog is the relatively small size and low content of muscle fibers of the regenerate. In the 2-week frog regenerate there are large numbers of muscle fibers, but most of them become vacuolated and disappear. We are currently investigating the correlation between this condition and a lack of innervation.

P. E. CONEN: There was a suggestion in the literature that a minute lesion in muscle could repair without scar tissue, but if it were a larger lesion there was repair by connective tissue. The connective tissue repair of a larger lesion could be avoided by adding some minced muscle to the lesion to achieve repair with muscle. Today, you have described your whole muscle experiments. Have you done any smaller replacements to try to heal smaller defects?

B. M. CARLSON: I have tried some experiments in which smaller defects were produced, but not so small as I suspect you have in mind. In early work I removed the distal half of the gastrocnemius muscle in the frog, replaced it with minced muscle and got quite poor regeneration. The filling of defects in muscles by minced muscle in the dog has been reported by Studitsky and Ignatieva (1961). Regeneration occurred within the defect, but replacement was not perfect.

D. ALLBROOK: The answer, Dr. Carlson, is apparently that if the muscle has intact endomysial tubes, all will be well. But if you have smashed up the tubes, it is a geometric problem that the muscle has to solve as well as the fibroblast problem.

B. M. CARLSON: This seems to be true. Experiments in which muscles have been shown to regenerate after interruption of the vascular supply demonstrate this quite well.

P. E. CONEN: You are going to need an endomysial plastic surgeon then.

D. ALLBROOK: The devascularization of muscle studies done by Le Gros Clark and Bloomfield (1945) and repeated by me (Allbrook and Aitken, 1951) in relation to reinnervation of the regenerated fibers shows that so long as the endomysial tubes are intact excellent regeneration resulted. Electron microscopy showed the importance of the basement membrane, external lamina or glycocalyx portion of the supporting tissue (Allbrook, 1962).

I think the pathologists here will support the assertion that if there is a toxic necrosis of muscle fiber, this is followed by a perfect fiber regeneration. I found the same thing in experimental viral myositis (Simbis virus) in mice. In both toxic necrosis and simple myositis, recovery is shown by a practically normal muscle histology. And yet one knew that the muscle was quite severely damaged a matter of 2 or 3 weeks before. Our experience with experimental nutritional myopathy in Perth is very similar. Kakulas showed a similar cycle of recovery in the Rottnest quokka. Only the fiber was affected. Perhaps the most elegant experimental demonstration of the importance of an intact endomysial tube and external lamina (basement membrane) for fiber reconstitution were the bat wing experiments we did at Makerere College in Uganda and which Dr. Church will be showing us later.

J. C. SLOPER: I can say, as a pathologist, that you are all rather speculating. Actual evidence of the fact that survival of the basement membrane is necessary for regeneration is still lacking, as far as I know. I think everybody agrees that you get good regeneration without basement membranes, but the point that Dr. Allbrook is making is a good one, and maybe it is much better if they are there.

A. R. MUIR: The early work of Le Gros Clark (1946) of turning a piece of the gracilis muscle through 90°, and then showing that the regeneration in that isolated portion also turned through 90° does emphasize the importance of the connective tissue framework.

G. PALADE: In the Studitsky-Carlson type of muscle regeneration, the cellular and fibrillar elements of the connective tissue of the minced muscle should be less affected than the muscle fibers (at the beginning and during the first days of the process). The surviving elements,

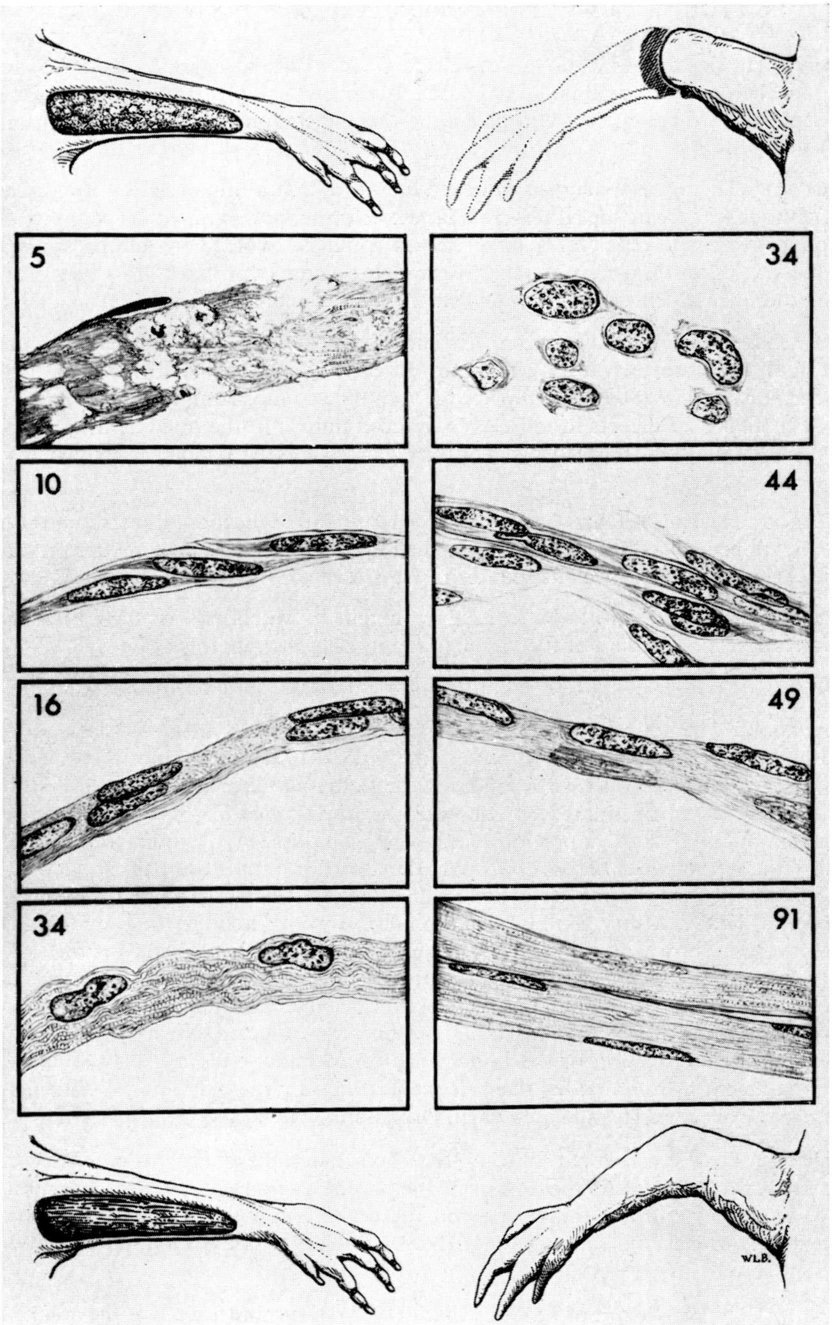

Fig. 53. Comparison of the course of muscle regeneration after mincing and after amputation in the mature axolotl. The left-hand column summarizes the morphology and time course of the principal cellular events occurring after the mincing of a muscle. The right-hand column summarizes the pattern of formation of muscle fibers in a regenerating limb. The numbers refer to days after mincing or amputation. (*from*: Carlson, 1970*c*)

i.e. basement membranes and collagen fibers, may act then as supporting and directing devices during subsequent muscle fiber regeneration, thereby accelerating the whole process. Since collagen fibers respond to mechanical forces by alignment along stress lines, regeneration in the presence of normal collagen should lead to a secondary regular alignment of muscle fibers which could mimic in part normal development. Collagen alignment may be defective in lathyritic animals in which, as noted by Dr. Aloisi, there is no 'proper' orientation of muscle fibers in the regenerating muscle.

B. M. CARLSON: That is a good system. This is related to observations which I have made on the regenerating Achilles tendon. Although the initially implanted mass of muscle is homogeneous throughout, the great proliferation of connective tissue in the area of the Achilles tendon actually seems to pinch off the regenerating muscle fibers.

E. ZAIMIS: Dr. Carlson, where do you stop the circulation?

B. M. CARLSON: Circulation is interrupted where the blood vessels enter the muscle. Hemostasis is obtained by pressure on the cut vessels. As a rule there is little bleeding during the operation.

J. ZELENA: Did I understand that what you really remove in the rat is not the gastrocnemius muscle but the gastrocnemius, plantaris and soleus muscles – all 3 of them? And what you get as a regenerate is only a 'bag' of muscle just fused together? Do you differentiate some individual muscles in it now? Is it not true that Studitsky claims that the more you remove the better the regeneration? Is it so that if you remove only one muscle you just get hypertrophy of the others?

B. M. CARLSON: In the rat, I remove 3 muscles, but the regenerate is a single structure. However, there are usually fairly prominent bands of connective tissue joining the proximal origins of all removed muscles to the regenerate.

With regard to hypertrophy, there is considerable hypertrophy of the biceps femoris muscle overlying a minced muscle regenerate. In the axolotl, mincing of the puboischiotibialis (p.i.t.) muscle causes obvious hypertrophy of the pubotibialis muscle, but yet in time individual muscle fibers in the regenerates attain normal size.

I would next like to describe regeneration of the puboischiotibialis muscle in the axolotl. This is a large strap-like muscle situated on the ventral surface of the upper segment of the hind limb. Sections taken at the fifth postoperative day show a predominance of degenerating old muscle fibers with blood cells scattered among them. Intensely basophilic spindle-shaped cells can be seen at the periphery of some of the degenerating muscle fibers. By 10 days (Fig. 26) the myoblasts have begun to fuse into myotubes. In contrast to frogs and rats, I have seen relatively few prominent basophilic cuffs in the axolotl, but occasionally they do occur. Myofibrils are seen at 15 days, and within a month (Fig. 27) one can find cross-striated muscle fibers. After the first month, maturation of the regenerated muscle fibers continues until they closely resemble normal ones (Fig. 28). Muscle regenerates in the axolotl always have more connective tissue than normal muscles, but they are closer to the original muscle than are frog regenerates. Additional details of this regenerative process are presented in Carlson (1970*b*).

The similarity of minced muscle regeneration in the axolotl to that seen in other forms is of particular interest because it indicates that in this species a given muscle can be regenerated by 2 different means. Dr. Hay (this volume, pp.3-24) has already discussed the formation of muscle from the blastema of the regenerating limb (epimorphic regeneration). The experiment outlined above shows that the same muscle can regenerate as a tissue (tissue regeneration). Of great interest to me is the time relationship between these processes. Figure 53 compares the time course and morphology of muscle fibers regenerating by the tissue and epimorphic modes. The left-hand side depicts the course of regeneration after mincing and summarizes the findings reported above. The right-hand side illustrates the sequence of appearance of muscle in the regenerating limb. By 34 days a well-developed blastema composed of a homogeneous-appearing population of cells is present in the limb, and no cells which could be called muscle are recognizable distal to the amputation surface. In another 10 days the regenerate is elongating and next to the cartilaginous skeletal primordia are groups of fusiform or spindle-shaped cells with very basophilic cytoplasm. They are all aligned parallel to one another and are very close together. I should stress that they often have no morphological continuity with the more proximal elements of the limb. I am assuming (and, I stress, only assuming) that these spindle-shaped cells arose from blastema cells looking like those depicted at the top of the right-hand column of Figure 53.

Cross striations appear by 49 days, and by 3 months the regenerate appears normal.

Thus there seem to me to be 2 basic types of regeneration of muscles, at least at the level on which I am working. There is the tissue regeneration of muscle, which has been found in all classes of vertebrates investigated up to now, and also the epimorphic regeneration of muscles seen during the postamputational regeneration of a limb. Further discussion of this point is given in the formal paper in this volume (pp. 25-37) and in Carlson (1970*b*, *c*).

H. Holtzer: Could we have some comments on Dr. Carlson's interesting presentation. I think he has covered a great deal of material which could bring together pathologists, morphologists, and cell biologists. As a cell biologist I do not see any difference on a cellular level between what happens during muscle regeneration in man and mouse, or in frog and salamander. Clearly there are differences in geometry and geography and this leads to differences in the time course of regeneration.

Also, there is the intriguing issue that the cut limb of a salamander regenerates, whereas ours does not; but this phenomenon concerns problems of morphogenesis, not of histogenesis, and is beyond the scope of this conference.

Considering the activity of a nucleus in a muscle fiber, and the activity of the muscle cytoplasm, I see no differences in the regenerative processes you have been studying. The theme is that mononucleated, presumptive myoblasts not synthesizing contractile proteins, are stimulated to replicate. Many of their progeny fuse to form multinucleated myotubes, which then produce striated myofibrils. Dr. Hay and others studying regeneration in the salamander contend that the original source of muscle fibers in the regenerate could be cartilage cells, skin cells, fibroblasts, or 'fragmented' muscle cells. I claim that neither she nor others who entertain this view have supplied compelling evidence supporting this concept.

B. M. Carlson: At the level of the individual muscle cell and with respect to its differentiation, I have not seen any noticeable light-microscopic differences between muscle fiber regeneration in tissue and epimorphic systems. The origin of the myogenic cells in neither case has been agreed upon by all. At the level of tissue organization, however, there are some distinct differences, and to me the key to the difference is the presence of the regeneration blastema in an epimorphic system. This is the most obvious qualitative difference between a limb-regenerating system and a muscle regenerating after mincing, and its presence could be related to both the differences in rates of the 2 processes and the morphology of the end products. Whether one believes that muscle in a limb regenerate arises from blastema cells or as a direct outgrowth from the muscles of the limb stump (my own observations would indicate that muscle arises from cells morphologically identical to blastema cells), there is little doubt that the presence of a blastema constitutes a delay factor in the differentiative phase of muscle regeneration. This delay period could well represent the time during which the myogenic cells receive morphogenetic information enabling them to aggregate into a morphologically perfect end product. In contrast, the myogenic cells in minced muscle regeneration are never associated with a blastema; they differentiate much more rapidly than do those in a regenerating limb and they do not form a morphologically perfect end product. To me these cells contain within themselves all the information they need to form muscle fibers, but they have not acquired that body of information or have not been exposed to the proper morphogenetic forces to form an anatomically perfect muscle.

Getting back to limb regeneration, I am still bothered by the question of their (muscle cells) origin if it is not the blastema. Muscle cells have to come from somewhere. It is nice to say that in certain cases they arise directly from the ends of the injured muscle in the limb stump without participating in the blastema, but one still must account for the muscles which form completely within the regenerate and which do not have any connection whatsoever with proximal cells.

There are, for instance, interosseus muscles in limbs. If you amputate a limb above the elbow, there will arise muscles whose origin is the ulna and whose insertion is the radius. Such muscles arise without direct reference to a more proximal muscle mass. In any explanation of the origin of muscles in limb regenerates, these muscles must be taken into account.

E. D. HAY: Here is a slide you can use to illustrate. This is a fully formed regenerate 8 months after amputation. The amputation was off to the left. None of these distal muscles are connected to the proximal (Fig. 7, F).

B. M. CARLSON: With reference to these muscles, we have some distinct differences from the situation in tissue regeneration. It is claimed, for instance, that tissue regeneration seems to require the continuity of the basement membrane around the muscle fibers. In the regenerating limb basement membranes are absent, yet the muscle fibers regenerate beautifully.

D. ALLBROOK: Can I just ask you – what you are really saying then is that you do not see any evidence of any difference at the cellular level between the mammals we have been looking at and your creatures?

H. HOLTZER: Absolutely none.

D. ALLBROOK: But, the end result is manifested differently – why?

H. HOLTZER: The regeneration of the entire limb obviously involves the coordinated activity of a large number of cell types. Events involving skeletal cells feedback and influence activities of myogenic cells which, in turn, influence regenerating nerve fibers and so on. Of course, I have nothing to say about why these processes occur in lower forms, and not in frogs and higher forms.

Having established my ignorance I would now like to establish my skepticism of at least some claims in the literature. In salamanders, as in higher forms, the precursor cells to those which will form muscle fibers in the regenerate are mononucleated, *presumptive* myoblasts. In higher forms they are now called 'satellite' cells, thanks to Dr. Mauro. In salamanders they may or may not exist under the basement membrane of the definitive muscle fibers. Alternatively, it would appear more likely, if Dr. Hay has really examined this as carefully as she claims, that such cells may simply be distributed among the connective tissue cells between the muscle fibers.

Lastly, it is worth pointing out that fusion of mononucleated, presumptive myoblasts to form multinucleated cross-striated myotubes holds for regeneration in Rhodnius and Diptera, and even for the striated cells in the umbrella of the jelly-fish.

B. M. CARLSON: I have never studied tail regeneration. Here again it is a matter of establishing levels of organization. I can definitely say that it will take longer to restore a given muscle after amputation of the limb than it will be after mincing. Why is there a delay and what cells are involved? This, to me, is one of the primary problems.

E. D. HAY: I think we are mixing up several issues, and making statements about the origin of the cells, with which I would not agree. The process that occurs in the formed muscle after amputation and after mincing is similar except that it is faster. But, there is a very definite difference between what you are calling epimorphic regeneration and the kind of regeneration that we are talking about in mammals. And, as we alluded to earlier, in the blastema there are no old basement membranes that guide the cells. The cells all have the same general morphology. They grow out, they proliferate, and then, without any of this scaffolding that we were alluding to earlier, they redifferentiate into the perfect muscle for the perfect place, and the limb is absolutely right. So, there is this problem of organization of the regenerate, which perhaps is the key difference. Now, I think we ought to consider that separately from origin of cells, a subject about which we still have a lot to talk.

J. C. SLOPER: I would imagine Dr. Holtzer for one admits that the scaffold is a good thing to have around probably. But, it is easy to show that where there is no scaffolding, you can still have very good muscle regeneration. I think this is the point Dr. Holtzer was making.

R. K. GILBERT: Why are you saying that you can have good regeneration without any scaffolding?

D. ALLBROOK: Surely the point is that in mammals in a totally destructive situation you get muscle regeneration all right but it does not do any good, whereas it does the salamander an awful lot of good; he has a nice little limb that he can flip around with. But the muscle regeneration does not do my rats or Dr. Sloper's humans much good.

A. R. MUIR: In the embryo, I think we would all agree that individual muscles differentiate, and are recognizable as adult muscles at a very early stage. This occurs at a stage when electron micrographs do not show clear basement membranes or connective tissue. I think this is the situation obtaining in the salamander limb. Hence the development of the blastema is more akin to the developing limb of the mammal than to regeneration in an established situation.

E. D. HAY: It is worth raising the issue as to whether or not that old scaffolding is inhibiting things in mammals and whether or not an excess of connective tissue leads to non-regeneration. There certainly is a change in ratio between amount of connective tissue per unit area of muscle, as you go from the salamander to the higher vertebrates. As the animals leave the water they get more and more connective tissue, and less and less capacity for muscle regeneration. Of course we all know there are other things happening, *e.g.* the number of nerves per unit area is also less.

P. E. CONEN: May we extend Dr. Holtzer's philosophy of cells, the species differences discussed and Dr. Hay's regeneration of whole limbs, to include the stage of development and effect of the surrounding milieu? Experiments have been described in which transplantation of the wing bud of the early-stage chick embryo to the peritoneal cavity resulted in development of a wing in the wrong place. However, if a later-stage transplantation of a wing bud was made it apparently developed into tissue appropriate to the new area.

H. HOLTZER: The statement that transplanting a piece of 'undifferentiated' embryonic tissue to another location 'transforms' cells in the graft, is not the most accurate way of describing what happens. For example, suppose you graft a piece of embryonic tissue containing 10^4 cells from region A to region B. In this piece of tissue there may be precursor cells to 5 different kinds of cell types. Now the 5 different kinds of precursor cells each undergo variable numbers of cell divisions. Left in site A, 2 lines of cells would normally have died out, leaving 3 types surviving and terminally differentiating. The same piece of tissue in site B, however, is subject to influences such that the 2 lines that would have died out are in fact stimulated to proliferate so that their progeny come to terminal expression in the new site. In this context the phrase 'transformation' is not satisfactory. Another possibility: the cells in the piece of tissue at the time of the original grafting were equivalent to the cells at sites A or B. Differences then in the fates of their *progeny* are differences gradually induced by influences in site B.

In either of the above cases it should be stressed that no transformation of cell type occurs. Cells 'half-way' to becoming cartilage or skin are not influenced by their new surroundings so as to switch their biosynthetic activity to the production of luxury molecules appropriate for muscle or lens. When responding to the influences of site B the cells in the graft were not committed to synthesizing any kind of terminal luxury molecule.

E. D. HAY: I might add, if you transplant a regenerating limb at the stage that we are calling an early blastema before there is any obvious redifferentiation of cartilage (that is the first tissue to form), it is resorbed. There were some earlier claims that would more or less fit into what Dr. Holtzer is talking about – that when the blastema is transplanted to the eye, limb cells transform into eye tissues, but it seems likely that they died.

By the time the blastema is large enough to be transplanted and firm enough to withstand this trauma, it is determined to form a limb-like structure (see Stocum, 1968*a*, *b*). Now you can do another kind of experiment: Glade (1963) has transplanted tail skin to the limb of the salamander. Under these conditions, the outgrowths which you get look more like tails than limbs, in fact you almost recognize some of the vertebral elements. It is a complicated experiment and if you try to think about it at the cell level, you can imagine the number of variables.

H. HOLTZER: Again I strongly differ with Dr. Hay's interpretation of the experimental data. The reports of Rose (1964) and others that tail muscle grafted to limb sites forms tail structures simply do not withstand critical inspection of the experimental data. A morphologically malformed limb is not an 'abnormal tail'. For those interested in this problem compare, not the conclusions, but the experimental data cited in Holtzer (1959) with those reported by the investigators cited by Dr. Hay.

P. E. CONEN: These new enlightenments, perhaps, may be applied to experiments with enucleation of sea-urchin eggs very early in development. These nuclei can continue cell development when they are transplanted into cytoplasm of eggs at later stages of development. But a nucleus from an older egg does not function properly when transferred to a very early urchin egg.

H. HOLTZER: The work on nuclear transplants you refer to involves the frog and most recently is largely the work of Gurdon. Nuclei from several kinds of somatic cells, including one neoplasm, when grafted into enucleated zygotes will often support reasonably normal development. These experiments are provocative in the present context. They suggest that the arch conservative of the cytoplasmic-nuclear relationship is in fact the cytoplasm, the nucleus being probably the more flexible metabolically. A nucleus, providing its divisions can be synchronized to that of the cytoplasm, can be re-instructed or re-programmed. Thus a nucleus from a gut cell, or from a Lucke tumor cell will yield descendant nuclei capable of sustaining the synthesis of chondroitin sulfate or myosin or cholinesterase depending upon the dictates of the cytoplasms they come to lie in. This, if you wish, is an instance of 'transformation'. Interestingly, it requires a good deal of mitoses.

G. PALADE: Considering for the moment only the production of muscle fibers during regeneration (because the whole limb system may be too complicated for proper analysis at the present time), the most important problem seems to be the origin of muscle precursor cells. According to Dr. Hay, in the regenerating limb of salamanders, the precursors are blastema cells which in turn derive, at least in part, from dedifferentiated muscle fibers by a process of syncytium fragmentation. In higher vertebrates, however, the precursor cells seem to form a distinct 'developmental reserve' represented by the satellite cells of muscle fibers.

The first question then is: what is known precisely about satellite cells in urodeles capable of limb regeneration? Do their muscle fibers truly lack such cells?

The second question concerns the nature of the cells which form the 'basophilic cuffs' described by Dr. Carlson in regenerating muscle fibers in amphibians as well as mammals. Can the morphological observations so far recorded shed any light about their general organization and their origin? Is there any evidence that indicates or suggests that the cuffs are formed by proliferating satellite cells or, conversely, by small cells budding off from degenerating muscle fibers? Perhaps Dr. Allbrook has some information bearing on this point.

E. D. HAY: I wonder if I could try to answer the first part about the salamanders. I thought of a couple of things I wanted to tell Dr. Palade in retrospect. You asked why we had not transplanted any muscle. In fact, we have followed labeled muscle by labeling with tritiated thymidine at a specific time in regeneration, and then fixing the limbs at daily intervals (Hay and Fischman, 1961). Under those circumstances, labeled cells left the muscle mass and appeared in the blastema.

We injected one group of animals at 5 days, another at 10 days, and another at 2 weeks. Then we fixed at daily intervals, and looked at what happened to an area like X in Figure 7, B (see p. 14). In an area of this sort that is replaced by blastema cells we counted the labeled cells at daily intervals. We could account in numbers for all the labeled blastema cells, allowing for one division every 2 days. Chalkley (1954) has also done a quantitative study relevant to origin of blastema.

Another point that I should have mentioned is that we know that regeneration is local from irradiation experiments (Butler, 1935; Butler and O'Brien, 1942). In other words, if you X-ray the limb and protect the local millimeter of tissue next to the amputation site, you will get perfect regeneration. If you do the reverse, and radiate just the last millimeter you will get no regeneration. So, there is no question of a reserve cell migrating in.

J. C. SLOPER: What method of radiation, and what dosage was used?

B. M. CARLSON: In the adult newt, 2000 roentgens is the customary minimum dose. In axolotls the dose is higher (4000 to 7000 roentgens). There are problems in larval forms, though, because after irradiation the limbs resorb.

E. D. HAY: There are other kinds of experiments of that nature. For example, you can label all the blood, and then look at the limbs and see there is no label in the blastema. So, regeneration is local and formed tissues such as muscle and cartilage are giving rise to the cells. Now the other issue you raised as to whether or not there are any satellite cells involved in salamanders is one that I cannot answer very well. I have not looked at the problem as extensively as Dr. Mauro and Dr. Allbrook. I have looked at salamander muscle without seeing any satellite cells; however, I shall ask them whether they have looked at limb muscle in salamander. As I said earlier, cells which label with tritiated thymidine have cleaved off the muscle fiber (Fig. 11, see p. 19). I cannot prove they did not derive from satellite cells.

One of the things we are trying to do is to label muscle ahead of time, not in a transplant, but by taking advantage of the fact that the cell has stopped making DNA. Then you have to follow it, but you have to do it with electron-microscopic autoradiography, so it is a long-term process. We are faced with many problems, I think.

G. PALADE: Are you anticipating a difference in labeling rates between the nuclei of muscle fibers and those of satellite cells (assuming that such cells exist in amphibians)?

E. D. HAY: If one adheres to the hypothesis of the satellite cell as a reserve cell stemming from embryonic cells and proliferating and giving rise to muscle during the whole process of differentiation, then the satellite cell, as I understand it, would be expected to be dividing during embryogenesis. Thus the label might become diluted completely in the satellite cell and not in the muscle fiber.

J. R. COLEMAN: Are they differentially labeled?

E. D. HAY: I do not know, I have not seen satellite cells in newt limbs and I do not think they are present in any large number here. So, my interpretation of the autoradiograph that I showed you (Fig. 11) is that *that* nucleus has become separated from the myofibrils of the formation of the plasma membrane. That is my interpretation here, and it is an interpretation which can also be applied to the mammal since it has not been ruled out for the mammal. Dr. Palade asked me if I would have any hope of labeling satellite cells if they are here, that is, preferentially labeling the muscle, not the satellite cell, and that was the reason why I made the comment about the satellite cell possibly existing in salamanders.

A. MAURO: Would you just clear up one point. Do you think the so-called blastema cells arise from the mature multinucleated cell, that is to say, that the nucleus gathers up cytoplasm and pops off as a cell, and forms a blastema cell? Is that what you think?

E. D. HAY: Yes, I really do, but I cannot prove it.

A. MAURO: That is not the point, I just want to clarify what you think.

J. C. T. CHURCH: Could we clarify a simple point here. Surely we define a cell, which is lying under the basement membrane and separated from the fiber syncytium by a double plasma membrane, as a satellite cell, without stating anything about its function under normal or pathological conditions. As far as our present discussion has gone, are we not defining such a cell as a satellite cell? In which case the last photograph (Fig. 11) was a very beautiful satellite cell.

E. D. HAY: That is all right as long as you indicate that you think it originated from a nucleus that was present in the formed muscle. That is the trouble.

G. PALADE: What is important here is Dr. Hay's statement that before regeneration there is no satellite cell population, or a negligible satellite cell population in salamander limb muscles.

H. HOLTZER: I am sorry, that has not been established.

G. PALADE: But this point has to be clarified because it is so important.

H. HOLTZER: I think what Dr. Church is saying – at least I hope this is what he is saying – is that the satellite cell in the salamander is not tightly apposed to the muscle fiber. Possibly it is not even buried underneath a common basement membrane. But the equivalent of the satellite cell exists between muscle fibers, or somewhere in the system. If this were true then the argument that a definitive muscle fiber fragments, giving rise to mononucleated cells becomes unnecessary.

J. C. T. CHURCH: I am only at present concerned with definitions. If you see a cell under a basement membrane and separated from syncytium by a double plasma membrane, that, by our present definition, is a 'satellite cell'. However, I should add, knowing something of mammalian systems and the sampling problems inherent in electron microscopy, that satellite cells might only be seen very infrequently. For instance, in the bat web muscle, there is only 1 satellite cell per 200 microns of muscle fiber.

E. D. HAY: But, you see, if you get this much labeling (Fig. 10, see p. 18), and you do not see any satellite cells when you look at a field like that in a non-regenerating muscle it is hard to believe that those are satellite cells that are labeling. But, when I look at a field like that at X in Figure 7, B under the electron microscope, so far, this is all I have found (Fig. 10). The nuclei are definitely separated off the fiber, but they are under the basement membrane. They are a part of the muscle no matter how we define a satellite cell, they are a part of the muscle – they are not fibroblasts in the endomysium, which is what Dr. Holtzer is trying to imply.

G. PALADE: Do you get unlabeled muscle nuclei in this situation?

E. D. HAY: Yes.

D. ALLBROOK: In my book then, what you have shown, is by definition a satellite cell.

E. D. HAY: Well, I do not think it came from what you call a satellite cell, I think it came from a muscle fiber.

D. ALLBROOK: I do not know where it comes from, it just looks like a satellite cell. What we are talking about is an entity which is under the basement membrane, and is separated from the main part of the muscle fiber by a double plasma membrane, and Dr. Hay showed that.

G. PALADE: That does not mean too much.

D. ALLBROOK: I agree, but we are talking about definitions.

J. C. SLOPER: But, you are Dr. Allbrook. Surely you have to admit at this stage that you are going to broaden your definition. It is an undivided cell, both heterogeneous in origin and source, and function.

P. E. CONEN: Dr. Allbrook should add to the definition a series of positive additional characteristics for identification, and a series of negative characteristics which are rarely seen in satellite cells and are more characteristic of other cell types. In other words, for each 'suspect' satellite cell, you should be able to 'score' the number of characteristics in favor of, and against, the identification. Some characteristics would be more heavily weighted in forming this core than others.

D. ALLBROOK: Perhaps that may come. But right now a definition has to be made. What Dr. Muir and I saw were cells which answered to that description. In postnatal muscle similar cells also had a lattice pattern of myofilaments developing in them. Well, rightly or wrongly we called them 'satellite cells' – because we did not know what else to call them (Muir *et al.*, 1965).

A. R. MUIR: I thought we actually said: "satellite in all other respects". I think we are going to have terrible difficulties in this Symposium unless we use words very carefully. The word 'satellite cell', which our chairman introduced, was given that specific meaning. Unless we stick to a definition (see p. 93), and do not prejudice the issue by what it becomes after injury and assumes 'pleomorphic forms', we are going to get hopelessly confused. If we take reserve cells outside the muscle fiber and tag the name 'satellite' on those, then I am afraid we are going to have a chaotic discussion.

But, in answer to Dr. Hay's point about the satellite cells in the salamander, Flood (1964) did show them in the trunk muscle of axolotl.

A. MAURO: Has Dr. Flood published an extensive paper? He wrote a brief note in the Proceedings of the Third International European Conference of Electronmicroscopy.

A. R. MUIR: That note is the one which I refer to, but it does include a convincing illustration.

E. D. HAY: The real issue is how frequent they are; I just could not see any. Then you get back to the problem of whether they are duplicated. It becomes a real quantitative issue, and I am not sure this can be solved by statistics. I think in the long run one will have to label the nuclei within the myofiber to make a convincing case. In both of these systems, either salamander or mammal.

J. C. SLOPER: Could not the photograph simply show the stage in a process of forming a blastema cell, which happens to look like a satellite cell? In this case, do we clarify or muddy the issue by calling it a satellite cell? If this tends to be a transit stage in the formation of the membrane which separates the blastema cell, I think we have a condition that looks like what you would find in satellite cells.

But, is it really realistic to define this transfer period in regeneration with this term?

D. ALLBROOK: I would agree, but what else can we do? When we look at a section, we are looking at frozen time.

B. M. CARLSON: We are talking specifically about the stability of the differentiated state, and if a satellite cell was never differentiated and did not produce any myofibrils or muscular units within the cell, this is not the same thing as a cell which had been surrounded by mature elements and is now no longer surrounded.

M. E. BARBERIE: Satellite cells, resembling those described in other vertebrates, are present in normal salamander muscle but are extremely rare. However, in injured muscle there is a slight increase in the number of cells that can be described as satellite cells. These cells could either be satellite cell progeny or nucleated fragments of muscle retained within the basement membrane of the fiber. In agreement with Dr. Hay, nuclei of injured salamander muscle fibers do become labeled with tritiated thymidine. In my material this occurs when they are surrounded by vesicles derived from sarcoplasmic reticulum. The vesicles are fusing and gradually separating cytoplasm muscle nuclei from the injured fibers.

J. R. COLEMAN: Are these cells closely apposed to fibers appearing frequently, or are they rare?

M. E. BARBERIE: The best estimate that I can make, and I hesitate to use it, is about one of these undifferentiated cell types for every 1000 muscle nuclei.

H. ISHIKAWA: Presently, I completely agree with Dr. Allbrook's idea and Dr. Barberie's idea on the definition of the satellite cell. I think here we should stick to the definition proposed by Dr. Mauro; or rather we stick to it at this moment, and then we can discuss those (satellite) cells. Otherwise we cannot communicate with each other.

J. C. T. CHURCH: I have noticed that the basement membrane, in addition to being around the fiber and satellite cell, sometimes invaginates itself to some extent between the satellite cell and the syncytium. This is seen in both transverse and longitudinal sections, to form what I would call an 'enclave', within which the satellite cell lies. Now a given transverse section may be at a level, at either end of the cell, where it appears as though the cell is entirely separated from the fiber by the basement membrane. But with serial transverse sections through the middle of the cell, the basement membrane disappears, and there is plasma membrane contact between the satellite cell and the syncytium.

I think that in this case we can still say that such a cell is a satellite cell, in that, though largely surrounded by basement membrane, it has, at least in part, a plasma membrane contact with the syncytium.

M. E. BARBERIE: Drs. Reger and Craig have also shown that in human muscle.

J. C. T. CHURCH: Are we in fact looking at 2 populations of satellite cells, those under a simple basement membrane, and those within an enclave?

J. R. FLORINI: Is there any hope of isolating satellite cells, and working with these? It seems to me that the muscle cannot really survive. They only have a bit of cytoplasm around them and therefore they will be less dense than isolated nuclei, and more dense than the rest of the cytoplasm.

E. D. HAY: The cell membrane can regenerate itself so you would not be able to prove you had not broken the fiber into many cells. If some of the cells came from nucleated muscle fragments, which formed new cell membranes, demarcating them from the sarcoplasm, you would not be able to distinguish them from the so-called satellite cells.

In fact, I think this is probably what normally happens in limb regeneration, because these nuclei in the myofibers are not degenerating. It is very hard to follow one or the other of those two populations. Either in a regenerating limb or in the kind of experiment you propose, where you chop up the muscle, all kinds of things could happen to your cell membranes.

A. MAURO: Could we just have one comment from Dr. Muir?

A. R. MUIR: When I was preparing the manuscript for this meeting, I thought I ought to write a definition of the 'satellite cell', and I was aware of Dr. Church's enclave with basement membrane. So, I thought that your definition might be slightly altered. The satellite cell of a striated muscle can be defined as a mononucleated cell whose cytoplasm does not contain myofilaments, and which is enclosed by and lies within the basement membrane component of the sarcolemma of the striated muscle fiber.

I think that, if we settle on some such definition, we will save ourselves a lot of time.

A. MAURO: Dr. Steen will now present his work on cell differentiation during salamander limb regeneration. This material, which was referred to by Dr. Elizabeth Hay, will supplement the presentation of Dr. Hay.

CELL DIFFERENTIATION DURING SALAMANDER LIMB REGENERATION*

TRYGVE P. STEEN

First, I would like to describe limb regeneration in the Mexican axolotl, *Amblystoma mexicanum*, a salamander which becomes sexually mature in the larval morphology. I work on the animal when it is small, around 40 millimeters long. When this size, the axolotl is analogous to larval *Amblystoma punctatum* in which Dr. Hay has studied finger regeneration, and it is considerably smaller and regenerates much more rapidly than the young adults Dr. Carlson talked about. I think that, rather than referring to the events of regeneration on the basis of time since amputation, it will be useful to think in terms of a sequence of descriptive stages, because the time necessary to reach a given stage of regeneration varies greatly with age as well as with kind of animal.

The flat stump remaining after amputation is immediately involved in a period of wound healing, including both wound closure and clean up. As these processes are completed, free cells are mobilized from the stump tissues and aggregate at the flat end of the limb to form a small mound-shaped bud, the mound-stage blastema. The free, mesenchymatous cells comprising the mound blastema arise by a strikingly complete morphological dedifferentiation of stump tissues, and they look like those shown in Figure 55. Cells are released from cartilage (see Fig. 59 for example) as well as muscle, events whose morphology Dr. Hay documented much more extensively during this morning's session. As the mound blastema elongates and the regenerate begins to assume a cone-like morphology (Fig. 54), regenerative activity shifts from cell mobilization to proliferation. Proliferation during the cone stage results in the formation of a very large, essentially homogeneous bud of cells (Figs. 55 and 56). It is impossible to assert that any given blastema cell has a particular origin within the stump tissues, either by means of morphological or by means of immunochemical techniques. In addition, it is impossible to say that a given blastema cell will differentiate into any particular cell type within the regenerate. When the distal end of the late cone blastema begins to flatten, we have the palette or paddle stage. The palette stage marks the beginning of a period of increasing emphasis on differentiation and morphogenesis. Cartilage begins to condense in the proximal regions of the blastema, and the cartilaginous condensations progressively form the elements of the limb skeleton (see Fig. 57). Digits become defined during later stages, and the limb musculature forms from the blastema cells surrounding the differentiating cartilage. Since a very large population of blastema cells forms before any differentiation takes place, the regenerating axolotl limb provides a substantial test of the differentiative properties of blastema cells derived from any given stump tissue.

The primary focus of my study was to determine whether the cells participating in amphibian limb regeneration redifferentiate only into their former cell types, exhibiting a

* The original work reported in this paper was supported by a Danforth fellowship; by USPHS developmental biology training grant T01 HD-00032; by NSF grant GB4265 to Dr. J. P. Trinkaus; and by The Muscular Dystrophy Associations of America, Inc. This paper was written during the author's tenure as a post-doctoral fellow of The Muscular Dystrophy Associations of America, Inc.

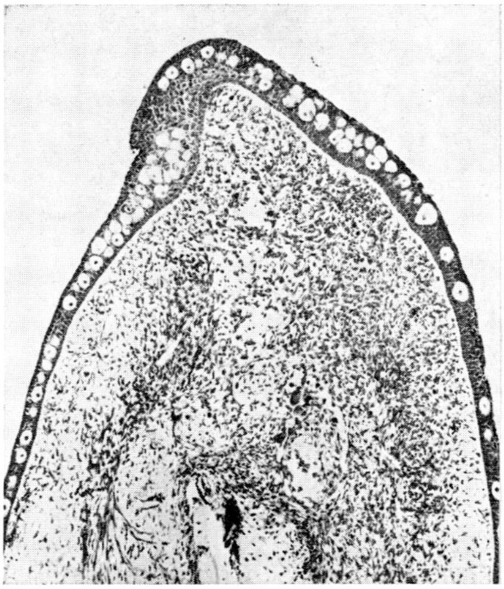

Fig. 54. In this early cone-stage blastema, the cut end of the humerus is apparent, as are some remnants of cartilage. The irregular contours of the humerus are due to the removal of matrix associated with cartilage dedifferentiation. A substantial group of blastema cells has aggregated at the tip of the stump. × 35.

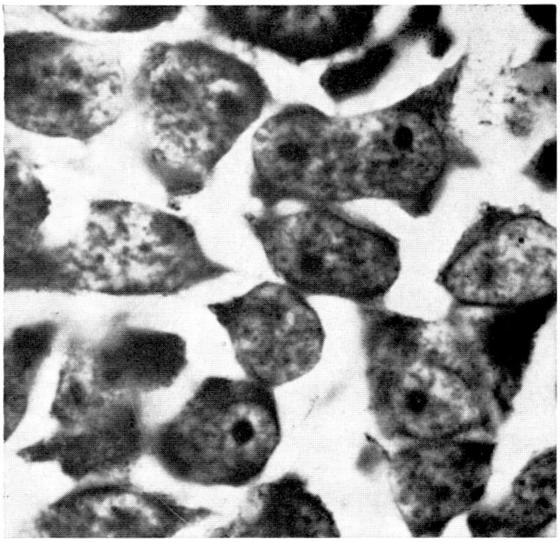

Fig. 55. Taken at much higher magnification, this illustrates part of a typical blastema. Like all blastema cells, these are morphologically indistinguishable and apparently interchangeable among one another. × 1200.

stability of determination, or whether they redifferentiate into different cell types, exhibiting metaplasia.

The latter of these alternatives is implied by the morphological evidence summarized in Figure 58. This work, therefore, shall define the arrow indicating morphological equivalency in terms of actual differentiative capacity. This work also applies to a question raised this morning: do cells in the stump degenerate or do they in reality dedifferentiate morphologically, contribute to the formation of the blastema, and subsequently to the tissues of the regenerate? These questions have been raised many times, but difficulties with labeling cells have foiled adequate answers. The crucial problems are developing the ability to dependably identify labeled cells and have the label maintained in spite of the proliferation involved in the process of regeneration.

Two different labeling methods were used in this study. One was triploidy. The method used to detect triploid cells depended on the presence of three nucleoli in a triploid nucleus, as opposed to the two nucleoli found in the diploid. The second label was tritiated thymidine, with labeled cells being detected in radioautographs. In order to obtain cells heavily labeled with thymidine, axolotls with regenerating limbs were given intraperitoneal injections of tritiated thymidine over a period of 4 days during the period of maximal proliferation – the cone blastema stage. Labeled tissues were carefully dissected free of other kinds of tissue and were transplanted into unlabeled limbs where their cells were available to participate in regeneration.

These labeling methods have a number of important advantages as well as some disadvantages for a study of this kind. The ploidy label is particularly advantageous. It is not affected by proliferation, because the label is reconstituted when the cells enter interphase after each division. One problem with using triploidy as a cell label is that spontaneous

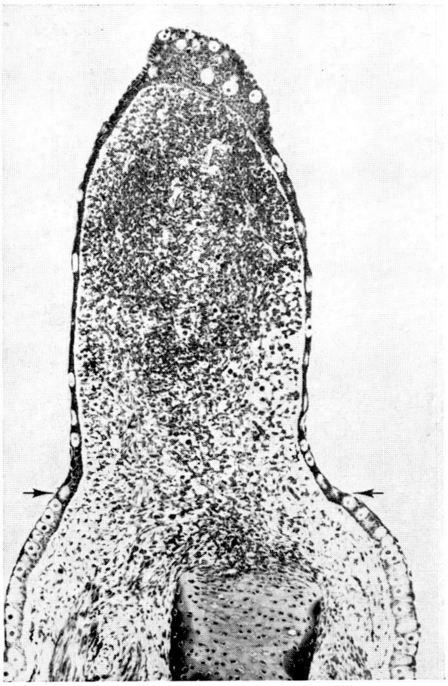

Fig. 56. In this late cone-stage blastema, the approximate plane of amputation is indicated by arrows. Note the disappearance of formed tissues for a significant distance proximal to the plane of amputation; these tissues dedifferentiated and were the source of the initial population of blastema cells. Clearly there is now a sufficient number of cells to form a new limb with little additional proliferation. × 35.

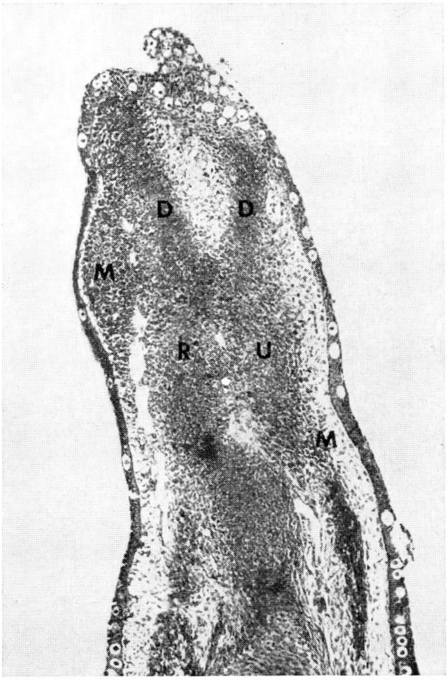

Fig. 57. The distal end of a palette-stage regenerate is shown. The central cells in the blastemal mass have begun to condense into the cartilaginous limb skeleton, the ulna (U), the radius (R), and the digital cartilages (D). In addition to the condensed cartilaginous elements, there are still many free mesenchymatous cells (M) remaining in the limb, available to form the rest of the limb tissues. × 35.

ploidy changes in cells of the host limb can cause ambiguous results. In the case of thymidine, dilution and transfer are substantial problems. The possibility that changes in nucleolar number significantly influenced my results was minimized by eliminating any cases containing cells with abnormal numbers of nucleoli. Control double label experiments enabled me to eliminate the problem of thymidine label transfer. Careful use of these labeling methods gave the results I am about to discuss.

UNIDENTIFIED SPEAKER: How well can you detect the nucleoli in triploid cells?

T. P. STEEN: In my preparations, nucleoli are distinct and obvious. Histological sections were stained with Azure B according to the method of Flax and Himes (0.25 milligram per milliliter of stain, pH 4.0, and 40°C). This method gives prominent, purple nucleoli in a basically green nucleus, and the nucleoli can be clearly identified because of this color difference. In addition, sections are cut at 12 microns, in order to maximize the number of triploid nuclei observable with three nucleoli in them. If you section a triploid regenerate at 12 microns, about 43 per cent of the cells will contain 3 nucleoli. Cartilage is of particular interest, and it is typical in this regard. If you reduce section thickness to 10, 8 or 5 microns,

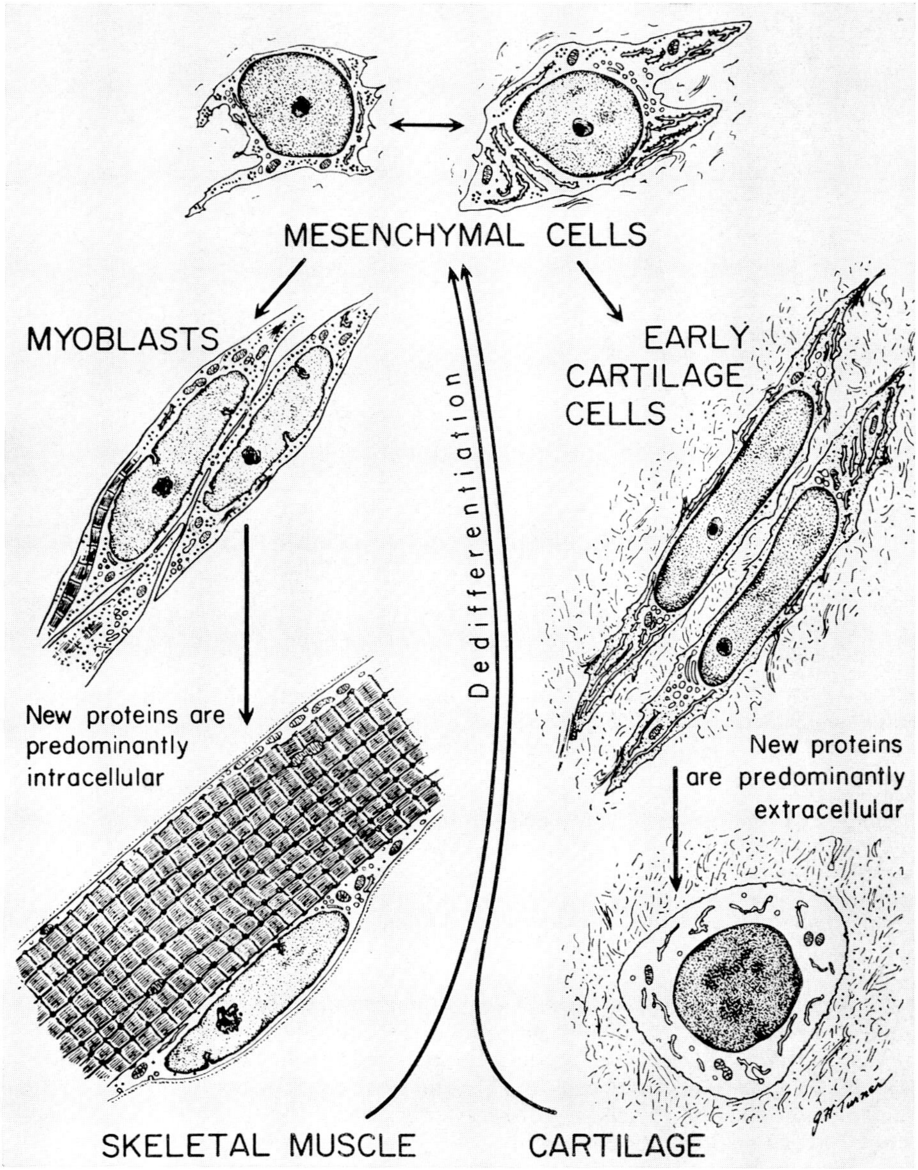

Fig. 58. The morphological changes occurring in cells that participate in limb regeneration are summarized in this diagram. Cells from skeletal muscle and cartilage dedifferentiate morphologically, forming mesenchymal (mesenchymatous) cells which contain no structural features to indicate their tissue of origin. As indicated at the top of the diagram, mesenchymal cells differ with respect to the amount of endoplasmic reticulum. The double-ended arrow also serves to emphasize the possibility that a cell originating from one tissue could possibly differentiate into another. Although morphological evidence raises fundamental questions about the stability of cell differentiation during regeneration, these questions must be answered by means of tracing labeled cells through the process of regeneration as was done in the present study. (*from*: Hay, 1962; copyright by The Ronald Press Co., New York, N.Y.)

the percentage of cells with 3 nucleoli goes down very radically. In fact, if you cut 4- or 5-micron sections, less than 1 per cent of the cells in a triploid limb would contain 3 nucleoli. In very thin sections, all you can detect are those cells which have virtually coplanar nucleoli.

UNIDENTIFIED SPEAKER: How well can you detect nucleoli in cartilage? The reason I ask about cartilage is that it is a notoriously poor cell in which to see nucleoli, because of heterochromatin.

T. P. STEEN: Problems with heterochromatin in cartilage nuclei were largely avoided by fixing experimental regenerates as soon as possible, since nuclei in young amphibian limb cartilage contain very little heterochromatin. Also, the high level specificity of Azure B stain enabled me to distinguish between the purple RNA of the nucleoli and the more green blobs of heterochromatin. The use of DNase to eliminate heterochromatin was considered but proved to be unnecessary. I should note that recently Marion Namenwirth has successfully used DNase to eliminate the very dense heterochromatin found in older cells. By this means she was able to identify triploid cells in older regenerates on the basis of nucleolar number.

UNIDENTIFIED SPEAKER: Cannot you add another parameter for the detection of triploid cells, the diameter of the nucleus. The nucleus of a triploid cell is $1\frac{1}{2}$ times the size of a diploid cell.

T. P. STEEN: Yes, you could possibly use nuclear dimensions for the detection of triploid cells, but there is a substantial problem with this particular cell property. The measurements from triploid nuclei overlap those of diploid nuclei so considerably, you can be sure about only a very small percentage of the cells at the extreme ends of the distributions. Many more cells can be very reliably identified on the basis of nucleolar number.

M. E. KAIGHN: Can you pick up the nucleoli in adjacent sections? Say a particular nucleus has 2, can you pick up the third one in another section containing part of that same nucleus?

T. P. STEEN: It would be conceivable to use that method, and the number of triploid cells detected would be increased. However, any error in the very difficult task of locating the other part of a given nucleus would reduce the reliability of one's identification of labeled cells. In order to be particularly confident in my identifications of triploid labeled cells, I counted only cells containing 3 nucleoli within a clearly definable nucleus located in only one section. I should note that you could never detect more than about 70 per cent of the triploid cells in a limb on the basis of nucleolar number, because nucleolar fusion also acts to reduce the number of triploid cells containing 3 nucleoli.

UNIDENTIFIED SPEAKER: How much of a problem do you have with the transfer of tritiated thymidine label in your experiments?

T. P. STEEN: In the case of cartilage grafts, control experiments gave essentially no evidence for transfer. However, when a soft tissue such as muscle was transplanted, control experiments showed a significant amount of low-level transfer. Differences in the amount of transfer are apparently related to the amount of cell death, which is probably greater for muscle than for well-protected cells like chondrocytes. Thus, any experiment involving the transplantation of labeled soft tissues will almost certainly result in significant amounts of transfer. However, control experiments such as the following one provide a basis for completely eliminating transfer cells from consideration by a safe margin.

In my control experiments to determine the existence of transfer, tritiated thymidine-labeled diploid cells were transplanted into triploid limbs. If there were no transfer, all radioactive cells should contain a maximum of 2 nucleoli. However, if transfer did occur, then some radioactive cells with 3 nucleoli should be found. The results from this experiment provided unequivocal evidence for low-level transfer. As a consequence, before I accepted a cell as labeled with tritiated thymidine, the count of silver grains over its nucleus had to exceed a minimum number. Further details, including the data for this transfer experiment can be found in Steen (1968).

In addition to labeling cells dependably, one must be sure that he is dealing with cells which have left their graft, have been free in the blastema, have proliferated, and finally have redifferentiated into one of the cells in the regenerate. Unfortunately, a single cell cannot be followed through all of these steps and thus be shown to participate fully in all major aspects of the process of regeneration. This is due to the limitations of using fixed material and of directly observing cells in living limbs. In this work, control experiments as well as careful observation of the slides containing fully developed experimental regenerates provide substantial assurance that labeled, grafted cells in fact do participate in these events of regeneration. This is critical, because cells in grafted tissue could exhibit an apparent stability of cell type simply by having the graft integrated into the regenerate as an intact piece. In order to verify that labeled graft cells do exist as free, undifferentiated elements in the blastema, regenerating limbs were fixed at different stages, with emphasis on mound and palette stages. These control regenerates contained labeled mesenchymatous cells in their blastemas (see Figs. 59 and 60). Although control limb blastemas do contain free, morpho-

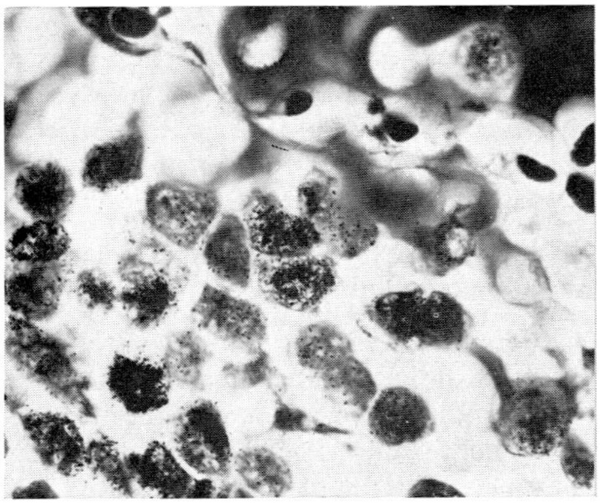

Fig. 59. This labeled cartilage graft is in an experimental limb fixed 6 days after amputation and transplantation. A group of cells has just been released from the graft and has not yet been incorporated into the blastema. × 400.

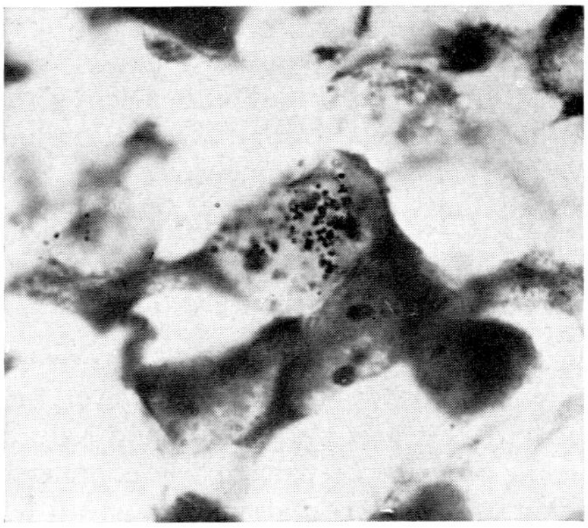

Fig. 60. The tritiated thymidine-labeled mesenchymatous cell shown here is free in a palette-stage blastema. It illustrates the dispersion of graft cells among blastema cells from other stump tissues. × 1000.

logically dedifferentiated cells of graft origin, one still needs to be sure that labeled cells in experimental regenerates were released from their grafts and free in the blastema. In the differentiated regenerates which furnished the evidence for this study, dispersion of labeled cells among unlabeled cells of the regenerate was used to infer that the grafted, labeled cells under study were released from the graft, were free as well as separate in the blastema, and were separate during their subsequent differentiation.

In experiments to study the stability of differentiation, cartilage is by far the best tissue to use. It is a homogeneous avascular tissue, which, when cleaned completely free of adherent cells, permits one to deal with a single cell type. All cartilage grafts were checked under a compound microscope to be sure of their tissue homogeneity and purity.

The contamination rate for control, fixed cartilage grafts like those labeled with tritiated thymidine, ran from 15/100 to about 3/10 of a per cent. The cells contaminating the cartilage

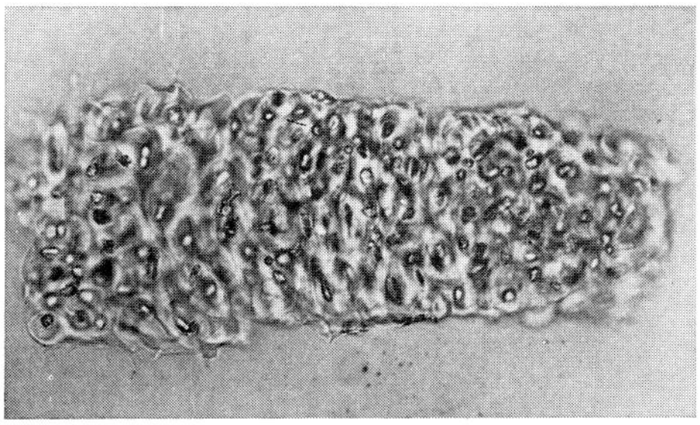

Fig. 61. Photograph of a cleaned, live piece of axolotl limb cartilage, taken just previous to transplantation. × 120.

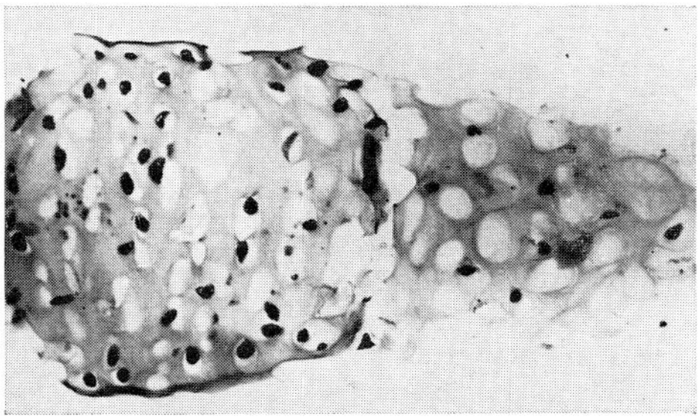

Fig. 62. Histological section of a piece of cleaned, control cartilage. × 100.

were primarily perichondrocytes; they were closely adherent to the surface of the matrix but were not completely enclosed within it. Even lower rates of contamination are possible when you clean cartilage that is within perichondral ossification. This was the situation for all triploid grafts, and the contamination rate was 1/10 of a per cent or lower.

When tritiated thymidine-labeled chondrocytes participated in regeneration, they predominantly redifferentiated in cartilage. As with all labeled cells I will talk about, they are located among unlabeled cells in a basically normal regenerate. In all experimental limbs, cells derived from the cartilage graft are found in all parts of the regenerated limb skeleton. See Figure 63 for an example of a stable radioactively labeled chondrocyte.

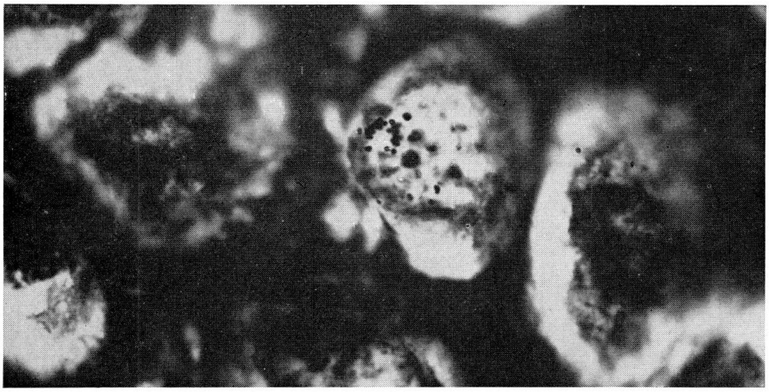

Fig. 63. The radioactively labeled chondrocyte in this Figure originated from a graft of cartilage containing cells labeled with tritiated thymidine. Since it is now found within the cartilage of a regenerate, this cell illustrates stability with respect to cell type. × 1200.

In this experiment, the use of labeled graft as well as labeled host limb cells provides a particularly effective control for the possibility of transfer of tritiated thymidine label. Of the labeled cells in these experimental regenerates, 72 per cent are chondrocytes. The remaining 28 per cent includes not only cells at the surface of the differentiating cartilage which are tabulated as perichondrocytes, but also the mesenchymatous cells. If you eliminate the mesenchymatous cells from consideration, because you cannot really tell what kind of

TABLE I

Cell types formed by tritiated thymidine-labeled limb chondrocytes that participated in the regeneration of triploid limbs

Graft number	1	2	3	4	5	Total
Chondrocytes	304	30	26	45	68	473
Perichondrocytes	8	11	0	0	0	19
Mesenchymatous cells	92	51	0	21	3	167
Limb totals	404	92	26	66	71	659

Each numbered column gives the data for one limb regenerate containing labeled graft cells.

cell they were going to become, the percentage of labeled cells that remained stable as chondrocytes becomes much greater, namely 96 per cent.

A comparison of grain counts over tritiated thymidine-labeled chondrocytes indicates that they typically divided at least 5 times between the time they were in the graft and the time when they were finally differentiating as cartilage in the regenerate. Based on a study of cells in sectioned, control pieces of cartilage, the initial grain count rate of grafted cells was 36 to 46 grains per day per nucleus (90 per cent confidence interval). In the results with cells that had redifferentiated after regeneration, the grain count rate was 0.57 to 0.71 grains per day per nucleus (90 per cent confidence interval).

TABLE II

Cell types formed by triploid chondrocytes that participated in regeneration in diploid limbs

Graft number	1	2	3	4	–	5	6	7	8	Total
Label observed	3N	3N	3N	3N	4N	3N	3N	3N	3N	–
Chondrocytes	68	6	6	485	0	58	75	71	111	880
Perichondrocytes	0	1	0	0	0	0	1	0	0	2
Joint connective tissue	0	0	0	7	0	0	1	0	0	8
Connective tissue	0	0	0	6	0	1	8	0	2	17
Muscle *or* connective tissue	0	0	1	2	1	0	0	0	0	4
Muscle	0	0	0	1	0	0	0	0	0	1
Mesenchymatous cells	0	0	0	4	0	0	0	14	9	27
Graft size	S	S	M	L	–	S	M	M	M	–
Limb totals	68	7	7	505	1	59	85	85	122	939

Each numbered column includes the results from a graft of cartilage transplanted into a different regenerating limb. Grafts 1 through 6 were derived from the humerus, graft 7 was from scapular cartilage, and graft 8 was from the coracoid. All numbers in this Table include cells that were actually observed. No correction has been applied to allow for the fact that all triploid cells cannot be detected.

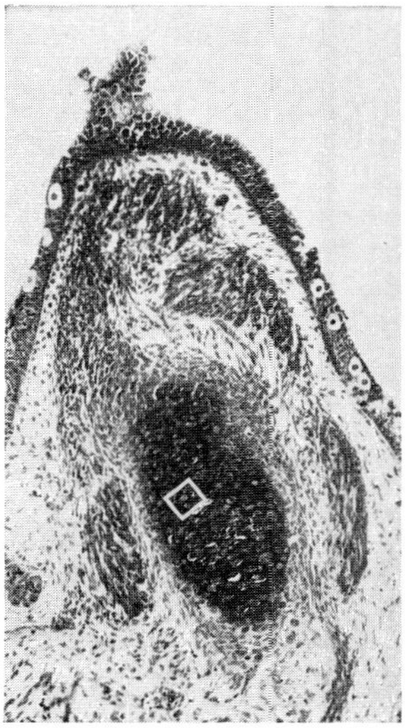

Fig. 64

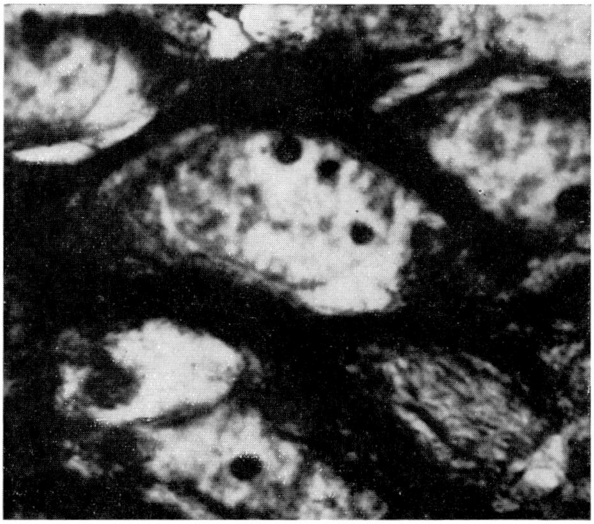

Fig. 65

Fig. 64. The portion of the humerus shown in this section contains 4 labeled cells that originated from a graft of triploid cartilage. The area included in the rectangle is shown at a higher magnification in Fig. 65. Cut on a diagonal through the upper arm, this section shows the typical histological features of a nearly completed axolotl regenerate. Differentiation of the cartilage is more advanced than that of the muscle, and a substantial volume of muscle is forming. This significant quantity of muscle indicates that blastema cells make a significant contribution to all cell types of the regenerate, not only to the skeleton, but also to the muscle. × 50.

Fig. 65. This differentiating triploid chondrocyte originated from a graft of limb cartilage and is now located in the humerus of the regenerate tabulated as number 1 in Table II. Since none of the adjacent cells contain 3 nucleoli, this cell must have been free in the blastema; yet it differentiated as cartilage, exhibiting its stability of cell type. × 1600.

In experiments using triploidy as a cell label, cells derived from cartilage were highly stable with respect to cell type (see Figs. 64 and 65 for illustrations and Table II for the data). In the results from triploid cartilage grafts (Table II), if you compare the chondrocyte row with all the rest, you will find that 93.7 per cent of *all* labeled cells found in these regenerates are chondrocytes. However, since undifferentiated mesenchymatous cells comprise 2.9 per cent of the labeled cells and cannot be placed into any particular category of differentiated cell, the percentage of cells going into cartilage can be adjusted. Thus, 96 per cent of the *differentiated* labeled cells in these regenerates are found in cartilage.

The remaining 4 per cent of the differentiated labeled cells not in cartilage can be explained in a variety of ways. The presence of the rare apparently labeled cells in muscle can be best explained as the occurrence of spontaneous polyploidy, since these cells were found primarily in parts of the limbs that were not involved in the formation of the regenerates. The various triploid connective tissue cells could represent: (1) cells about to differentiate into cartilage (a particularly good possibility for the perichondrocyte category and some of the cells in the joint connective tissue category); (2) the occurrence of spontaneous polyploidy (a particularly good possibility for cells in the muscle connective tissue category); (3) connective tissue contaminants of the graft (an unlikely possibility, because the exterior of the grafts were so thoroughly scrubbed with hair loops that no free cell should have been able to survive); (4) limited metaplasia of the chondrocytes (the only remaining possibility for the cells

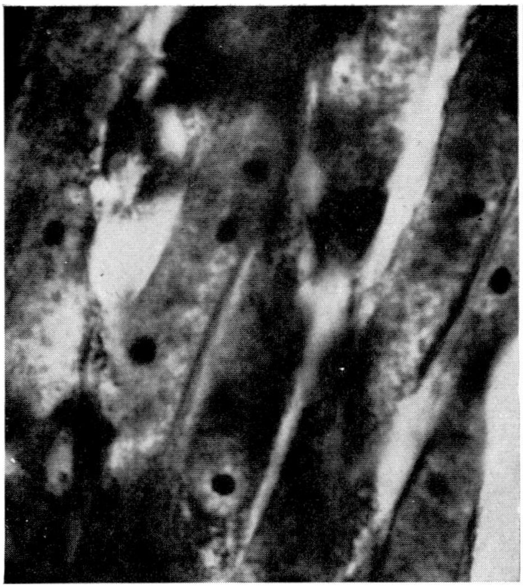

Fig. 66. The differentiating muscle cell containing 3 nucleoli originated from a muscle graft. It illustrates a relative stability of cell type. × 1300.

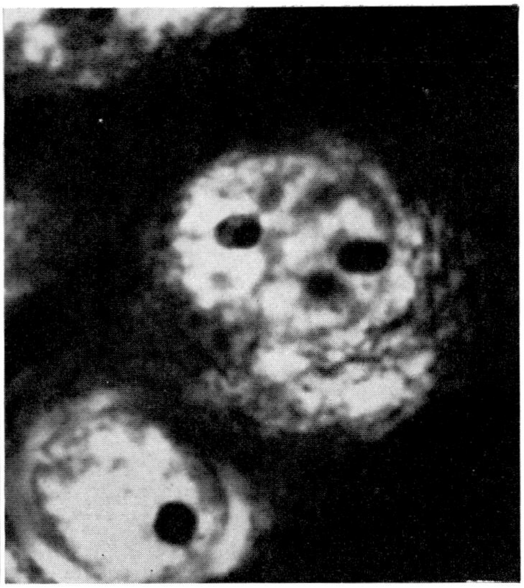

Fig. 67. This triploid chondrocyte originated from the same whole muscle graft as the cell illustrated in Fig. 66. Illustrating the incorporation of cells from muscle into regenerate cartilage, this cell provides an example of metaplasia which took place in the course of regeneration. × 1600.

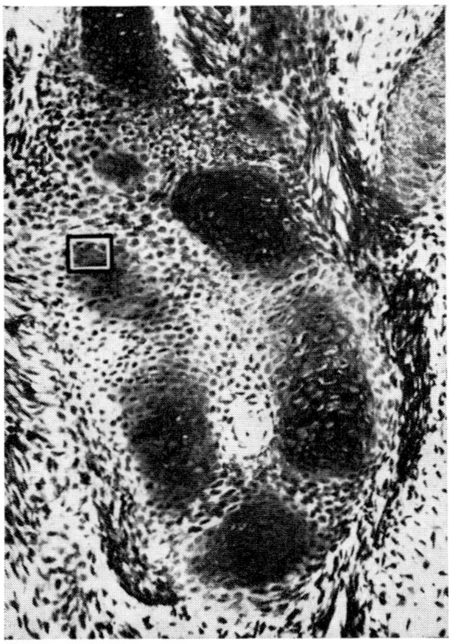

Fig. 68. The distal end of the regenerate tabulated as number 4 in Table III is shown here. This limb received a graft of whole limb muscle containing cells labeled by both triploidy and tritiated thymidine. The area indicated by the rectangle is within one of the carpals and is shown at higher magnification in Fig. 69. × 60.

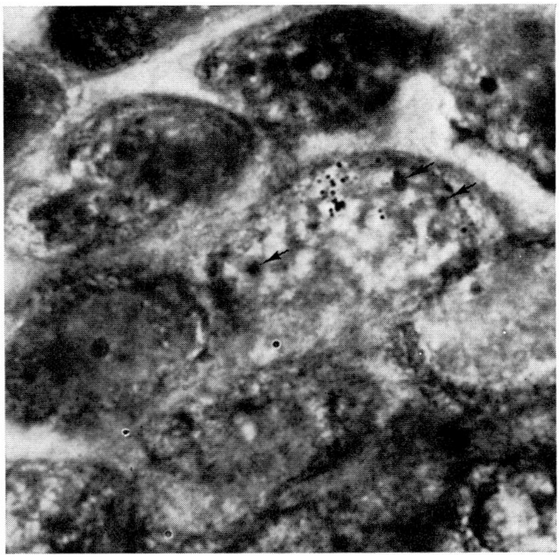

Fig. 69. The chondrocyte nucleus of primary interest in this Figure is doubly labeled. Its 3 nucleoli are indicated by arrows, and some of the 32 silver grains over this nucleus are also visible. The simultaneous presence of both labels in a cell such as this provides particularly strong evidence for the incorporation of cells from muscle into cartilage. × 1300.

not fitting into any of the above categories). Thus, between 1.5 and 3 per cent of the labeled, differentiated cells in these limbs could represent limited metaplasia of chondrocytes into some other connective tissue. However, this possibility of limited metaplasia should not obscure the impressive degree to which chondrocytes retain their cell type during normal limb regeneration.

Study of the role of cartilage in regeneration also gave some preliminary insights into the contributions of other tissues to the normal limb regenerate. In the cartilage graft experiments, all of the cartilage in the limb stump was labeled, but yet the regenerated cartilage contained only a very low percentage of labeled cells. Clearly much of the cartilage in the regenerate was derived from other tissues. Since muscle is a major contributor of cells to the blastema, it is a very likely source for these cartilage cells.

Labeled whole limb muscle was taken from either the upper arm or the proximal forearm and was carefully cleaned so it was free of tendon, excess connective tissue, and cartilage. After formation of the regenerates containing the labeled whole muscle grafts, there were many labeled cells in the muscle, in the connective tissues, in the cartilage, and even among the remaining mesenchymatous cells at the limb tip. The labeled mesenchymatous cells are a clear indication that the muscle contributed free cells to the blastema. Results from these experiments are illustrated in Figures 66 through 69 while the quantitative data are presented in Table III.

TABLE III

Contribution of labeled limb muscle to cartilage in limb regenerates

Graft number	1	2	3	4			5			Total*
Label observed	H^3	H^3	3N	$3NH^3$	H^3	3N	$3NH^3$	H^3	3N	
Chondrocytes	35	59	6	12	76	6	5	40	4	194
Muscle, connective tissue, or mesenchymatous cells	42	140	16	0	54	–	–	–	–	252
Limb totals	77	199	22	12	130	6				446

Grafts 1 through 3 contained cells labeled by only one method. Grafts 4 and 5 utilized doubly labeled cells, requiring multiple columns for a tabulation of the results. See text for explanation.

* Totals include cells from grafts 1 through 4 only.

Including results from both labeling methods used in this study, Table III provides documentation of the significant extent to which cells from muscle can contribute to cartilage during limb regeneration. In limbs 1 through 4, host limb cartilage was removed to make room for the muscle grafts. In limb 5, the muscle graft primarily replaced host muscle, and most of the host limb cartilage remained in the stump. In limbs 4 and 5, the graft nuclei were doubly labeled, containing 3 nucleoli as well as tritiated thymidine. Some labeled cells in the resulting regenerates contained both labels ($3NH^3$ columns). However, either of the labels could be lost from any given nucleus. For instance, when nucleolar number was reduced by sectioning or nucleolar fusion, a labeled cell may be identifiable from radioactivity only. Such cells are tabulated in the H^3 column. On the other hand, if the thymidine label became diluted beyond detection, then a cell can still be identified by the presence of 3 nucleoli and be tabulated in the 3N column. Unfortunately, whole muscle is a heterogeneous mixture of cell types. As a consequence, the metaplasia observed in these experiments could either be a limited transformation of connective tissue into cartilage, or it could be a substantial transformation of cells derived from a myotube into cartilage.

The ambiguity of these results from the study of muscle can be clarified somewhat by studying the participation of fin connective tissue in regeneration. Tail fin connective tissue provides an easily isolated, essentially pure source of loose connective tissue, and when a small piece is transplanted into a limb, it furnishes cells that will participate in the formation of a limb regenerate. Although limbs containing grafts of tail tissues are generally abnormal in appearance, normal limb structures will also form when a small graft is used. Two preliminary cases which received small grafts of tail fin connective tissue are available for study. Although both are abnormal in external appearance (see Fig. 70 for an example), their distal ends contain both carpal and metacarpal elements. In both cases, tritiated thymidine-labeled triploid cells were transplanted, and labeled cells were later found in essentially all of the tissues in the regenerate. The substantial contribution of cells to cartilage is of particular interest in this context (see Fig. 71 and Table IV), since it implies that the connective tissue component of muscle contributes cells to cartilage. The contribution of fin connective tissue cells to the myotubes of limb regenerate muscle represents an example of very significant metaplasia (see Fig. 72 and Table IV). It will be well worthwhile to pursue this latter observation further, especially with the use of 2 labeling methods.

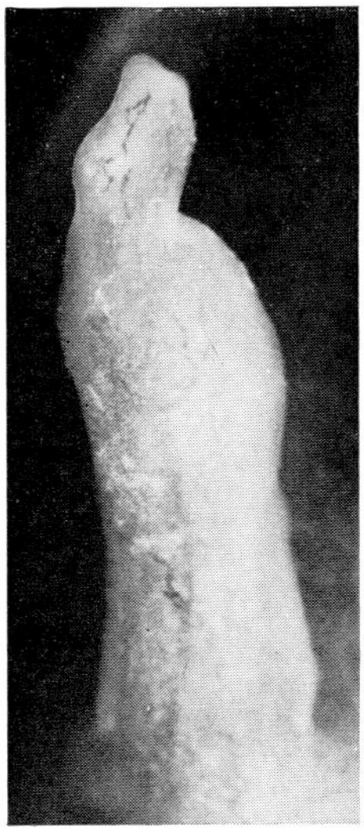

Fig. 70

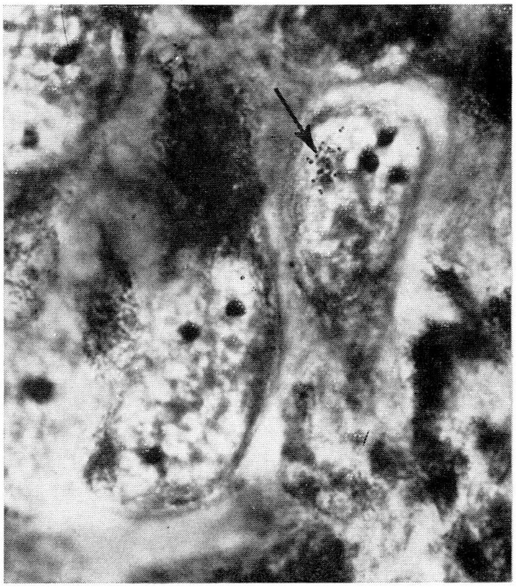

Fig. 71

Fig. 72

Fig. 70. Photograph of the limb tabulated as number 1 in Table IV. Cells from the fin connective tissue graft in this regenerate have contributed to the cartilage (see Fig. 71) as well as to the normal limb connective tissues. × 25.

Fig. 71. The 2 labeled, differentiating chondrocytes in this photomicrograph came from a tail fin connective tissue graft. One cell nucleus contains 3 nucleoli, and the other contains 3 nucleoli as well as tritiated thymidine – note the silver grains (arrow). × 1000.

Fig. 72. A labeled, triploid muscle nucleus derived from a fin connective tissue graft. × 900.

The number of cell types formed by fin connective tissue cells reveals a striking diversity of capacity. Consequently, these results imply that the connective tissue component could be responsible for the diversity of cell types formed from whole limb muscle. It should be emphasized, however, that additional information about the cells derived from the myotubes of whole muscle awaits the completion of current experiments in which clones of labeled muscle cells will be transplanted into regenerating limbs.

TABLE IV

Cell types formed by labeled tail fin connective tissue that participated in limb regeneration

Graft number	1			2			Total
Label observed	$3NH^3$	H^3	3N	$3NH^3$	H^3	3N	
Chondrocytes	2	8	10	3	4	25	52
Perichondrocytes	0	0	0	1	5	3	9
Connective tissue	1	7	5	3	60	12	88
'Muscle' connective tissue	0	0	0	0	18	1	19
Muscle *or* connective tissue	0	0	0	0	26	1	27
Muscle	0	0	0	0	0	2	2
Limb totals	3	15	15	7	113	44	197

Both of the cases in this experiment utilized doubly labeled cells, and as a consequence multiple columns are required for a tabulation of the results. See text dealing with the muscle results for an explanation. It should be noted that some localized areas of loose connective tissue were not tabulated in these results, because they contained a high percentage of labeled cells, which implied that they were unintegrated graft tissue.

UNIDENTIFIED SPEAKER: Would you kindly define a blastema cell in your own words?

T. P. STEEN: In morphological terms, a blastema cell or mesenchymatous cell is a cell with an enlarged nucleus, prominent nucleoli, and very scanty, highly basophilic cytoplasm (see Fig. 55). A blastema cell cannot be uniformly defined in terms of its potential or its capacity for differentiation.

UNIDENTIFIED SPEAKER: Does this mean that the blastema cell population is already sequestered? In other words, do you have sub-groups of cells, so the population of cells in the blastema is structured?

T. P. STEEN: There is at least 1 definite sub-population of cells in the normal limb blastema, because cartilage gives rise to blastema cells which can give rise only to cartilage. Therefore, every cell in the blastema apparently is not complete in its capacity to differentiate into any limb tissue. However, some blastema cells must still maintain the capacity to differentiate into a variety of cell types, as is apparently the case for cells of muscle and fin connective tissue.

H. HOLTZER: I would question 3 central dogmas in the literature on amphibian regeneration. I submit that (1) only the progeny of chondrocytes and fibroblasts contribute to the blastema, (2) only chondrocytes and fibroblasts and related connective tissue cells are derived from the blastemas, and (3) multinucleated muscle fibers do not fragment or dedifferentiate into mononucleated cells which, after replicating as pluripotential cells, then differentiate into muscle, or chondrocytes or fibroblasts of the regenerate. These views differ considerably from those usually expressed (*e.g.* Nicholas, 1955; Rose, 1964; Hay, 1962, 1965, 1966, 1968; Goss, 1969).

The bulk of the literature on amphibian regeneration involves one 'experiment': amputation is followed by fixation on successive days and the 'stump', 'blastema' and 'regenerate' analyzed histologically. Two considerations limit the interpretations of this material: (1) absence of sharp boundaries delimiting stump, blastema and regenerate in space or time, and (2) difficulties of reconstructing a dynamic sequence from fixed histological sections. For example, immediately following amputation, the transected muscles retract unequally millimeters proximal to the distal surface of the skeletal stump. Consequently, early degenerative and regenerative processes in muscle tissue take place in pockets remote from early degenerative and regenerative processes in skeletal tissues. Only days later, after considerable proliferation and cell migration, do cells from the muscle tissues overlap cells from the skeletal tissues. By then the early decisive mitotic activity so crucial to regeneration is over.

In skeletal tissues, 24 to 48 hours after amputation, degenerative and regenerative processes occur side by side. In regions where mucopolysaccharide matrix is being resorbed, many cells are in mitosis. Chondrocytes liberated from their matrix are induced to re-enter the mitotic cycle (for comparable behavior of avian chondroblasts see Abbott and Holtzer, 1966; Holtzer and Abbott, 1968; Chacko *et al.*, 1969) and their progeny constitute the major source of blastema cells. Equally evident is the mitotic activity in perichondrial cells and surrounding connective tissue cells; they also contribute to the blastema. To my knowledge there is no *critical* experiment demonstrating that a nucleus once part of a muscle fiber, enters the blastema as part of a mononucleated cell, then to differentiate into a cartilage cell.

Weiss demonstrated that prior removal of skeletal elements in the stump did not preclude the formation of skeletal elements in the regenerate. These experiments suggest that after removal of skeletal elements the perichondrial cells and fibroblasts in the stump comprise a reservoir for blastema cells. The capacity of amphibian perichondrial cells and amphibian fibroblasts to be induced to form chondrocytes is matched by avian and mammalian perichondrial cells and fibroblasts (Holtzer, 1968; Holtzer and Abbott, 1968). Neither amphibian, nor avian, nor mammalian chondrocytes or fibroblasts readily dedifferentiate to give rise to progeny that differentiate into muscle cells.

In spite of the report of Hay (1952) exploiting nucleolar markers of polyploid cells, there is no reliable evidence that cells of the skin contribute to the blastema or that they differentiate into any cell type but skin cells. Similarly there is no evidence that endothelial cells of blood vessels or pigment cells or Schwann cells contribute to the blastema.

Hay cited the work of Thornton (1942) and Chalkley (1954) who counted mitoses in muscles of the stump. The key issue, however, is what kinds of cells in muscle tissues display mitotic figures, and in the work of Steen what kind of nuclei incorporate labeled thymidine? Are they: (1) nuclei within a multinucleated muscle fiber; (2) nuclei of presumptive myoblasts or satellite cells; (3) nuclei of fibroblasts; (4) nuclei of endothelial cells? etc. There is no compelling microscopic evidence, either on the light- or electron-microscopic level, that nuclei within multinucleated muscle fibers, before or after the reconstitution of a new cell membrane, synthesize DNA, divide and yield viable, functional mononucleated cells. Until this sequence is demonstrated the assertion that functional muscle nuclei contribute mononucleated cells to the blastema remains unproven. A more likely identification of the cell types undergoing mitosis in muscle tissue would be: (1) presumptive myoblasts or satellite cells (Mauro, 1961; Ishikawa, 1966; Church, 1970); (2) fibroblasts; (3) endothelial cells; (4) Schwann cells, etc. There is no electron-microscopic study scrutinizing the kinds of cells undergoing mitosis in regenerating limbs. Similarly there are no data in the literature to contra-indicate the suggestion that nuclei released from injured salamander muscle degenerate as do their counterparts from avian and mammalian muscle (Okazaki and Holtzer, 1965; Holtzer *et al.*, 1969).

Implicit in the data, though often explicitly denied in the discussion of papers in this field, is the fact that cells of the blastema only differentiate into chondrocytes and associated connective tissues. This has been unequivocally demonstrated in the regenerating salamander tail (S. Holtzer, 1956; Holtzer, 1959) and limbs. Young limb blastemas, free of stump muscle, implanted into the dorsal fin of host salamanders display only cartilages and connective tissues in the resulting regenerate (Holtzer *et al.*, 1954). Implants of older 'blastemas', taken when the distally regenerating myogenic elements girdle the proximal portion of the blastema display structures consisting of cartilage and muscle.

Two other experiments (Holtzer, unpublished) demonstrate that blastema cells of both limb and tail differentiate only into chondrocytes and connective tissue cells and that muscle of the regenerate is not derived from blastema cells. The first experiment is the converse of

Weiss: mainly, radical extirpation of muscle proximal to the plane of amputation. The second experiment is the converse of Goss: mainly, prior grafting of supernumerary muscle masses so that the stump at the time of amputation has a larger mass and larger number of muscle fibers than do normal limbs. In the first experiment – no muscle proximal to the amputation plane – poor to good skeletal elements and fibroblasts were present in the regenerate (see Holtzer, 1959); muscle was either totally absent or at best sparsely represented. In the second series – excess groups of muscle in the stump – no effect was observed on rate or morphology of the regenerated skeletal cells. The supernumerary muscle masses, however, formed irregular bundles of muscle which jutted into the regenerate. These results are comparable to Goss' experiments of supernumerary skeletal elements elongating into the regenerate as supernumerary skeletal elements.

When tail muscle, grafted either into the limb or into the dorsal fin, is transected, it regenerates only as muscle and connective tissue. The addition of spinal cord with such grafts results in the induction of vertebral cartilages (S. Holtzer, 1956; Holtzer, 1959). This inductive interaction is assumed to involve fibroblasts among the trunk muscle cells. This interpretation would be in accord with the observation that avian and mammalian fibroblasts associated with 'irritated' muscle, bladder, liver and other tissues may yield a progeny of typical chondrocytes provided they are first induced to undergo mitosis (Holtzer, 1968). There is no evidence that chondroblasts or fibroblasts yield progeny capable of forming or being incorporated into muscle cells (Okazaki and Holtzer, 1965, 1966).

The data in the recent papers of Stocum (1968*b*) and Namenwirth (1968) are all consistent with the above interpretations and suggest that amphibian blastema cells are no more pluripotential than are homologous cells from higher forms. The unique feature of amphibian regeneration is on the morphogenetic rather than cytogenetic levels.

In conclusion it is worth reiterating that: (1) with the exception of the inductive role of the regenerating spinal cord or ependyma (Holtzer, 1959; Simpson, 1970), tail and limb regeneration involve basically similar events on the cellular level, and (2) the conventional contention that myogenic cells yield progeny that contribute to the blastema and eventually differentiate into chondrocytes is not only unproven, but, in my opinion, highly improbable.

E. D. HAY: I think this is a very serious question of terminology getting back to the Chairman's opening remarks. I think Dr. Holtzer is defining the blastema in a different way from the rest of us. The early blastema extends proximally and is composed of a mixture of cells derived from muscle and connective tissues (Fig. 7, B). The blastema outgrowth, it is true, first forms cartilage. But even if you use the term blastema only to refer to the outgrowth of cells at the end of the limb you still cannot say that it does not form muscle, when it does.

It was shown earlier (Fig. 2, E) that muscle appears in the blastema (Thornton, 1938). You cannot say that the blastema does not form muscle unless you have a different definition. Here is a newly forming muscle, this is all from the blastema (Fig. 7, F).

H. HOLTZER: You showed a lovely picture at low magnification with the blastema on the extreme right and the 'stump' on the left. Cells in the blastema divide, but at the same time among the muscles in the stump many division figures can be observed. These cells – the mononucleated cells – from between the degenerating muscle fibers undergo many cell divisions and later will migrate into the area dominated by the blastema cells originally derived from the cartilage cells or the connective tissue cells. I submit this interpretation would account for why Dr. Steen does not find labeled cartilage cells yielding a progeny which populate a muscle fiber.

E. D. HAY: But labeled muscle appears in the blastema when transplanted there (Steen, 1968; Namenwirth, 1968). Where does it come from? It has to come from the blastema. The blastema is entirely a local affair (Butler, 1933; Butler and O'Brien, 1942). Cells are not migrating in from elsewhere, not in the case of limb anyway.

T. P. STEEN: I would like to add the following in reply to Dr. Holtzer's comments. Although the technique of using triploidy as a cell label can be tricky and in spite of the fact it has caused many problems in the past, it is important to emphasize that my present use of triploidy as a cell marker differs from these past uses in many important ways. As was brought out in our earlier discussion, the use of 3 nucleoli as the criterion for the identification of triploid cells is much more reliable than other cell properties that could be used and have, in fact, been used. Identification of triploid cells is made particularly certain by my use of Azure B to specifically stain the RNA of the nucleoli in a manner distinctively different from the DNA of the surrounding nucleus. In addition, for the cells considered in my data, only nuclei within a single section were used, further strengthening the certainty of my identifications of labeled cells. Finally, when both triploidy and tritiated thymidine are used as labels within the same cell, I do not think there can be any reasonable doubt that such a cell is labeled.

With this reminder about labeling techniques in mind, I should assert that there is quite reasonable evidence from studies of labeled cells showing that blastema cells of regenerating limbs do more than just make cartilage. In fact, blastema cells give rise to all the internal tissues of the limb regenerate, except for nerves and perhaps Schwann cells. Since all of my evidence comes from regenerating limbs, I shall not say anything about the blastema cells of regenerating tails, except that we clearly need some studies using labeled cells in that system too. Two different tissue sources, limb muscle and fin connective tissue, provide cells which can give rise to essentially all of the internal tissues of limb regenerates. Cells from both tissues give rise to chondrocytes, perichondrocytes, general loose connective tissue, muscle connective tissues, as well as the myotubes of limb muscle. Both Tables III and IV in my presentation support this assertion, although they were made with a different purpose in mind. In addition, the work of Namenwirth (1970*b*) provides particularly strong evidence that blastema cells can give rise to all the internal tissues of a limb regenerate. X-rayed limbs, which could not otherwise regenerate, formed nearly normal regenerates after receiving grafts of either triploid whole skin or triploid whole muscle. Since the regenerative capacity of the host limb tissues was completely destroyed by the X-ray treatment, internal tissues of these experimental regenerates were completely made up of triploid, labeled graft cells. The triploid cells were identified in the same way I did, with the added use of DNase to eliminate heterochromatin from their nuclei. When triploid whole skin was transplanted, excellent regenerates formed, and they contained a full complement of normal, triploid limb tissues. Since small numbers of muscle and connective tissue cells remained on the inner surface of the skin grafts, these results do not apply to the capacity of any particular cell type, but they clearly show the diverse capacities of the blastema cells. In the case of cells derived from whole muscle, Dr. Namenwirth concluded they can differentiate into cartilage, perichondrium, joint connective tissue, subepidermal fibroblasts, muscle fibroblasts, and myoblasts.

The conclusion that blastema cells can form all the internal tissues of the regenerated limb, except nerves and perhaps Schwann cells, is supported not only by the above-mentioned studies of cell behavior done with labeled cells *in situ*, but also by studies in which blastemas were transplanted to the fin or were isolated *in vitro*. Pietsch (1961*b*) transplanted blastemas devoid of all stump tissues to the fin where they gave rise to cartilage and some muscle. Although muscle differentiation was limited, it occurred in cases from almost all experimental groups. These results are difficult to interpret in any greater detail, because the status of differentiation in the transplanted blastemas cannot be judged. This is because the blastemas under study were referred to only in terms of days since amputation, none of the stages used were illustrated, and descriptive stages of regeneration were not referred to. In a recent study of morphogenetic properties of the limb blastema, Stocum (1968*b*) provides very valuable information about the differentiative capacities of the component cells. The 39 cases where whole cone-stage blastemas were transplanted without stump tissues are of particular interest, because blastemas of this stage are homogeneous buds of undifferentiated cells (see Figs. 55 and 56). Since freedom from differentiated stump tissues was verified by histological study of operational controls, these cases provide a particularly critical test of what cell types can be formed by blastema cells. These cone blastemas were able to form essentially normal limb regenerates after transplantation to the fin. Cartilage and other connective tissues differentiated, and in 90% of the cases apparently normal muscle formed. In addition, when cone- and palette-stage blastemas were isolated free of stump tissues and allowed to develop *in vitro*, they gave rise to cartilage, associated connective tissues, and muscle (Stocum, 1968*a*). It should be noted that only 7.7 per cent of the cone cases and 45 per cent of the palette cases gave rise to muscle in these cultures. Nevertheless, the differentiation of muscle in these cultures is highly significant, because the absence of muscle in isolated blastemas of these stages was verified by histological study of operational controls, and the development of these blastemas took place in the complete absence of stump muscle. These results, therefore, provide significant additional evidence that the blastema of a regenerating limb is able to form all the major internal tissues of the regenerate.

H. HOLTZER: It has been known for over 30 years (Nagotte) that ectopic cartilage will form or be induced in a variety of tissues, including muscle. Most investigators in this area propose that the induced cartilage is derived from fibroblasts that were stimulated by the 'injured' muscle to proliferate. This sequence of fibroblasts, proliferation and, eventually, cartilage might be termed 'metaplasia'. This is not what I am talking about. To me metaplasia in the current context refers to the proposition that muscle nuclei – that is nuclei once within a muscle fiber – somehow get out of the fiber, replicate as mononucleated cells and then some of these differentiate into chondrocytes.

T. P. STEEN: In your comment about getting pure muscle, you are primarily referring to the difficulties of obtaining pure axial muscle from older, more mature urodeles, and I agree, that is a very difficult problem. However, whole limb muscle can be dissected out, in a very pure state, quite easily. In my experiments, the distal end of the forearm was removed before taking any muscle from the limb. This precaution makes it impossible to accidentally include any of the very small cartilages of the carpals or metacarpals in the muscle graft. After removal from the major elements of the limb skeleton, the muscle transplanted in these experiments was carefully dissected free of tendon and excess connective tissue. Clearly, it is impossible to remove all connective tissue by direct dissection. Grafts of whole muscle are therefore primarily a mixture of connective tissues and myotubes. A detailed knowledge of what these elements do awaits results from cloning experiments. However, some cells in whole muscle do undergo metaplasia, since it contributes cells to the limb regenerate cartilage. In addition, cells from fin connective tissue undergo metaplasia as they definitely contribute to limb regenerate cartilage and as they quite probably contribute to the myotubes of regenerated muscle.

H. HOLTZER: The question is: the cells that go into that induced ectopic cartilage, did they come out of a myotube or out of the muscle cell and produce a chondroblast, or was it the connective tissue, which once again I for one would never quibble with, that made cartilage? My proposition was that the only things that go into the blastema are connective tissue elements, and the only thing that is going to come out will be cartilage.

Within that limit, that is the kind of metaplasia I would expect to see.

T. P. STEEN: I cannot agree that the blastema of a regenerating limb only makes cartilage. However, I should note that the metaplasia observed in the labeling experiments done to date, both mine and those of Dr. Namenwirth, can be explained in terms of an interchange among connective tissue cells. Connective tissue from either skin or fin can clearly contribute to dermis, loose connective tissue, joint connective tissue, perichondrium, muscle connective tissue, and cartilage. Cartilage is fundamentally stable as a cell type (Namenwirth, 1970*a*;

Steen, 1968), but under some conditions it can contribute to any of the connective tissues of the limb in a limited way (Namenwirth, 1970*a*, *b*).

Also, the diversity of cell types that whole muscle can form in a regenerate could simply be ascribed to its connective tissue component, with the muscle cells themselves remaining fundamentally stable. Completion of this picture requires evidence from the behavior of the individual cellular components of muscle. What can muscle connective tissue cells do? What are myoblasts capable of differentiating into when in the blastema? Answers to these questions await results from my current work in which labeled clones of cells from muscle will be transplanted into limb stumps so the differentiative capacities of their cells can be evaluated after participation in limb regeneration. At the very least, therefore, present results show that limited metaplasia takes place during limb regeneration. An unequivocal demonstration of more dramatic metaplasia, such as myoblast into some connective tissue component, or the reverse, awaits results from cloning experiments.

A. MAURO: May we cut off further discussion on these matters until later. Dr. Muir and Dr. Church would like to present their material on the satellite cell.

THE STRUCTURE AND DISTRIBUTION OF SATELLITE CELLS

ALAN R. MUIR

A mononucleated cell enclosed within the striated muscle fibre was described by Mauro (1961) who chose the name *satellite cell.* An almost concurrent report by Katz (1961) on the ultrastructure of the muscle spindle showed the presence of similar cells related to the intrafusal fibres. So these authors share the original discovery of one of the few cells, as distinct from cellular organelles, which have been revealed by electron microscopy. The crucial feature of the satellite cell is its intimate relationship to the sheath of the parent striated muscle fibre and so it is necessary, before defining this relationship, to discuss the components of this sheath.

THE SHEATH OF A STRIATED MUSCLE FIBRE (Fig. 73)

If the complexities of the terminology are overlooked, the observations of various authors (Robertson, 1960; Mauro and Adams, 1961) on the structures present at the surface of muscle fibres can be summarized as follows:

1. Bounding the sarcoplasm, there is a membrane consisting of 2 electron-opaque laminae separated by a clear layer to make an overall thickness of $<$100 Ångströms. After many preparative techniques, this structure has asymmetry with the outer lamina being less dense than the inner one (Sjostrand, 1963). It is this membrane which is invaginated to form caveoli and the transverse tubular system (Franzini-Armstrong and Porter, 1964). Because of its resemblance to the membrane covering all types of cell this part of the sheath is called the plasma membrane (syn: unit membrane, plasmalemma, cell membrane).
2. An amorphous layer 300 Ångströms thick, which is separated from the outer lamina of the plasma membrane by a gap of 100 Ångströms, is called the basement membrane (syn: glycocalyx, ectolemma, basal lamina).
3. In the intercellular spaces between skeletal muscle fibres, collagen and reticulin fibres weave a delicate sheath around each fibre and this is referred to as the endomysium.

Considerable semantic problems have arisen in relating this fine structure to the earlier nomenclature, as it is difficult to correlate the findings by different techniques. The review of Bennett (1958) clarifies the terminology and, in particular, shows that the sarcolemma, as described by Bowman (1840), consists of the plasma and basement membranes. These 2 components are mutually adherent and together they survive drastic physical and chemical treatment as a transparent tube, which is structureless at light-microscope resolution. Draper and Hodge (1949) report the presence of fine reticular fibres in their study of shadowed sarcolemmal tubes; these could represent adherent endomysial components but it is possible that the basement membrane has a finely woven reticular framework, which escapes recognition in thin sections as it is embedded in an amorphous mucopolysaccharide matrix. Certainly the construction of the basement membrane creates a layer, which can remain intact when the contained fibre atrophies or is damaged (Birks *et al.*, 1959; Allbrook, 1962; Church and Noronha, 1965).

The limits of the endomysium cannot be sharply defined as its reticular and collagenous fibres blend with those which permeate the whole muscle. Mauro and Adams (1961) distinguish a layer of collagen fibres close to the basement membrane, with an outer layer

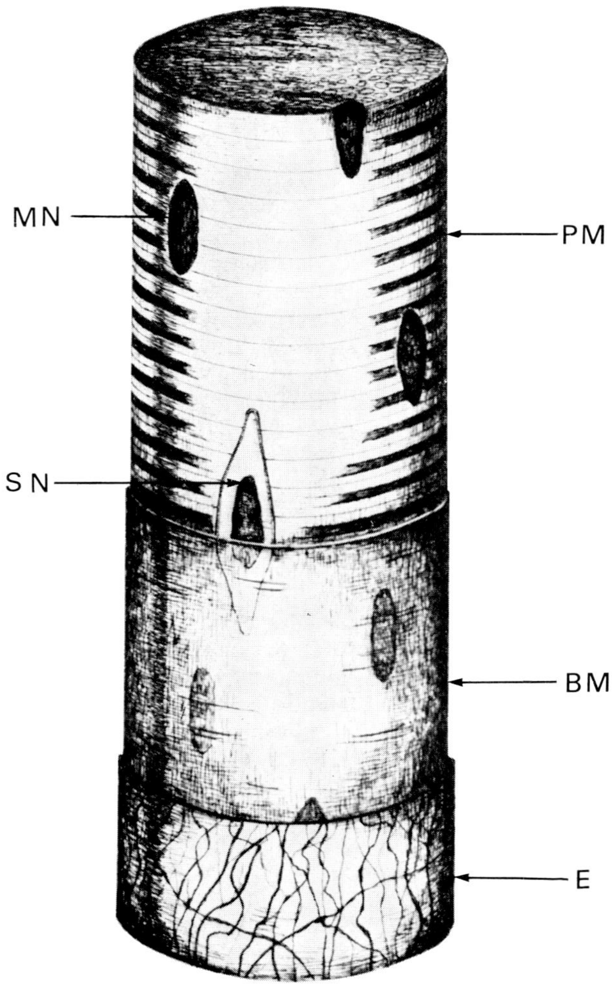

Fig. 73. Diagram showing the relations of a satellite cell (SN) to the plasma membrane (PM), basement membrane (BM) and endomysium (E) of a skeletal muscle fibre. The number of myonuclei (MN) and the length of the segment are drawn in proportion.

of finer filaments which show no periodicity. Components of the endomysium can be demonstrated by conventional light-microscopic methods; reticulin can be impregnated with silver and the connective tissue mucopolysaccharides of both endomysium and basement membrane can be stained by the PAS technique (MacConnachie *et al.*, 1964). The resolution of the instrument prevents a distinction being made between these 2 layers.

DEFINITION

The satellite cell of striated muscle can be defined as a mononuclear cell, whose cytoplasm does not contain myofilaments, and which is enclosed by or lies within the basement membrane component of the sarcolemma of a striated muscle fibre. This definition does not exclude cells which are partially or completely separated from the muscle fibre plasma membrane by extensions of the basement membrane, providing that the basement membrane forms a complete investment on their outer surfaces.

IDENTIFICATION

The only absolute criterion for identification is the location of satellite cells as defined above. As the combined basement and plasma membranes are less than 500 Ångströms thick, it is evident that positive initial identification depends on electron-microscopic studies. Their inclusion within the basement membrane distinguishes them from endothelial cells, pericytes, macrophages, and Schwann cells associated with muscle innervation and endomysial fibroblasts, but does not differentiate them from myonuclei in the muscle fibre syncytium. The latter distinction depends on the demonstration of the plasma membrane surrounding the satellite cell.

In some species and muscles which have been studied by light microscopy, the satellite cell nuclei resemble myonuclei sufficiently to be distinguished from the slender long dense nuclei of fibroblasts and endothelial cells and from the large vesicular nuclei of macrophages. In other muscles, a satellite cell nucleus may be separated from a myonucleus in paraffin or plastic embedded specimens by the clumped chromatin and the absence of a nucleolus (Venable, 1966*b*). These nuclear characteristics are shared by the adjacent fibroblasts and endothelial cells and so identification depends on location within the basement membrane. It is evident that light-microscopic quantitation of satellite cells will contain some errors, but with skill and wise choice of material the error may be admissible. These remarks apply to control adult muscle and it will introduce confusion if other than electron-microscopic criteria are used for their identification in experimental or developmental material where variations occur in the structure of these cells and their nuclei. It is worth noting that the literature on skeletal muscle prior to 1961 appears not to contain references to 2 populations of nuclei (Meigs, 1932; Barer, 1948; Walls, 1960).

DISTRIBUTION

After satellite cells had been demonstrated by Mauro (1961) in frog skeletal muscle, many investigators must have confirmed their existence in other skeletal muscles. Published reports (see p. 94) show a wide distribution in adult muscle:

Satellite cells have not been observed in mammalian cardiac muscle (Stenger and Spiro, 1961; Muir, 1965) and the brachyuran cardiac muscle containing them is structurally the same as, and functionally resembles, vertebrate skeletal muscle. Red and pale muscle fibres contain a complement of satellite cells (Muir *et al.*, 1965) and they are present on intrafusal fibres (Katz, 1961; Karlsson *et al.*, 1966; Landon, 1966). On intrafusal fibres, satellite cells are concentrated in the compact sensory zones, but they are distributed randomly along the length of ordinary extrafusal fibres. They are no more numerous at the myoneural or myotendinous junctions.

Estimations of the frequency of satellite cells lack precision because they must depend on small electron-microscopic samples. The proportion of the nuclei, within the basement membrane, which belong to satellite cells has been variously given as 10 per cent (Muir *et al.*,

Phylum	*Class*	*Species*	*Muscle*	*Author*
Chordata	Mammalia	Human	—	Laguens (1963), De Recondo *et al.* (1966), Shafiq *et al.* (1967)
		Human	Gastrocnemius	Ishikawa (1966)
		Pig	Biceps femoris	Muir (1970)
		Rat	Sartorius	Palade, quoted by Mauro (1961)
		Rat	Tongue	Palade, quoted by Mauro (1961)
		Rat	Intrafusal fibres	Landon (1966)
		Mouse	Levator ani	Venable (1966*b*)
		Guinea-pig	—	Birks, quoted by MacConnachie (1962)
		Bat	Pectoral, web, tongue etc.	Muir *et al.* (1965), Church (1968)
	Aves	Fowl	Ext. hallucis	(Fig. 75)
	Amphibia	Frog	Tibialis anterior	Mauro (1961)
			Sartorius	Mauro (1961)
			Ileofibularis	Mauro (1961)
			Intrafusal fibres	Katz (1961), Karlsson *et al.* (1966)
		Axolotl	Trunk	Flood (1964)
	Marsipobranchii	Myxine	Trunk	Flood (1964)
Arthropoda	Brachyura	Crab	Cardiac	Midzukami (1964)
	Insecta	Rhodnius	Abdominal wall	Toselli and Pepe (1968)

1965), 4.8 to 5.8 per cent (Venable, 1966*b*) and 4 per cent (Flood, 1964, quoted by Venable, 1966*b*). Allowing a mean nuclear content of 80 per millimetre length, these figures mean that there is 1 satellite cell contained within each 125 to 250 microns length of a skeletal muscle fibre.

SHAPE

The satellite cells are fusiform with their long axes directed along the muscle fibres. They reach 25 microns in length, but are less than 5 microns wide and 4 microns deep. They often nestle into a depression made by the plasma membrane of the muscle fibre, so they do not distort the cylindrical shape of the enclosing basement membrane.

STRUCTURE (Fig. 74)

The nuclei of satellite cells are often similar to, and may be indistinguishable from, the myonuclei in the associated muscle fibre. Muir *et al.* (1965), Flood (1964), Midzukami (1964) and Laguens (1963) could notice no significant differences, but Ishikawa (1966) and Venable (1966*b*) observed that the chromatin material of the satellite cell was less evenly dispersed and that this clumping was accentuated by aldehyde fixation. Church (1968) draws attention to the presence of nucleoli in the myonuclei, to a high nucleo-cytoplasmic ratio and to the more fusiform shape of satellite cell nuclei. Most authors agree that the satellite cells are mononucleate, with the double nuclear membrane forming an uncrenated envelope with nuclear pores. Species differences undoubtedly occur as the mouse satellite cell nuclei measure $13 \times 4.5 \times 2.8$ microns compared with $16 \times 7 \times 2.5$ microns myonuclei (Venable, 1966*b*), while the nuclei of satellite cells in crustacean cardiac muscle are only 5 to 7 microns long when the ordinary nuclei measure 7 to 9 microns (Midzukami, 1964). In avian muscle the satellite cell nuclei are noticeably flattened, 2.5×1 microns in transverse section com-

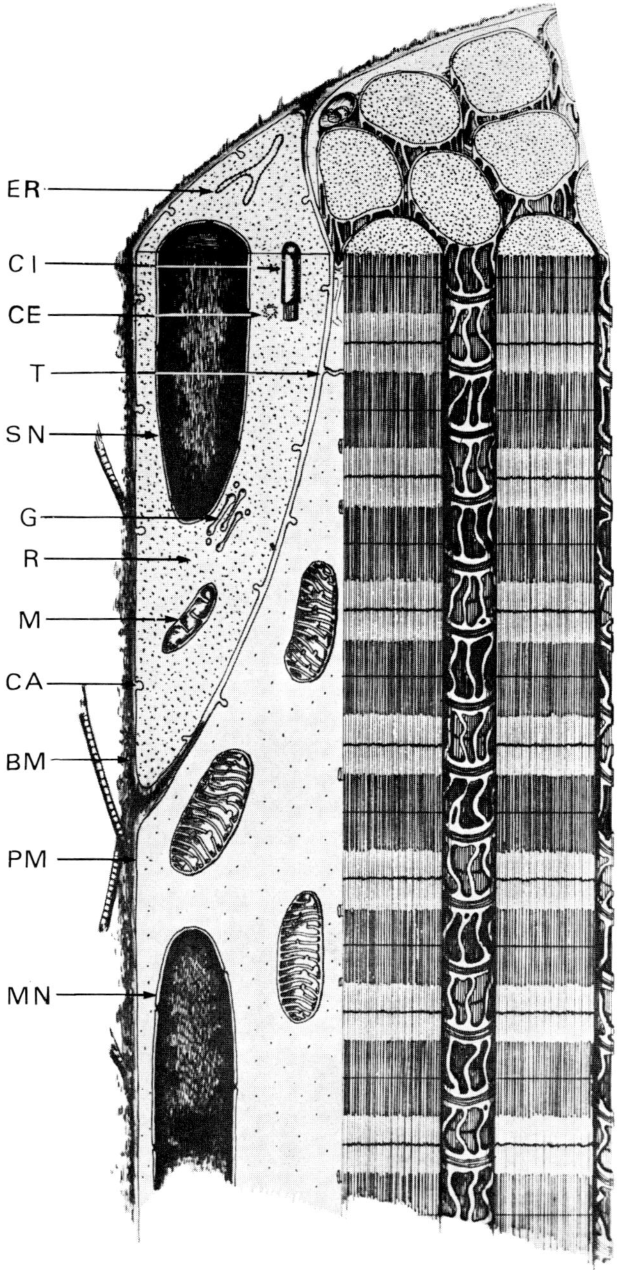

Fig. 74. Diagram summarizing the fine structural features of a satellite cell: BM, basement membrane; CA, caveolus; CE, centriole; CI, cilium; ER, endoplasmic reticulum; G, Golgi apparatus; M, mitochondrion; MN, myonucleus; PM, plasma membrane; R, ribosomes; SN, satellite cell nucleus; T, T-tubule opening into the cleft between the satellite cell and muscle fibre.

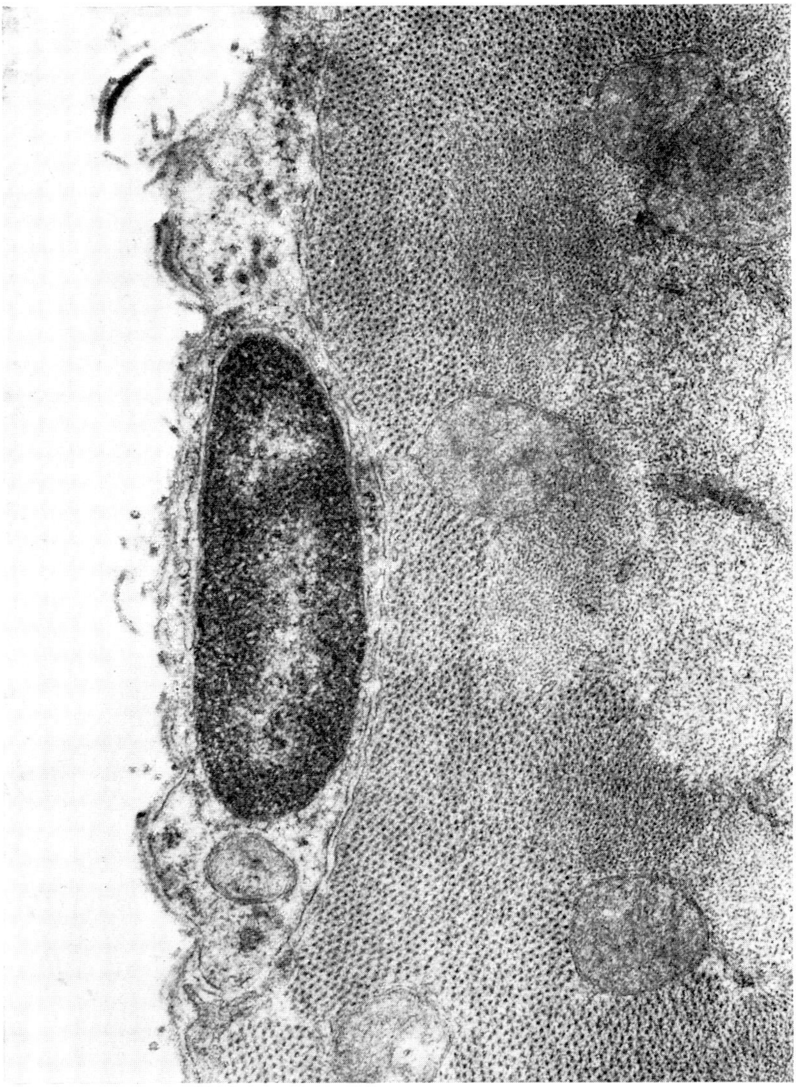

Fig. 75. Satellite cell on a white muscle fibre from the extensor hallucis longus muscle of an adult fowl. Osmium fixation. Araldite embedding. × 22,000.

pared with the almost tubular myonuclei, 3 × 3 microns, most of which are embedded in the myofibrillar material even of pale muscle fibres.

Flattened membranous sacs forming the Golgi apparatus lie close to the nucleus and occasionally pairs of similar sacs are scattered through the cytoplasm. A pair of centrioles is present either deep or superficial to the middle of the satellite cell nucleus. In bat muscles, one of these centrioles gives rise to an atypical cilium with 9 × 2 + 0 filaments which project in a membrane-lined cavity for a few microns (Muir *et al.*, 1965; Church, 1968), but this feature has not been noted by other authors on other species. It is perhaps noteworthy that centrioles have never been illustrated in the sarcoplasm of the multinucleated skeletal muscle fibre.

Very few mitochondria are present and they are smaller and contain fewer cristae than in the adjoining muscle fibre.

The cytoplasmic matrix contains numerous ribosomes and some vesicles, but these are only rarely combined to form isolated sacs of rough-surfaced endoplasmic reticulum. Glycogen has not been identified and it is noticeably absent from the satellite cells of amphibia where the adjacent sarcoplasm is heavily infiltrated (Flood, 1964).

If myofilaments are present in a cell contained within the sarcolemma, this cell is not encompassed by the definition given above. This appearance is common in developing muscles and may precede longitudinal splitting by basement membrane invasion and so account for the increase in number of muscle fibres (Boyd, 1960). A similar arrangement could be observed in adult muscle if the tapering end of one multinucleated fibre is transected. Only De Recondo *et al.* (1966) report the presence of microtubules and they also note non-myofibrillar filaments, which are especially abundant in the satellite cells associated with atrophied fibres. Lysosomes and other dense bodies have not been reported in satellite cells.

The plasma membrane on the internal and external aspect of the cell carries a large number of pinocytotic caveoli; these are also evident on the underlying muscle fibre plasma membrane. Coated vesicles (Fawcett, 1965) are not present although they are frequently seen on mononuclear cells and myotubes in developing and regenerating muscle. The transverse tubules of the muscle fibre open onto the surface membrane where it is apposed to the satellite cell (Midzukami, 1964). Some indentation of the satellite cell by the plasma membrane of the fibre is noted by Flood (1964), but most micrographs show the internal satellite cell plasma membrane running a straight course separated by some 150 to 1000 Ångströms of extracellular space from the muscle plasma membrane. Only in brachyuran cardiac muscle (Midzukami, 1964) are there specializations of the 2 plasma membranes to form desmosomes and close junctions.

The plasma membrane of the external surface is applied to the basement membrane with an intervening clear gap of about 150 Ångströms. The material of the basement membrane may penetrate the intercellular gap on the inside of the satellite cell (Venable, 1966*b*; Church, 1968). This penetration does not form a complete capsule for the satellite cell, but its presence may indicate some restriction of satellite cell migration.

Continuity between the satellite cell cytoplasm and the adjoining sarcoplasm is described by De Recondo *et al.* (1966), who suggest that some satellite cells apparent in transverse sections are merely pimples growing from the surface of the parent fibre. Similarly, in regenerating muscle, Reznik (1968*a*) describes myonuclei passing through a satellite cell morphology by the development of separating membranes. Obviously, these suggestions cannot be excluded on morphological evidence, but a simple experiment shows that the normal complement of satellite cells is osmotically independent of the adjoining muscle fibre (Muir *et al.*, 1965).

MITOTIC DIVISION

The existence of centrioles in satellite cells indicates the capacity to divide mitotically, but this has not yet been demonstrated in control adult or growing muscle. After colchicine administration to 3-week-old rats, Shafiq *et al.* (1968) were able to show an electron micrograph of an arrested division of a satellite cell. In adult pig biceps femoris, Muir (1970) shows a telophase, but this muscle also shows evidence of regeneration in other fibres. It is likely that the dividing nuclei, which seem to be within the muscle fibre in growing rat muscle examined by light microscopy, are truly divisions of satellite cells (MacConnachie *et al.*, 1964).

COMMENTS

If it is assumed that the cells forming the basophilic myoblasts are derived from satellite cells, the problem of their activation remains. It could be due to the change in their chemical or thermal environment at injury, or possibly a release from the physical restrictions of the intact fibre enclosed by the basement membrane. A cell could not be better placed to detect such physical or chemical changes than one occupied by the satellite cells which are stuck on the inside of the basement membrane.

The relationship between the satellite cell and the muscle fibre must be a subtle one, because in growing muscle there is an increase in muscle nuclei and the evidence of MacConnachie *et al.* (1964) and Shafiq *et al.* (1968) indicates that this is due to satellite cell proliferation. So this implies that an enlarging muscle fibre can, in some way, trigger satellite cells to divide when its nucleo-cytoplasmic ratio is falling.

A useful analogy can be drawn to the only other multinucleated syncytium in a mammal. The syncytiotrophoblast of the placenta is the only other tissue which contains more than, say, 100 nuclei within a single plasma membrane envelope. There is a cytotrophoblast, which is composed of mononucleated cells and related to the basement membrane in the same fashion as the satellite cell of muscle. In the placenta there is good evidence that the syncytial nuclei are derived from the mononucleated cytotrophoblast cells (Enders, 1965). There is also evidence of intercellular connections, such as desmosomes, lying stranded in syncytial cytoplasm after cellular fusion. A similar relationship obtains in skeletal muscle, but the absence of specialized cell junctions between satellite cell and myotube excludes the possibility of similar evidence for fusion.

SUMMARY

1. A morphologically discrete population of mononuclear cells is enclosed within the basement membrane of multinucleated striated muscle fibres obtained from various vertebrate and invertebrate species.
2. These enclosed satellite cells are fusiform in shape and their cytoplasm contains the usual organelles, but no specializations indicating phagocytic or secretory activity.
3. The presence of centrioles suggests that the satellite cells can divide mitotically.

J. H. Venable: In the satellite cell shown diagrammatically as Fig. 74 you have a sarcoplasmic space between the myonucleus and the myofibrils. In my studies in rats and mice I have never seen such a big space as you have shown in the drawing. I found that a space occurred less than once in every 200 muscle nuclei, but I wanted to point out that such a space as you have is greatly exaggerated.

A. R. Muir: I agree. In order to get everything in the satellite cell, the subsarcolemmal space in the muscle fibre has been enlarged. Also, the nucleo-cytoplasmic ratio of this satellite cell gives the impression of too much cytoplasm. The only excuse is that only part of the nucleus is represented in the diagram.

J. H. Venable: I bring this point, because you can identify satellite cells on 3/4-micron sections by light microscopy as these have a narrow separating space.

Unidentified speaker: With regard to possibly identifying satellite cells under the light microscope, now that you bring this up, I note that in looking in axolotl muscle and attempting to decide whether a given nucleus is within or without the cytoplasm, I use the criterion that a muscle nucleus does not have any basophilia associated with it, whereas occasionally there are cells which are very intimately associated with the sarcoplasm, which have adjacent to their ends, with azure B staining, very intense basophilia. It is intensely purple just like the nucleoli. Perhaps what I was actually observing was a satellite cell in which your large concentration of ribosomes will show enough basophilia with azure B.

A. R. Muir: The concentration of ribosomes varies considerably. I hope I am not confusing the established criteria for identification which will pick up the satellite cells by light microscopy. What I find hard while looking at them is to believe that any light-microscope criterion will distinguish them all. Any criterion you use would need to be checked on that particular nucleus from the thin section to the thick and this is a tedious and difficult matter.

J. H. Venable: In the pig muscle which you refer to is there necrosis and regeneration?

A. R. Muir: Yes.

J. H. Venable: Did it have any trichinosis?

A. R. Muir: No. It had a weight of 190 pounds and was a perfectly healthy specimen of the Belgian breed of Pietrain pig. I draw attention to this matter to pose the question: Have we seen the division in satellite cells in apparently normal muscle, and then certainly found some regeneration occurring? I wonder if an inherent regenerative process explains the differences in regeneration rates which were mentioned this morning. In adult muscle, you may have a low rate of spontaneous death and replacement, and this may affect the number of satellite cells which are in division.

P. E. Conen: Had the pigs been running around their paddock smashing into things?

A. R. Muir: No. In a given section, the muscle fibres will be at all stages of regeneration. They are widely separated from each other and also they are distributed in most of the muscles of the body. I found them particularly commonly in the ventral serrate. So I do not think that physical trauma has initiated the regeneration.

Unidentified speaker: It occurred to me that one way of controlling those satellite cells is how closely they communicate through things like tight junctions. One good way of controlling such activities would be coupling and uncoupling the communication. One could easily imagine, for instance, that ordinarily the satellite cells are coupled to a multinuclear muscle fibre. Damage to a muscle fibre could cause uncoupling.

A. R. Muir: All I can say is that I have never seen tight junctions between the satellites and the muscle fibre.

Unidentified speaker: You mentioned that one of the features of the satellite cell is the association of the centriole of the nucleus. I wonder what evidence there is that muscle fibre nuclei have associated centrioles.

A. R. Muir: As stated in the manuscript, I have never seen or heard of a description of a centriole in a multinucleated muscle fibre. In young developing heart muscle I have seen them and there is a report by Manasek (1968).

E. D. Hay: There has to be a difference because these are single cells. Heart muscle cells continue to proliferate even while they are making myofibrils.

Unidentified speaker: There is a long time when you see the formation of myofilaments and myofibrils in multinucleated cytoplasmic tissue, and sometimes a centriole can be picked up.

Unidentified speaker: Can I get back to the question of the light-microscope identification of satellite cells. You pose an impossible question, Dr. Muir. As you know, I think the evidence for it is that if you apply the criteria that Dr. Venable suggested in his paper, which is a highly chromatinic nucleus with a little halo around it, looking at it in thin sections of plastic-embedded material, the numbers of cells that you see in a given muscle come out at about the same number as you see when you do electron microscope counts, so that this suggests to me the fact that you can in those particular muscles, at any rate, identify all of them under the light microscope, and this simplifies our job considerably.

A. R. MUIR: I go a long way with you but you could get the same figures by including a few endomysial fibroblasts and losing a few satellite nuclei that happen to resemble myonuclei.

CELL QUANTITATION IN REGENERATING BAT WEB MUSCLE

JOHN C. T. CHURCH

I should like to show you first a few pictures of the fruit bat *Eidolon helvum* (Kerr), because I think it is of general interest.

Figure 76 shows fruit bats roosting in trees. They roost in large colonies. A particularly well-known colony, within a mile of the Medical School at Kampala, Uganda, contains between 250,000 and 500,000 animals.

Figure 77 shows a female bat in captivity, with a youngster at the breast. The 2 mammary glands are in the axilla. You can see how relatively large this youngster is. Young are born at about 1/4 to 1/5 of the adult weight.

Figure 78 shows a dead adult fruit bat, laid out to show its general morphology. The adult wing span is about 3 feet. The part we are most interested in is the main web (W). Notice that the leg is as much part of the flight apparatus as the wing, extension of both leading to maximal stretching of the webs.

These bats are vicious animals, and difficult to handle, which is perhaps the reason why they have not been used before as experimental animals. Consequently, we first had to devise a way of handling them in the laboratory. We designed a simple apparatus consisting of a perspex base board on which the bat is extended supine (Fig. 79). The wings are held with foam-lined batons (B), so that the main webs lie on the perspex board. The feet are held with rubber bands passed through holes in the perspex. In this way the bat can be very adequately immobilized, particularly when anaesthetized. I might add that these animals have a range of characters: some are more docile like this one, and some perverse. For long-term experiments it is wiser to choose a docile one! This particular animal is being used for a wound healing experiment; a full thickness portion of the right web, 2 square centimeters, has been excised about 2 weeks previously, and goes on in fact to heal completely about 4 weeks after injury.

We have studied the main web (Fig. 80) in some detail. It is, I would suggest, a unique tissue, consisting of dorsal and ventral skin layers with a thin sheet of connective tissue between them, in which is a most remarkable arrangement of small skeletal muscles (M), lying mostly in parallel. The main neurovascular supply to the web (NV) comes in at the elbow, and crosses the muscles, to extend obliquely down to the distal corner of the web, where there is a smaller group of muscles, diminishing in size, the smallest being about 1 centimeter long.

The longer muscles, which I designate the 'ulnar' group, are 8 to 10 centimeters long in the adult, and end in little elastic tendons which are attached to the ulnar border of the forearm at one end and the free edge of the web at the other. Each muscle has its own neurovascular supply, branching from the main axis. The muscles contain 500 to 800 fibers in transverse section. Individual fibers, however, are quite short, with a mean length of about 10 millimeters. The fibers are distributed randomly along the muscle, and this seems to be a distinctive feature of these muscles. We had hoped initially that we would find the end-plates in the centre part of the muscles, but this was not so; they are scattered in groups along the course of the muscles.

Fig. 76

Fig. 77

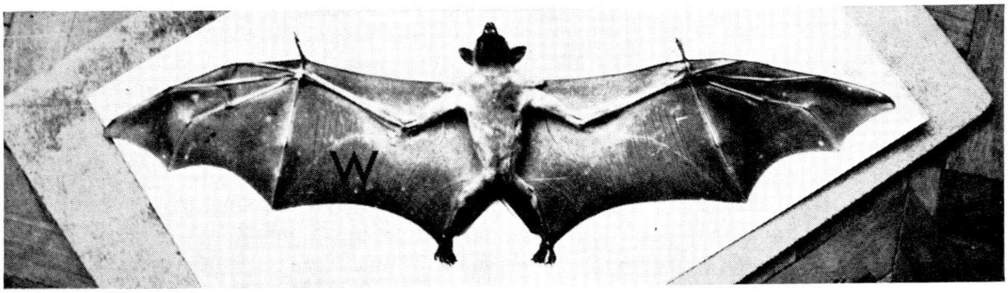

Fig. 78

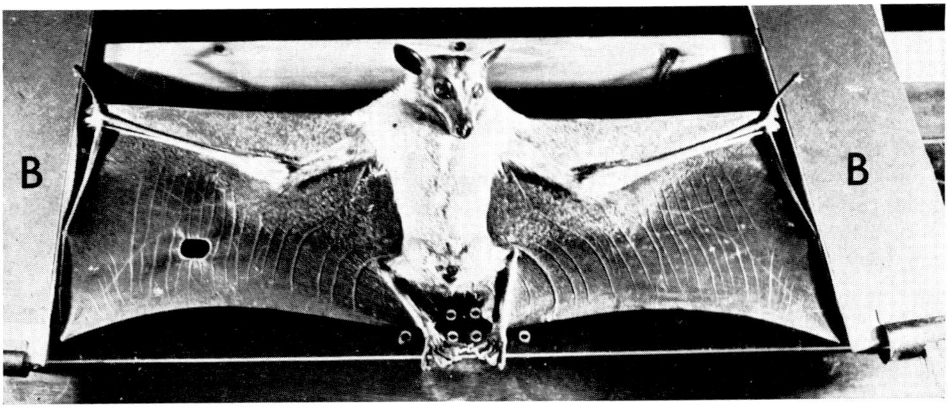

Fig. 79

Fig. 76. Fruit bats, *Eidolon helvum* (Kerr), roosting in eucalyptus trees.

Fig. 77. Adult female fruit bat in captivity, with youngster at the breast.

Fig. 78. Adult fruit bat, extended to show its general morphology. W, the main web.

Fig. 79. Adult fruit bat, extended on a perspex frame, for experimental use. B, plastic foam-lined batons holding the wings.

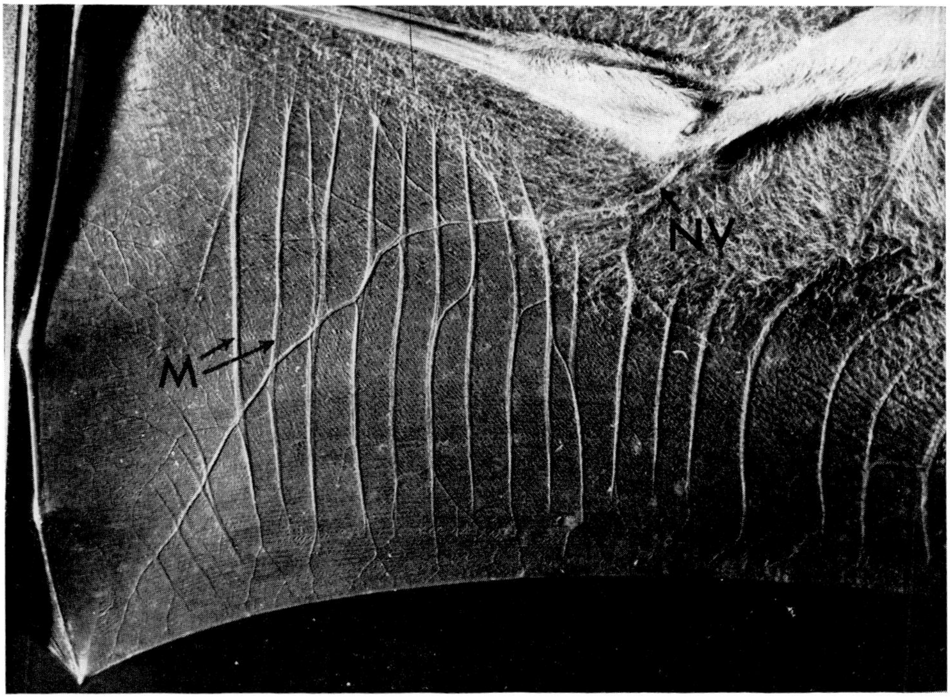

Fig. 80

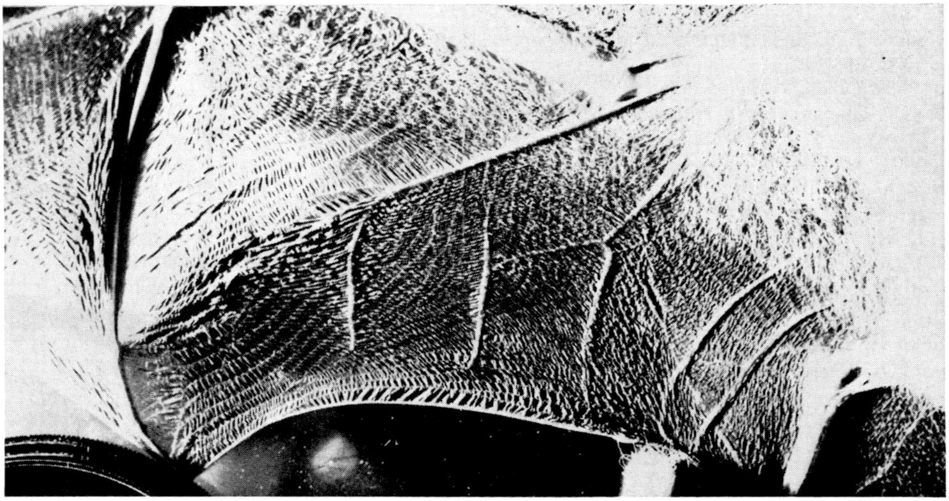

Fig. 81

Fig. 80. The right main web of an adult fruit bat, showing the web muscles (M), and their neurovascular supply (NV) entering the web at the elbow. × 0.6.

Fig. 81. The right main web of a vespertilionid bat, showing the web muscles, one of which is labeled (M). × 1.2.

The next picture (Fig. 81) is to illustrate that most bats, perhaps all, have web muscles. Though well developed in the megachiroptera, such as the East African fruit bat *Eidolon helvum* (Kerr), they are nonetheless present in the microchiroptera, and would certainly be present in various species in the United States. Their size and distribution varies, of course, from species to species. In this one, a vespertilionid bat from Uganda, there are several isolated little muscles, the central one (M) presenting an excellent neuromuscular preparation for *in vivo* study.

Low-power light microscopy of a web muscle in transverse section (Fig. 82) shows little but the centrally placed neurovascular bundle in the muscle, and the dorsal (D) and ventral (V) skin on either side of it. At higher magnification (Fig. 83) individual muscle fibers can be made out. With the limited number of fibers, their quantitation in transverse section of the whole muscle is eminently feasible. High-power light microscopy (Fig. 84) reveals further cell detail: myelinated axons (A), myonuclei (My), capillaries (C), endomysial fibroblast (EF), and satellite cells (S). These 3 pictures (Figs. 82-84) are of a section 1 to 2 microns thick cut from araldite-embedded material. Consecutive thin sections for electron microscopy were then cut from the area shown in Figure 84. Figure 85 shows some of the cells seen in Figure 84. The satellite cell (S) is enlarged in Figure 85, to show the adjacent plasma membranes (PM) between it and the fiber syncytium, and the basement membrane around both (BM). By working in this way we have gained considerable experience in the recognition of the different cell types in bat web muscle, and in particular in distinguishing satellite cells from myonuclei and endomysial fibroblasts. I must slightly disagree with Professor Muir here, that at least in the bat web muscle there are distinctive morphological characteristics, seen both in light and electron microscopy, between the majority of satellite cells and myonuclei, so that they can be distinguished from each other. Grouping Figures 82 to 86 together demonstrates how one can correlate light- and electron-microscopic appearances.

UNIDENTIFIED SPEAKER: On the right corner, down (Fig. 84), you have a satellite cell. What is the cell just above?

J. C. T. CHURCH: An endomysial fibroblast (EF, Figs. 84 and 85).

UNIDENTIFIED SPEAKER: At the electron-microscope level, you can recognize 2 different types of cells, but in the light micrograph they just look the same. How can you say that one is a satellite cell and the other a fibroblast?

J. C. T. CHURCH: I agree that these 2 cells (S and EF, Fig. 84) look similar in their shape and in their density with the staining used. But we can distinguish them in light microscopy by other criteria. For instance, endomysial fibroblasts have extracellular spaces on either side of them, containing collagen, whereas satellite cells will, in the majority of cases, adopt the contour of the fibers against which they lie, as does the one shown (S, Figs. 84-86).

UNIDENTIFIED SPEAKER: But sometimes you may have a fibroblast which appears like a pocket in the fiber.

J. C. T. CHURCH: This is possible, though it is not a feature of the bat web muscle. I am concerned here with the accuracy with which one can recognize satellite and other cells in light microscopy, in order to quantitate them. We could go into further detail of the various criteria used for their recognition. One, for instance, is the shape of the cytoplasm. The cytoplasm of these fibroblasts is often drawn out into long protrusions, extending between the muscle fibers. Satellite cell cytoplasm is more compact. One cannot build a story around 1 picture, of course.

S. A. SHAFIQ: What about the nucleus on the left side in Figure 84?

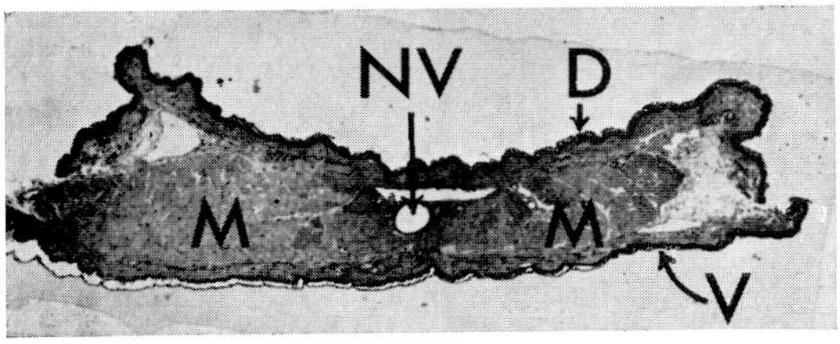

Fig. 82

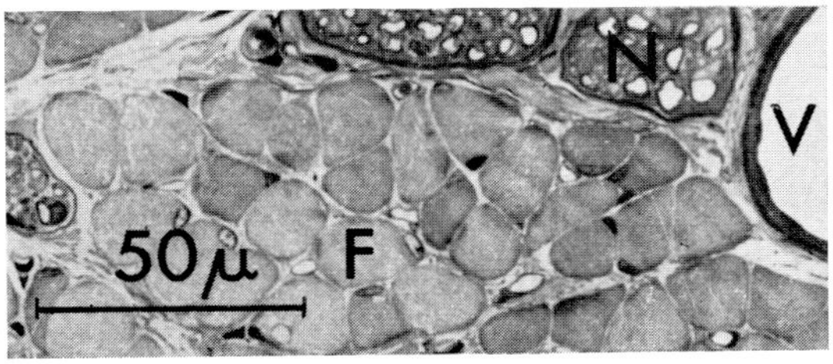

Fig. 83

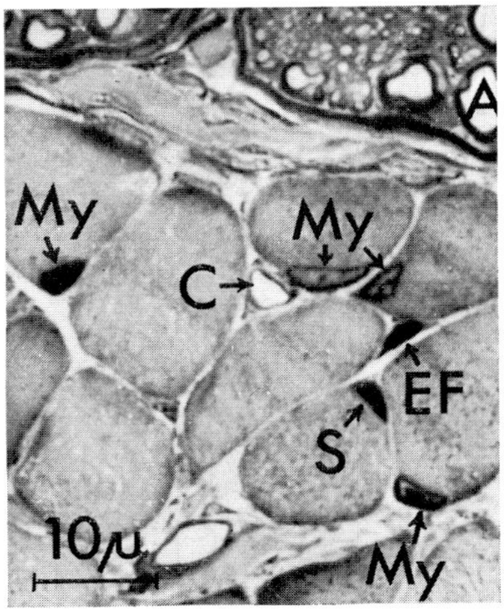

Fig. 84

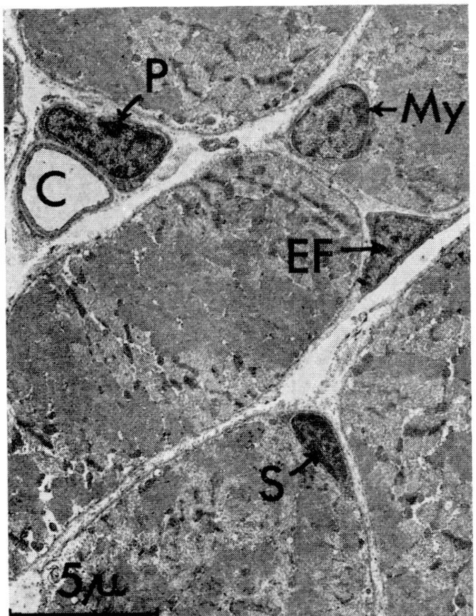

Fig. 85

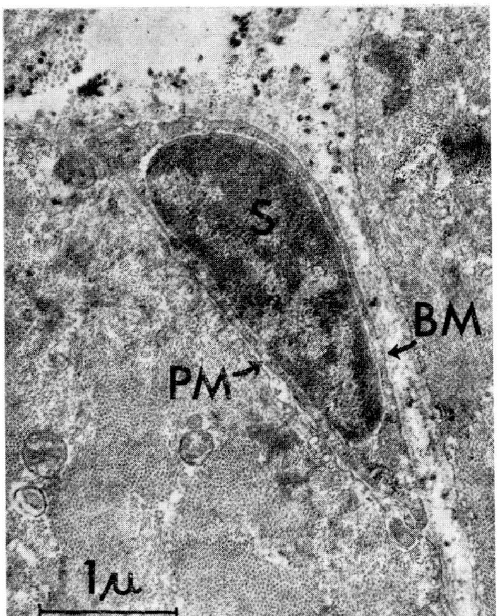

Fig. 86

Fig. 82. Light photomicrograph of part of a web, containing a web muscle (M) in transverse section, showing dorsal (D) and ventral (V) skin, and the neurovascular supply (NV) to the muscle. × 46.

Fig. 83. The centre part of Fig. 82 enlarged, to show part of a vessel (V), nerve (N), and muscle fibers (F). × 450.

Fig. 84. Part of Fig. 83 enlarged, to show myonuclei (My), an endomysial fibroblast (EF), a satellite cell (S), capillary (C), and myelinated axon (A). × 1,250.

Fig. 85. Low-power electron photomicrograph of part of Fig. 84, showing the same cells; satellite cell (S), endomysial fibroblast (EF), a myonucleus (My), capillary (C), and a pericyte (P). × 3,600.

Fig. 86. The satellite cell (S) of Fig. 85 enlarged, to show the adjacent plasma membranes (PM), and the basement membrane (BM) enclosing both satellite cell and fiber syncytium. × 16,300.

J. C. T. CHURCH: I think the nucleus (My) on the left in Figure 84 is another myonucleus. Though it is rather densely stained, it contains a nucleolus and does not have a rim of cytoplasm around it. Detail is lost in the photography – more detail can be made out in oil-immersion high power, with focus control.

UNIDENTIFIED SPEAKER: How frequently are the endomysial fibroblasts in such close association as that satellite cell?

J. C. T. CHURCH: I do not know. I would assume that it is not very frequent. In the bat web muscle it is very rare.

Fig. 87

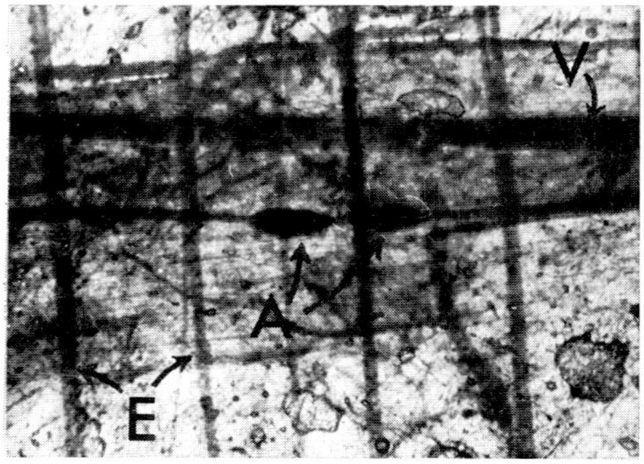

Fig. 88

Fig. 87. Three types of apparatus used to make crush lesions to the bat web muscles. × 0.21.

Fig. 88. Part of a web, transilluminated *in vivo*, showing elastin strands (E), a vein (V), and small aneurysms (A) in the artery, indicating the site of a previous crush lesion. × 17.

To continue, Figure 87 shows the types of apparatus used to inflict small crush lesions to the bat web muscles. The crushing block is of different sizes. That used for most experiments was 1 × 8 millimeters. By placing it accurately over a desired point, and placing a weight on it for a given time, one can deliver a standard crush lesion, which is of course repeatable many times. One of the great advantages of the web is that it can be transilluminated, so that a lesion can be studied *in vivo* (Fig. 88). These lesions can be identified by the fusiform aneurysms which develop in the arteries (A) of the crushed segment of muscle. These aneurysms are due to damage to the smooth muscle fibers of the arteriolar wall, which are disrupted in similar fashion to the striated fibers of the web muscles.

In low-power light microscopy, in longitudinal section (Fig. 89), a crush lesion is quite distinctive, with disruption and destruction of the sarcoplasm, the damage being maximal at either end of the lesion. Higher magnification (Fig. 90) shows the gross disorganization of the sarcoplasm (Sa), with the basement membranes (BM) distorted but still intact. These can also be seen in transverse section (BM, Fig. 91). Macrophages (M) appear within 24 hours of injury, and are found both between and within the damaged fibers.

UNIDENTIFIED SPEAKER: Is that a satellite cell, dead in the centre of Figure 91?

J. C. T. CHURCH: The small structure (S?, Fig. 91) could be the tail end of a satellite cell. In looking at many of these lesions soon after injury, we sometimes saw intact satellite cells lying within empty basement membrane tubes, or in tubes containing grossly disorganized sarcoplasm. This corroborates Professor Muir's work, when he showed by a simple osmotic

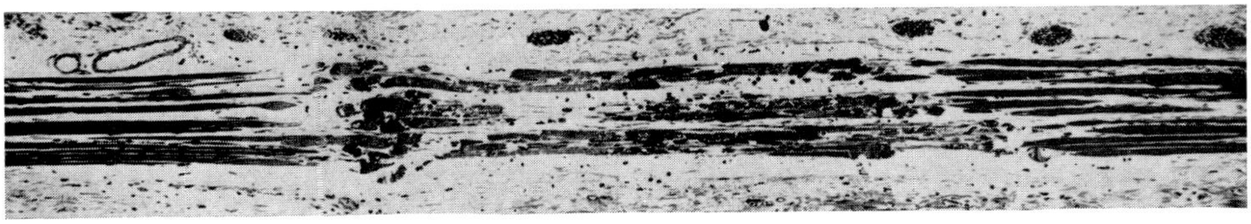
Fig. 89

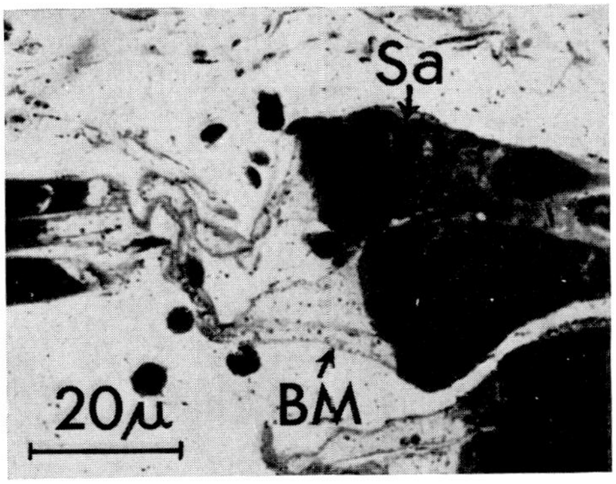

Fig. 90

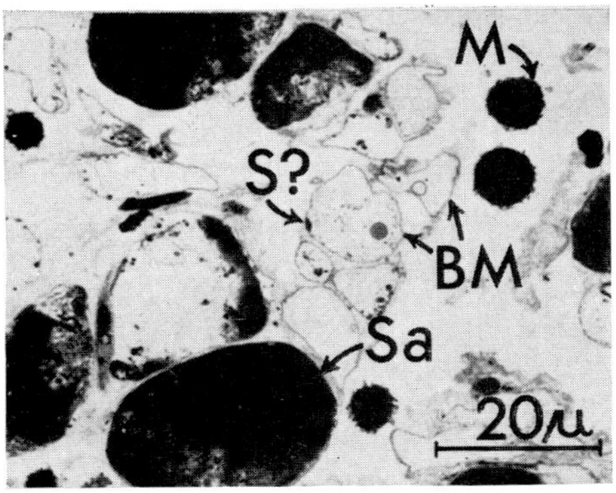

Fig. 91

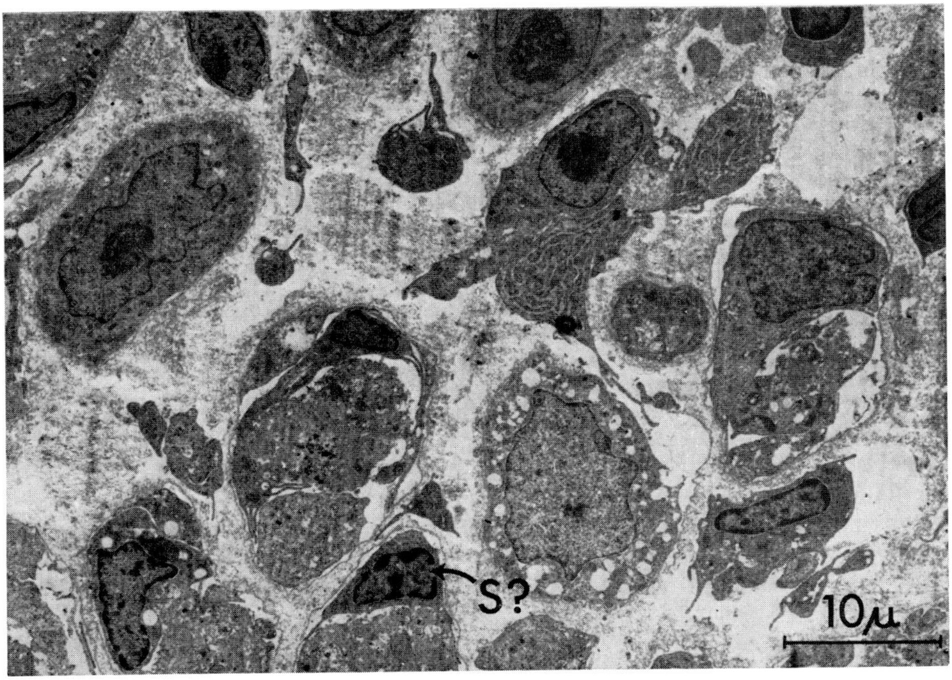

Fig. 92

Fig. 89. Longitudinal section of a crush lesion to a web muscle, 2 days after injury. × 59.

Fig. 90. Severely damaged muscle fibers in a crush lesion, in longitudinal section, showing the disorganized sarcoplasm (Sa), and empty basement membrane tubes (BM). × 720.

Fig. 91. The same subject as Fig. 90, but in transverse section, showing disorganized sarcoplasm (Sa), basement membrane tubes (BM), macrophages (M), and possibly the end of a satellite cell (S?) lying under a basement membrane. × 800.

Fig. 92. Low-power electron photomicrograph of crushed muscle fibers in transverse section, 6 days after injury, showing the pleomorphism of the cells, one of which (S?) has certain features of a satellite cell. × 2,000.

experiment in the bat web that the satellite cells are distinct from the syncytial cytoplasm, in their response to osmotic insult. It is interesting that the forces which are seemingly strong enough to destroy sarcoplasm in this way do not disrupt the smaller satellite cells, or dislodge them from their situation beneath the fiber basement membrane.

Within the first few days after injury, there is tremendous cytological activity in these crushed muscles. Certain cells, such as polymorphonuclear leucocytes and macrophages, can be recognized, but by about 6 days after injury numbers of presumptive myoblasts have appeared, and there is such pleomorphism (Fig. 92) that accurate cell identification is impossible. We could say that some cells are somewhat similar to satellite cells (S?, Fig. 92), in that they have scanty cytoplasm, nuclei of similar appearance, and lie under the basement membrane, but beyond that we cannot go.

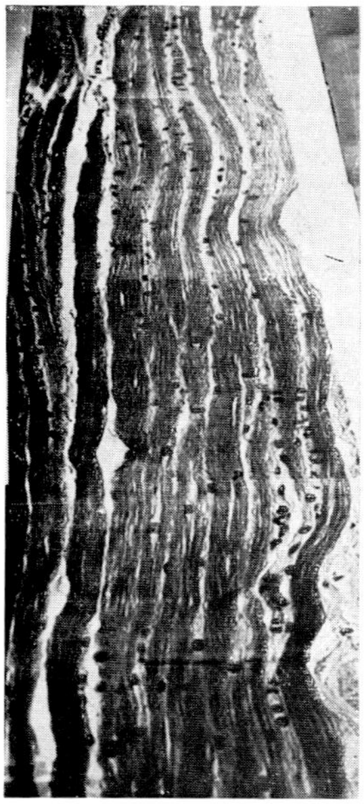

Fig. 93

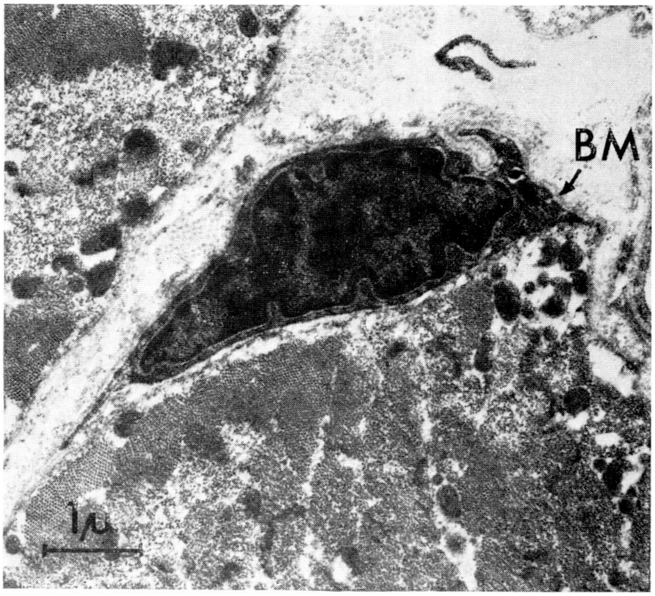

Fig. 94

Fig. 93. A crush lesion in longitudinal section, 4 weeks after injury, mounted as a montage. This is viewed very obliquely to demonstrate the continuity of muscle fibers through the previously crushed region, which occupies about the middle 1/3 of the Figure.

Fig. 94. An electron photomicrograph of muscle fibers in transverse section, from a region crushed 8 weeks and 4 weeks previously showing the well-regenerated fibers, and a satellite cell (S) lying under the basement membrane (BM). × 12,200.

The next picture (Fig. 93) is perhaps a bit difficult to understand. This is a montage of many photographs of the muscle region that was previously damaged, showing the extent of recovery after 4 weeks. The montage is viewed very obliquely to demonstrate that fibers pass right through the previously crushed region, which occupies about the middle 1/3 of the Figure.

Figure 94 shows a satellite cell (S) from the centre of an injured region, crushed 8 weeks and 4 weeks previously. By this time, despite the double injury, the muscle fibers have again regenerated sufficiently for the syncytium to be restored to near-normal appearance, and for the satellite cells to be recognizable.

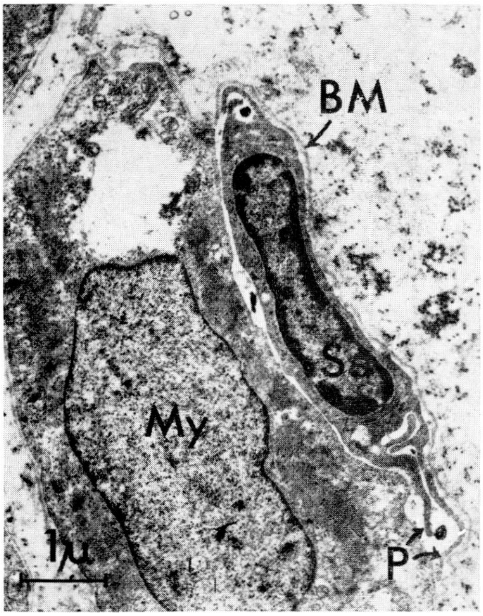

Fig. 95

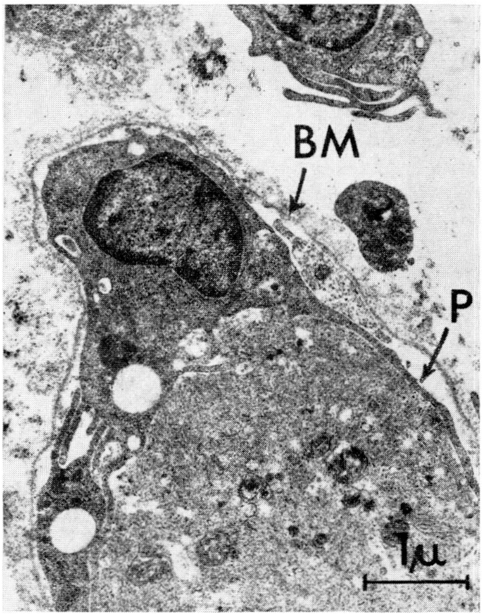

Fig. 96

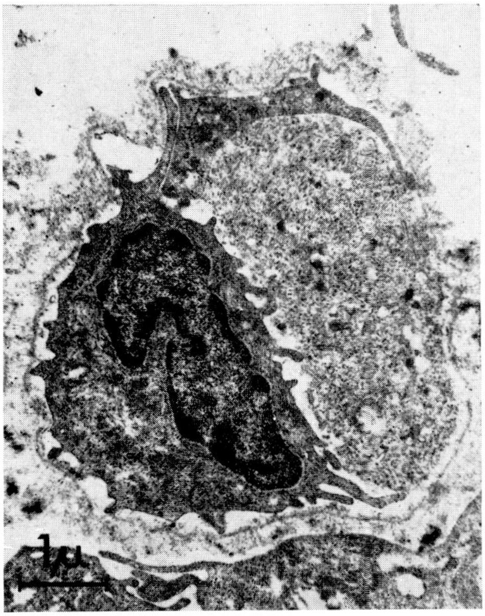

Fig. 97

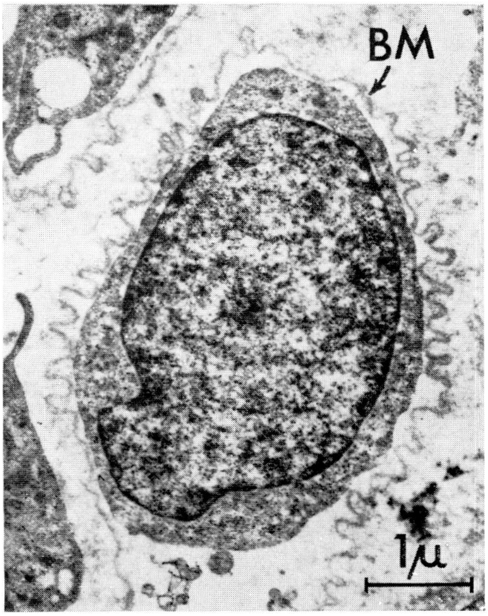

Fig. 98

Fig. 95. An electron photomicrograph of a crushed muscle fiber in transverse section, 6 days after injury, showing a large myonucleus (My), and a satellite-like cell (S?) semidetached from the basement membrane (BM), and throwing out cytoplasmic projections (P). × 10,100.

Fig. 96. A subject similar to Fig. 95, showing a cell with rounded nucleus and long cytoplasmic protrusions (P) lying within, but well separated from, the basement membrane (BM). × 14,800.

Fig. 97. A subject similar to Fig. 96. × 10,000.

Fig. 98. A cell, probably a myoblast, lying within the collapsed basement membrane tube (BM) of a damaged fiber, 6 days after injury. × 12,300.

The next 4 pictures (Figs. 95-98) are of crushed muscle fibers in transverse section 6 days after injury, during the time when there is so much pleomorphism. These pictures illustrate again the relentless question of cell identification, which is so pertinent to all that we are discussing in this conference. The smaller cell in Figure 95 (S?) has a number of the features of a satellite cell. But it is semidetached from the basement membrane, and is throwing out projections which are not seen in 'resting' satellite cells. I interpret this as evidence of mobility of the cell. It is lying against a damaged fiber containing a large nucleus (My) and somewhat disorganized myofibrillar material.

The cell in Figure 96 is different again. It is well separated from the basement membrane (BM), and is throwing out longer projections (P). Its nucleus is more rounded, and its cytoplasm contains various inclusions. The sarcoplasmic material, against which it is lying, is very disorganized. Is this cell a macrophage? Or is it an altered satellite cell?

The cell in Figure 97 is somewhat similar to that in Figure 96. Yet the nucleus is indented, and the cytoplasmic projections are not so extensive. We have taken many micrographs of these cells, and there is a complete range of morphology from cells which have the now well-established features of a satellite cell, through cells of the type shown in Figures 95 to 97, to cells of a more 'primitive' form, such as that shown in Figure 98. In this picture

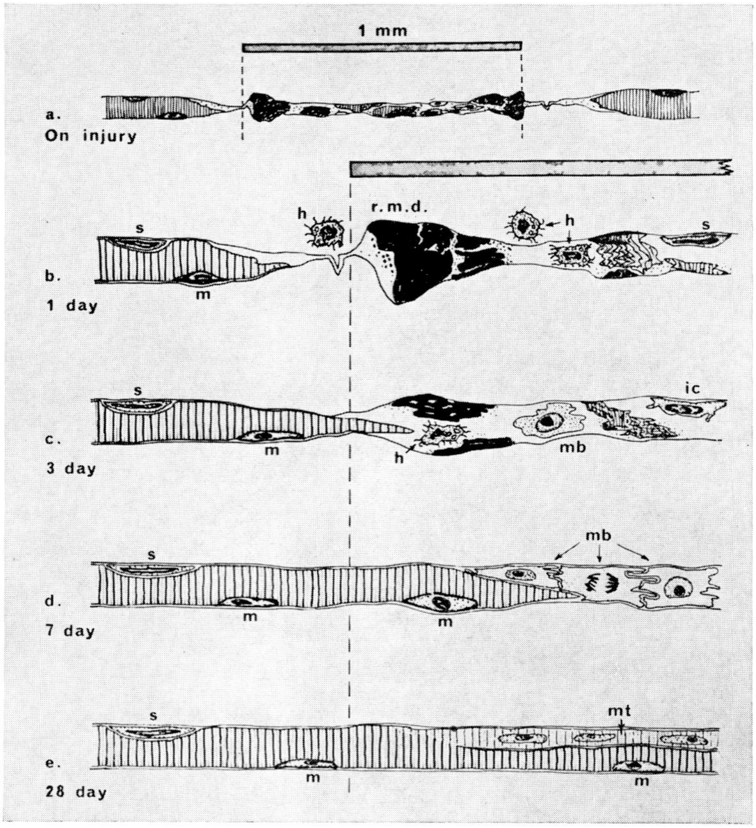

Fig. 99. Diagram of the crush lesion showing consecutive stages of regeneration in a single muscle fiber. h = histiocyte; ic = 'intermediary cell'; m = myonucleus; mb = myoblast; mt = myotube; s = satellite cell; r.m.d. = region of maximal damage.

the basement membrane (BM) can be seen to be well folded, following the disappearance of the sarcoplasm, and the space is filled by a cell containing a large nucleus which is more like a myonucleus in type. I would call this cell a myoblast.

The process of regeneration in these crush lesions is summarized in Figure 99. By 24 hours after injury, histiocytes invade the area to remove debris. By 3 days, there are no 'typical' satellite cells in the injured fibers but many cells of the 'intermediary' type. I do not know what they are becoming of course, but the inference is that they are becoming myoblasts, which are seen in increasing numbers from this time on. Some of these are seen in mitosis, and the mitotic figures are always orientated in the axis of the fiber. Myoblasts fuse to form myotubes, which fuse with the sprouting ends of the surviving fibers coming in from both sides of the lesion, so that by 4 weeks many fibers are continuous again through the lesion, and the muscle looks normal enough a few days later.

A small point not illustrated in Figure 99 is that by 28 days after injury satellite cells (S, Fig. 94) can again be recognized in the previously damaged region and, furthermore, they are more numerous than before.

The small size of the web muscles, and the remarkable regeneration which occurs by 4 weeks after crush injury, enables the quantitation of these regenerated lesions. I am not aware of any previous attempt to quantitate a whole muscle, following injury. I would not attempt to quantitate before 4 weeks after injury because of the cellular pleomorphism prior to that time.

TABLE V

The numbers of muscle fibers, myonuclei, satellite cells and endomysial fibroblasts, counted in serial transverse sections (1-21) at 250-micron intervals through a segment of muscle crushed 13 weeks previously

Transverse sections	Number of fibers	Number of myonuclei	Number of satellite cells	Number of endomysial fibroblasts
1	643	251	42	37
2	660	250	26	33
3	647	270	33	33
4	673	277	26	35
5	687	237	24	35
6	671	272	33	32
7	683	254	24	66
8	701	257	25	75
9	704	251	20	57
10	704	252	38	65
11	692	262	41	103
12	704	283	42	73
13	690	263	54	68
14	693	194	36	81
15	683	208	28	59
16	695	298	29	64
17	696	233	31	47
18	669	242	31	46
19	674	224	20	50
20	686	245	28	31
21	680	262	23	38

The counts of muscle fibers, myonuclei, satellite cells and fibroblasts, in serial transverse sections through a portion of muscle injured in the region indicated (lesion), are given in Table V, and in histogram form in Figure 100. Sections were taken at 250-micron intervals along the muscle, so that at least 4 of them would go through the previously crushed region. This particular muscle was sectioned 13 weeks after injury.

The number of fibers remains virtually unchanged through the lesion, with, if anything, a slight increase. You will recall that the fibers in these muscles are short, with a mean length of 10 millimeters, and overlapping, so that there is a slight variation in fiber number anyway.

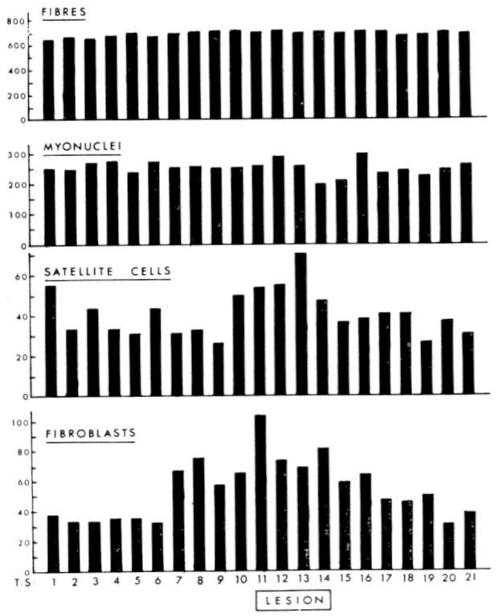

Fig. 100

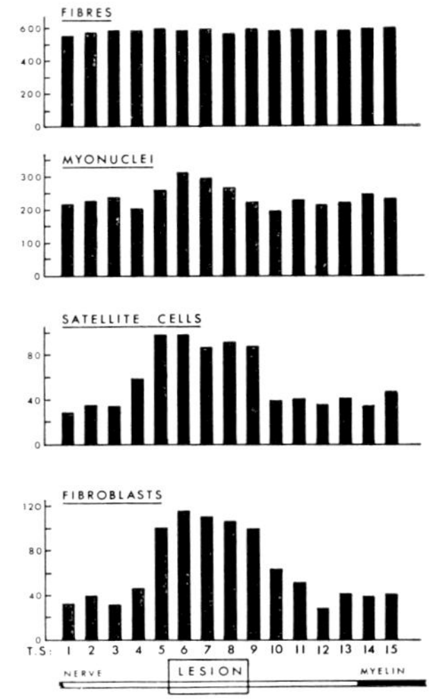

Fig. 101

Fig. 100. A histogram of the number of muscle fibers, myonuclei, satellite cells, and endomysial fibroblasts, seen in serial transverse sections (1-21) cut at 250-micron intervals along the muscle, 13 weeks after a crush injury to the region indicated (lesion).

Fig. 101. A histogram of the number of muscle fibers, myonuclei, satellite cells and endomysial fibroblasts, seen in serial transverse sections (1-15) cut at 250-micron intervals along a muscle, crushed 8 weeks and 4 weeks previously at the site marked (lesion).

The myonuclei show a slight reduction in number in sections 14 and 15, but with increase in number on either side, in sections 12 and 16, so that there is little overall change in number. The satellite cells are increased in number in the lesion, and this is matched by a marked increase in the endomysial fibroblasts extending for about 1 millimeter on either side of the lesion.

The interpretation of these results is that there is complete regeneration of the damaged muscle fibers, with restoration of the formed elements – myonuclei, myofibrils, satellite cells, fibroblasts – and, if anything, an exuberant cellular response.

The webs are so accessible, and given structures can be recognized *in vivo* with sufficient accuracy, for repeated injury to be made to a selected part. Thus repeated crush lesions can be made to the same segment of a web muscle. Table VI and Figure 101 give the results of serial counts through a muscle crushed 8 weeks and 4 weeks previously. Again, muscle fiber number remains constant through the injured region. Myonuclei show a definite increase

TABLE VI

The numbers of muscle fibers, myonuclei, satellite cells and endomysial fibroblasts, counted in serial sections (1-15) at 250-micron intervals through a segment of muscle crushed 8 and 4 weeks previously

Transverse sections	Number of fibers	Number of myonuclei	Number of satellite cells	Number of endomysial fibroblasts
1	550	217	22	32
2	568	224	27	39
3	577	237	26	31
4	577	202	45	46
5	594	257	75	100
6	575	312	75	115
7	579	294	66	110
8	560	266	70	106
9	588	221	67	99
10	572	195	30	63
11	591	228	31	51
12	573	214	27	28
13	575	219	31	41
14	578	245	26	39
15	594	233	36	41

in the injured area. Satellite cells are markedly increased, and fibroblasts even more so. An additional feature in this case was that the nerve in the muscle could be followed through the lesion, and was noted to contain myelinated axons up to the point shown (between sections 13 and 14). Beyond that, axons had regenerated along the Schwann-cell tubes but were not yet myelinated. This gives an indication of internodal length, which is probably 1.0 to 1.5 millimeters in these muscles.

Once again, the interpretation of these results is simple, that regeneration is complete with the formed elements being more numerous than normal. The implication is, of course, that under these conditions, far from having limited potential for regeneration, skeletal muscle can withstand repeated injury, and still recover completely. I must reiterate that these studies were only undertaken in the 1- to 3-month period after injury. It would be of some interest to continue these studies for longer terms to see if the cell populations eventually return to normal numbers.

H. HOLTZER: I would like to know more about increases in the number of satellite cells. Your system lends itself to a variety of intriguing experiments. A satellite cell divides, producing 2 daughter cells. Will both of them differentiate into replicating or non-replicating satellite cells? Will both daughter cells differentiate into myoblasts capable of fusing and producing myofilaments? Will one daughter remain a replicating cell, the other differentiating into a postmitotic myoblast? Your data on increasing cell number would, in fact, indicate an overshoot. You are producing more mother cells in the population. This, in turn, might mean that satellite cells themselves *cannot* fuse, this being a property only displayed by their progeny. From this it might be deduced that there are functional subclasses of satellite cells – those that replicate to produce more replicating cells and those that replicate to produce non-replicating myoblasts with the property to fuse and form myofilaments. All this is quite similar to the way we view myogenesis in the embryo (Holtzer *et al.*, 1958; Okazaki and Holtzer, 1965; Holtzer *et al.*, 1969; Holtzer and Bischoff, 1970; Bischoff and Holtzer, 1970).

J. C. T. CHURCH: I should like to develop the idea that there are subpopulations of satellite cells. If satellite cells are incapable of fusion, yet can give rise to daughter cells which are destined to fuse, there would be a time when 'non-fusing' and 'fusing' cells were in the satellite cell position, that is lying under the basement membrane, and they might appear very similar in their general morphology at that time. This would increase the 'satellite' cell count.

In addition, there is the question again to what extent one cell type can become another, and change its function. If we accept that a histiocyte remains a phagocytic cell wherever it is, then we could equally say that a satellite cell is always non-phagocytic, or conversely that a phagocytic cell in muscle is not a satellite cell. This is not to deny, though, that a satellite cell could take on phagocytic function. A Schwann cell, for instance, takes on phagocytic function during Wallerian degeneration of the axon. In the early stages after skeletal muscle damage, there is a great deal of damaged sarcoplasm to be removed, and phagocytic cells are the predominant cell type seen. Undoubtedly some of these cells are tissue histiocytes that have entered the damaged region, but some could also be the resident satellite cells that have survived the injury, and taken on phagocytic properties.

UNIDENTIFIED SPEAKER: I never did see phagocytosis *per se*. Did you?

J. C. T. CHURCH: The pictures shown (Figs. 95-98) are selected micrographs showing cells which are not evidently engaged in phagocytosis, though there are some peculiar inclusions in the cell in Figure 96. However, histiocytes containing phagocytic vacuoles are frequently seen in these damaged muscles. I think we can stipulate ultrastructural criteria for the identification of satellite cells, histiocytes, myoblasts, and other cells, but difficulties arise when a given cell appears to have features of more than one cell type, and the supposition then is that, at least with certain cell types like satellite cells and myoblasts, they are interchangeable.

E. D. HAY: What happens to the nuclei of the muscle fibers themselves? You showed one in which the cytoplasm was 'degenerating' but the nucleus was vesicular, an active-looking healthy nucleus. Can you rule out that some of these muscle nuclei become satellite cells?

J. C. T. CHURCH: Of course, I cannot.

E. D. HAY: One would merely have to form a new cell membrane around it.

J. C. T. CHURCH: In the crush lesions which I have described there is clearly a range of damage, maximal at the 2 ends of the crushed segment of muscle, where there is complete disorganization of sarcoplasm and I doubt if even satellite cells survive. Within a week or so following injury all the damaged sarcoplasm in the crushed segment is removed, essentially by phagocytosis, and there is no syncytium containing myofibrillar material. However, some myonuclei are not disrupted (*e.g.* My, Fig. 90), and these could retain some cytoplasm around them, and continue as viable cells. The design of these crush experiments was to give complete disruption of sarcoplasm without disruption of basement membranes and connective tissue architecture, so that if regeneration were possible it could proceed to full recovery. This it clearly did, with, if anything, an exuberant cellular response in the first 3 months after injury.

UNIDENTIFIED SPEAKER: If you have a minute, I would like to ask one more question. You can see the histiocytes in some cells, and then you say basement membrane was preserved with no disruption. Can you accept the idea that histiocytes penetrate the basement membrane?

J. C. T. CHURCH: This is a good question. We know that certain cells must move through basement membranes. A 'dust' cell in the lung alveolus is one such. Lymphocytes penetrate between the cells of epithelia, and most if not all epithelia have well-developed basement membranes. If the basement membranes of crushed muscle fibers are intact, as we have said, and yet histiocytes are seen within them, either these histiocytes are derived from cells previously existing within the basement membranes, or they have invaded the area and penetrated the basement membranes. I believe motile cells can so alter their shape that they can pass through small apertures by assuming an hourglass shape. Dr. Walker (1963) showed that, though leucocytes invade an injured area of muscle, they do not contribute to the regeneration of the fibers.

UNIDENTIFIED SPEAKER: How long are your fibers?

J. C. T. CHURCH: As indicated earlier, the fibers in the bat web muscles are quite short, with a mean length of 10 millimeters, lying randomly distributed along the muscle, which is itself 8 to 10 centimeters long. As the crush lesions are only 1 millimeter in width, across the muscle, individual fibers will be damaged in at least 1 millimeter of their length, but the injury will be at any point along a given fiber, depending on its position in the muscle. This means that the majority of fibers will be crushed somewhere along their length, and will have a relatively normal segment on either side of the crushed segment.

UNIDENTIFIED SPEAKER: What happens to those segments on the left and the right of the crushed region?

J. C. T. CHURCH: As studied in longitudinal and transverse sections, evidence of injury can be seen extending about 1 millimeter on either side of the crushed region, but beyond that the muscle appears normal.

M. REZNIK: You said there was an increase in the number of satellite cells. Do you assume this is due only to mitotic division?

J. C. T. CHURCH: Yes.

M. REZNIK: Did you try to estimate what should be the duration of the mitotic cycle to obtain this increase in number? What should be the time of DNA synthesis, G2 and mitotic division to produce this increase?

J. C. T. CHURCH: The quantitation described (Tables V and VI, Figs. 100 and 101) was undertaken after single injury 3 months before, and double injury 2 months and 1 month before; 4 weeks is adequate for the increase in cells seen. I know nothing about the possible rapid increase in satellite cells within a day of injury, but I assume that satellite cells are capable of moving into an injured area, just as leucocytes do, to give a rapid increase in cell number which need not be only due to mitotic division of the cells in the injured area.

M. REZNIK: The study I conducted was over a 24-hour interval, in which time you can see a very great increase in cell numbers. Mitotic division by itself cannot explain the increase in satellite cells.

J. C. T. CHURCH: If we consider a fiber, injured in its central region, but containing satellite cells along its length, and we then assume that numbers of these cells in and around the injured region go into mitosis, then some of the daughters could migrate along the fiber to give a population increase in the injured part without a parallel mitotic rate there. There is evidence that cells adjacent to an injured area are stimulated to divide, with some daughter cells migrating into the lesion.

M. REZNIK: But such an increase in only 1 or 2 days?

J. C. T. CHURCH: This is a valid criticism, but can you offer another suggestion for the origins of these cells?

M. REZNIK: Just suppose your population of satellite cells in division are a lot of histiocytes.

J. C. T. CHURCH: We all act on presuppositions. My theme is that the satellite cell is the stem cell of skeletal muscle, distinct and different in behaviour from histiocytes or other cells. Someone else could suggest equally well that the satellite cell has nothing to do with myogenesis, but is a separate cell line with a specific function, perhaps some supportive role, but never dividing to produce myoblasts. If these cells divide following injury, it would only be part of a non-specific response to any injury, to replenish lost cells of that population.

UNIDENTIFIED SPEAKER: Is there a technique to tell the difference between a satellite cell and a histiocyte?

J. C. T. CHURCH: There are excellent techniques for the identification of histiocytes, though we have not applied them in the bat web work. In electron microscopy, a histiocyte is readily identified by its phagocytic vacuoles, whereas a satellite cell can be strictly defined as a cell which has no phagocytic properties.

UNIDENTIFIED SPEAKER: I wonder how much of your cellular population after the injury reflects the response to damage of the vessels and supportive connective tissue.

J. C. T. CHURCH: With a tissue as accessible as the bat web, a whole range of traumatic insults, physical or chemical, could be applied, to produce desired results. Our desire was to achieve maximal damage to cells without disruption of connective tissue. As shown in Figure 88, vessels can be observed after injury and, with high-power microscopy, capillaries and their contained cells can be seen. There was no haemorrhage in these crush lesions, indicating to us that the capillary basement membranes were not disrupted. I would add that, if our presupposition is correct that the satellite cell is a stem cell for skeletal muscle, and if it only functions best within a fiber basement membrane, then we have in the web crush lesions a situation which will allow it to function optimally.

UNIDENTIFIED SPEAKER: You obviously have damaged the vessels on the right side. Ideally you should destroy the contractile substance selectively, leaving the basement membrane intact.

J. C. T. CHURCH: I have tried ultrasonic vibrations, as a physical insult to the bat web. I placed an ultrasonic probe above the web, with a droplet of water between it and the web, but vibrations sufficiently strong to scatter the water caused no effect in the web. I was attempting to disrupt the sarcoplasm without any distortion of the connective tissues. I believe this sort of technique could be developed, and some sort of physical or chemical agent be used to give this result.

UNIDENTIFIED SPEAKER: There was a viral probe and an anti-malarial probe.

J. C. T. CHURCH: This is a good suggestion, and well worth trying in the bat web. We have inoculated a mycobacterium (*Myco. buruli*) into the web muscles, and it causes specific lesions (Church and Griffin, 1968), but with marked necrosis of muscle fibers, inflammatory response and loss of basement membranes.

UNIDENTIFIED SPEAKER: Could I ask Dr. Church to return to the point that surprised me, namely, that the whole fiber did not degenerate from injury? What happens? Does the fiber grow from both ends?

J. C. T. CHURCH: We can distinguish between several phenomena here. Firstly, though individual fibers are severely injured somewhere along their length, they do not degenerate except in the region damaged. Rather, their surviving ends sprout into the injured region. This must be accomplished by the addition of new sarcomeres at the sprouting end as sarcomere length remains the same in normal and regenerating muscle. Then there must be a critical point in the ratio between the myonuclei and the volume of sarcoplasm they subserve. If hypertrophy – the increase in cytoplasmic volume without increase in nuclear number – can occur when functional demands on a muscle are increased, then I would say that the same process could occur longitudinally when part of the fiber is deficient following injury. But in either case a limiting factor would be the nucleo-cytoplasmic ratio. Further sprouting in regenerating muscle can only then occur if more myonuclei are recruited, by the addition of myoblasts to the syncytium. This could occur, however, at any point along its length, not necessarily only at its sprouting end. As myonuclei increase in number in this way, they would redistribute themselves along the sprouting syncytium.

Another phenomenon is the fusing together of myoblasts within a damaged region. The myotubes so formed then fuse with sprouting fibers coming into the damaged region (Fig. 99, e).

UNIDENTIFIED SPEAKER: Then one postulates that one multinucleated fiber will fuse with another multinucleated fiber?

J. C. T. CHURCH: Correct.

UNIDENTIFIED SPEAKER: In your slide (Fig. 99) showing a late stage of regeneration one can see an almost fully regenerated fiber with a myotube in it. Where does it come from and what is the significance of its presence at this late stage?

J. C. T. CHURCH: We saw occasional myotubes in the otherwise fully regenerated muscle fibers 4 weeks after injury. I think this just indicates that some fibers, perhaps the most severely injured ones, take longer to recover.

May I make one final comment. We have indicated that there is plenty of evidence that myoblasts fuse with each other, that myotubes fuse with more mature fibers, or even that sprouting fibers will fuse when they meet in the centre of an injured area; yet the actual process of fusion has yet to be convincingly demonstrated, at least in electron microscopy. Fusion is initiated "by the recognition of homotypic cells" as Dr. Holtzer (1966) has shown. I believe that, in this sense, satellite cells are heterotypic, as they do not fuse with other cells.

A MODEL FOR MYOGENESIS USING THE CONCEPT OF THE SATELLITE CELL SEGMENT

JOHN C. T. CHURCH

INTRODUCTION

Satellite cells, lying between the basement membrane and the syncytium, are a well-established feature of skeletal muscle fibers (Mauro, 1961; Muir *et al.*, 1965). These cells are recognizable from the time that basement membranes become visible around early myotubes in embryonic muscle (Ishikawa, 1966; Church, 1969). They have been seen in a variety of normal adult muscles, and in muscle spindles (Landon, 1966) and have been studied in various myopathies (Laguens, 1963; Shafiq *et al.*, 1967) and in experimental lesions (Church *et al.*, 1966; Venable, 1966*b*). It is generally concluded that they are 'stem' cells or 'reserve myoblasts', capable of giving rise to daughter myoblasts, both in developing and regenerating muscle. Though they are probably a regular feature of all skeletal muscle, much is still unknown of their form and function.

THE SATELLITE CELL SEGMENT

Morphological descriptions of skeletal muscle should now include the satellite cells, which appear to be regularly spaced along the fibers (Muir *et al.*, 1965). The smallest morphological unit of skeletal muscle could be described as a single satellite cell and its adjacent segment of fiber syncytium, containing a proportionate number of myonuclei. This would be a 'satellite cell segment' (lower part of Fig. 102), the whole fiber being made up of a number of such 'segments'.

The ratio between satellite cells and myonuclei would probably vary for different species, or different muscles in a given individual, as for instance the size of a motor unit does, so that the size of the 'satellite cell segment' would vary accordingly. This ratio has been calculated for some muscles (Muir *et al.*, 1965; Venable, 1966*b*; Church, 1970). The accurate quantitation of satellite cells presents its own problems, as they are usually placed well apart. Also they can only be identified with certainty in electron microscopy, and this necessitates a limited sampling. However, as studied in the bat web muscles, they can be recognized sufficiently well, in transverse sections 1 to 3 microns thick from araldite-embedded material, for quantitation in light microscopy (Church, 1970).

The ratio of satellite cells to myonuclei in the bat web muscles is about 1 to 8, or 12.5 per cent. This is not constant, and can rise to 35 per cent or more in regenerating muscle, after crush injury. In normal muscle, the 'satellite cell segment' is about 200 microns in length. The fibers of these muscles have a mean length of about 10 millimeters, so that a single fiber would be made up of about 50 'segments'.

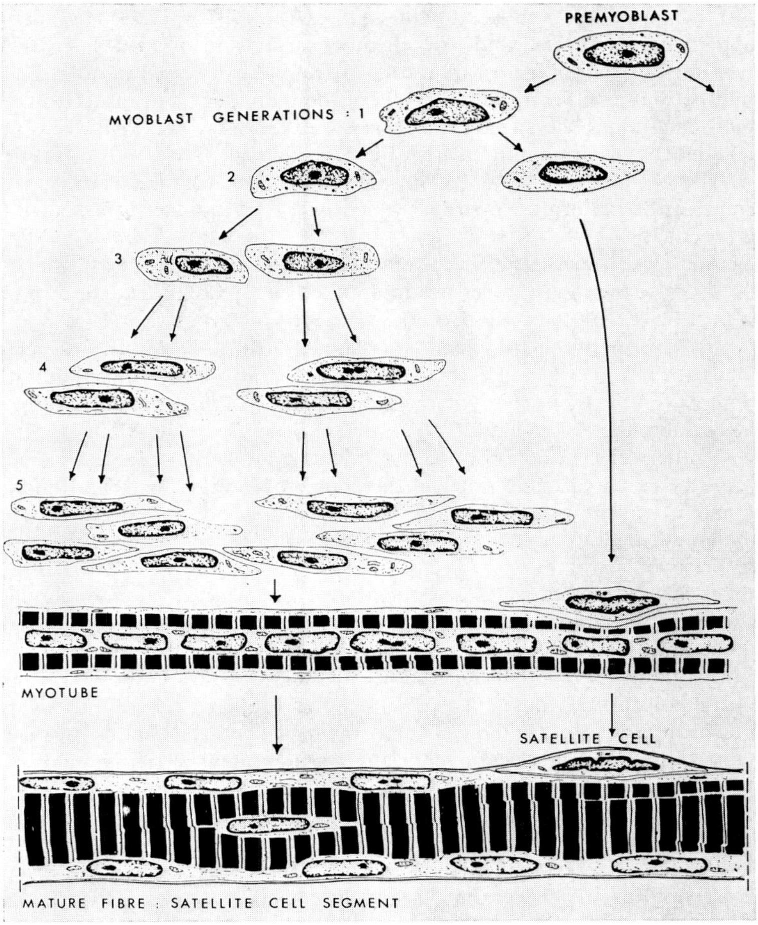

Fig. 102. A diagram of a suggested model for myogenesis, whereby a 'premyoblast' gives rise to a 'satellite cell segment'.

A MODEL FOR MYOGENESIS

Whatever the ratio of satellite cells to myonuclei, it must represent the result of genetic and environmental control mechanisms. If satellite cells are 'reserve myoblasts' as suggested, the ratio assumes a greater significance, and a model for myogenesis should be attempted, fitting the available data. Such a model should derive satellite cells and myonuclei from one 'stem' cell line; should allow for an almost geometric progression in the increase in number of myonuclei, despite the fact that once incorporated into a syncytium they divide no more; should allow for the appearance of differentiated muscle from an early stage; and should allow for the presence of satellite cells throughout development, some remaining in fully mature muscle. Ideally, a model should also be applicable to both developing and regenerating muscle, and should permit genetic programming of each stage.

When cells divide, there are at least 3 types of progeny. (1) Daughter cells may be identical with the parent cell, a division of an (A) → (A) + (A) type. (2) One daughter may be dissimilar to the parent cell, a division of an (A) → (A) + (B) type. (3) Both daughters may be dissimilar to the parent cell and to each other, a division of an (A) → (B) + (C) type. These differences of progeny from parent cells are referable not to their total DNA content, which would be the same for all diploid cells of the individual, but rather to the 'readable' DNA, which determines the specific biochemical expression of a given cell, and thus its behaviour. 'Genetic programming' is then the process whereby a series of specific genes or 'operons' (Harnden, 1968) are 'readable' for a given cell, determining its own specificity, and also determining the pattern of 'operons' which will be 'readable' in its daughters, giving them their specificity.

The suggested sequence of events in the development of a 'satellite cell segment' is depicted in Figure 102. A 'premyoblast' gives rise to another 'premyoblast' (not shown), and a 'first generation myoblast', by a division of the (A) → (A) + (B) type. The 'first generation myoblast', by a division of the 3rd type, in this case (B) → (C) + (A), gives a 'second generation myoblast' (C), and a satellite cell (A). The 'second generation myoblast' divides 3 times, by simple (C) → (C) + (C) type division, to give 8 similar daughter cells, which fuse to form a myotube, the myonuclei of which divide no more. The satellite cell, which does not fuse with other cells, is incorporated into the fiber, under its basement membrane. This process would be repeated many times in the course of development, and 'fifth generation myoblasts' might become admixed before fusion, but without alteration of the basic ratio between them and satellite cells. By designating the satellite cell an (A) cell, it is equivalent to another 'premyoblast'.

This model fits the prerequisites set out above. As with any model though, possible modifications must be considered.

It is possible that satellite cells are only set aside at the 'fifth generation myoblast' stage, as one of the 8 daughters of that generation. If these cells are homotypical, all (C) cells, their fusion or non-fusion would then have to be controlled by other factors, presumably from their environment. Holtzer indicates that myoblast fusion is 'initiated by the recognition of homotypic cells' (Holtzer, 1966), rendering a purely environmental control of fusion unlikely. On the other hand, a more complex 'genetic programming' would have to be postulated to ensure that 1 cell in 8, after 3 cell divisions, was heterotypical, and this again would be unlikely.

As there is much evidence that satellite cells are themselves 'reserve myoblasts', peculiar in that they do not fuse with other cells, it must be considered whether they are any different from the 'premyoblasts' or 'first generation myoblasts' in the model. If they were identical with the 'first generation myoblasts' this would mean that division of these cells would be of the (A) → (A) + (B) type, that already suggested for the 'premyoblast'. This would simplify the model, so that the 'first generation myoblast' becomes the earliest cell. But if this were so, then with each on-going division only one 'stem' cell could be formed, whereas the 'second' to 'fifth generation' myoblasts would steadily increase in number, rendering the ratio incompatible with that known for satellite cells and myonuclei both in developing and adult muscle. This would indicate that 'first generation myoblasts' and satellite cells must be distinct. The satellite cells could however be identical with the 'premyoblasts' as already suggested. Then, by a step-wise division of 'premyoblasts', (A) → (A) + (B), and 'first generation myoblasts', (B) → (C) + (A), there is a steady increase in the (A) cell population, which are 'premyoblasts' or satellite cells, together with the regular production of (C) cell 'clones', destined to form myotubes.

If the 'premyoblast' is identical with the satellite cell, then they would presumably be exposed to the same environmental control mechanisms for mitosis, and the satellite cell would not necessarily stop dividing, as depicted in Figure 102. It must be asked, therefore, what its progeny would be if it went directly on into mitosis. At least 1 of its daughters would have to be a satellite cell if its own 'segment' is to retain one. It is unlikely that it would divide to give 2 satellite cells, as the predominant need is for myoblasts to furnish myonuclei in the developing syncytium. It is likely therefore that it gives a satellite cell and a myoblast, by an (A) → (A) + (B) type division, that which the 'premyoblast' is 'programmed' to do in the model. This again would indicate that satellite cells and 'premyoblasts' are identical, or, put more simply, that satellite cells are the 'stem' cells of skeletal muscle.

It is possible that the 'threshold' for mitosis is different for the different cell types suggested in the model. If the 'threshold' is higher for the earlier cells, their mitotic rate would be slower, and this would skew the model in favour of 'fifth generation myoblasts' appearing earlier, but would not alter the suggested 'genetic programming'.

In the later stages of development, and during regeneration, most mononuclear cells are seen in the satellite cell position, lying under the basement membrane. Thus the various cell types depicted in the model could be morphologically very similar. However, there is evidence for variation in the morphology of 'satellite' cells, and the implication is that, with further study, it will be possible to divide these cells into subpopulations.

With the present limitation of knowledge about satellite cells, the concept of the 'satellite cell segment', and its relation to myogenesis, must remain somewhat conjectural. However, it is in keeping with current concepts of myogenesis, and is applicable to both developing and regenerating muscle.

SATELLITE CELLS AND FIBER NUCLEI IN MUSCLE REGENERATION*

SAIYID A. SHAFIQ

I would like to show some slides which bear on the problem of the role of satellite cells and muscle nuclei in regeneration. These preparations are from the tibialis muscle of mouse which was induced to regenerate following either a crush injury or a localized heat shock.

Figures 103-105 illustrate the general histological features of the injured region during the first 2 days after injury. As noted in previous studies (Price *et al.*, 1964; Shafiq and Gorycki, 1965; Sloper and Pegrum, 1967; Reznik, 1968*b*) the injured regions became sealed off from the healthy parts of fibers by contraction clots (Fig. 103), showed localized necrosis and became infiltrated by the inflammatory cells (Figs. 104, 105). After the first two days (during which a good deal of the necrotic debris appeared to be removed) there was evidence of pronounced regenerative activity in the injured regions. The role of satellite cells in these regions, however, could not be studied effectively because the plasma membranes disintegrated in the necrotic areas and the satellite cells became difficult to distinguish from other infiltrating cells. We, therefore, tried to get around this difficulty by concentrating on the uninjured regions (up to a distance of about 1 millimeter from the crush lesion) of the muscle. In these regions the general architecture was well preserved, the fiber plasma membranes were intact, and, fortunately, there was also some regenerative response which could be studied with reference to normal morphology of muscle fibers.

The regenerative response in the uninjured regions consisted of the appearance of clusters of new cells which then proceeded to various stages of myogenesis. The new cells appeared to be derived from mitotic activity in the periphery of muscle fibers; these mitoses could be readily accumulated by colchicine treatment (Figs. 106 and 107). Several such mitotic figures were examined by electron microscopy. Invariably they appeared to be in cells which in ultrastructural features corresponded to satellite cells (Fig. 108) – and never inside the muscle fibers themselves. Similarly, when the cluster of cells on the periphery of the fibers (Fig. 107) was examined by electron microscopy, they also, like the satellite cells, had their own separate plasma membranes. At 3 to 4 days after injury different stages of myofibril formation could be seen in this peripheral cluster of cells; 2 stages, namely the appearance of a large number of ribosomes and the formation of myofilaments are illustrated in Figures 109 and 110. So it seemed that during regeneration the satellite cells became activated to undergo mitoses and produced cells (myoblasts) which proceeded to different stages of myogenesis.

Healthy regions adjacent to the crush lesion (in which the plasma membranes were intact) were also examined for changes in fiber nuclei and the surrounding cytoplasm. In the perinuclear sarcoplasm rough-surfaced endoplasmic reticulum and free ribosomes were found (Fig. 111). Around other nuclei myelin figures (probably derived from mitochondria) were common (Fig. 112). In some cases filaments (which appeared to be rudimentary myofilaments) were also present (Fig. 113). The fiber nuclei themselves appeared enlarged and vesicular and had prominent nucleoli (Fig. 113). We could find no convincing evidence

* Supported by grants from Muscular Dystrophy Associations of America.

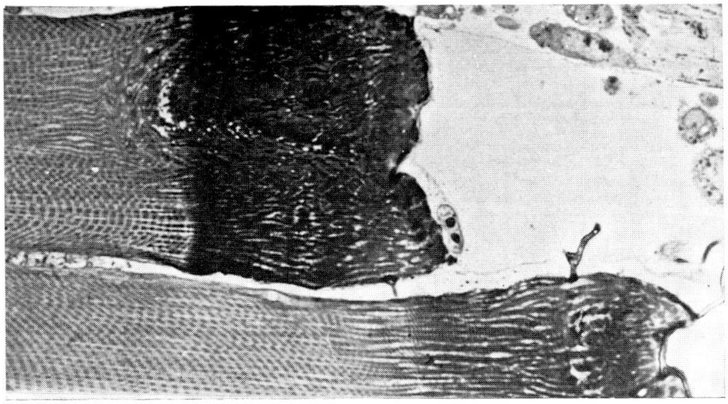

Fig. 103

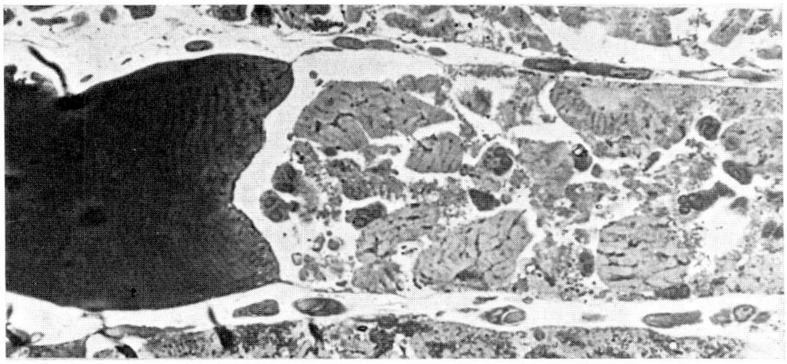

Fig. 104

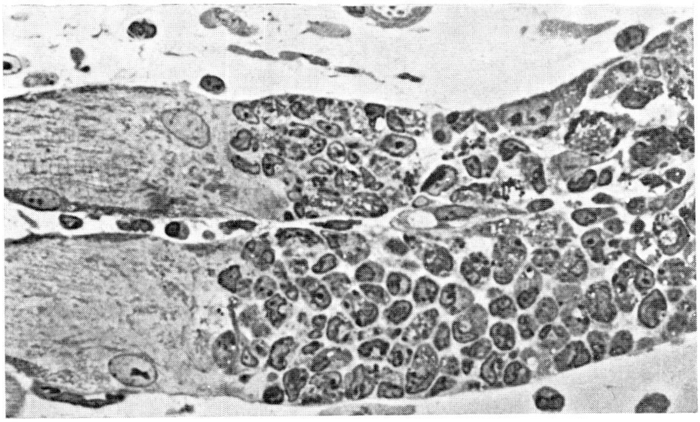

Fig. 105

Fig. 103. Photograph showing contraction clots at 1 day after injury to the tibialis muscle. The parts of fibers on the left of the contraction clots have normal morphology. × 416.

Fig. 104. Necrotic part of a fiber at 2 days after injury. The degenerating myofibrils are in irregular clumps. × 416.

Fig. 105. Necrotic parts of two other fibers at 2 days after injury. The degenerating areas are heavily infiltrated with phagocytic and other inflammatory cells. The contraction clots have disappeared. × 416.

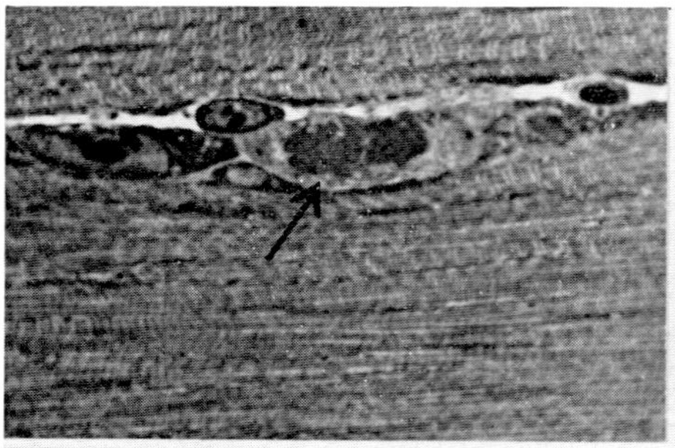

Fig. 106

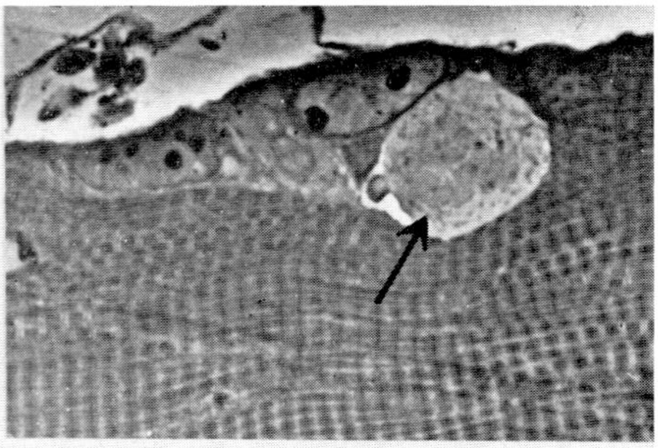

Fig. 107

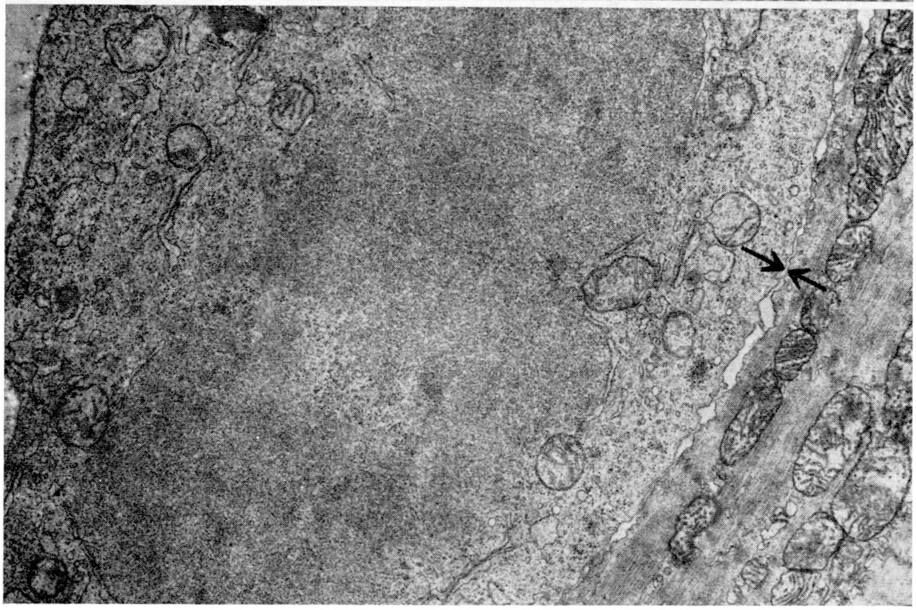

Fig. 108

Fig. 106. A mitotic figure in the uninjured portion of a muscle fiber adjacent to the crush lesion, 2 days after injury. × 1040.

Fig. 107. Mitotic figure and 2 other cells in the peripheral part of a muscle fiber (with normal morphology) adjacent to the crush lesion, 3 days after injury. × 1040.

Fig. 108. Electron micrograph showing mitosis in a satellite cell, 2 days after injury. The apposed plasma membranes of the satellite cell and the muscle fiber are indicated by arrows. × 12,000.

that these changes in the fiber nuclei and in the perinuclear cytoplasm were related to formation of new cells. On the contrary, it looked as though many of these vesicular nuclei in fact moved from the periphery into more internal parts of fibers and finally into the so-called sarcoplasmic outgrowths or buds (Figs. 114 and 115) whose development has been described in several previous studies (*e.g.* Clark, 1946; Shafiq and Gorycki, 1965).

In summary, therefore, it seems that during regenerative response both the satellite cells and fiber nuclei are 'activated', the former to give rise to myoblasts and the latter to start development of the sarcoplasmic buds.

J. R. Florini: Are not those awfully large nucleoli in the nuclei of the injured fibers?

S. A. Shafiq: Yes, the nuclei which have become internalized appear much larger than the

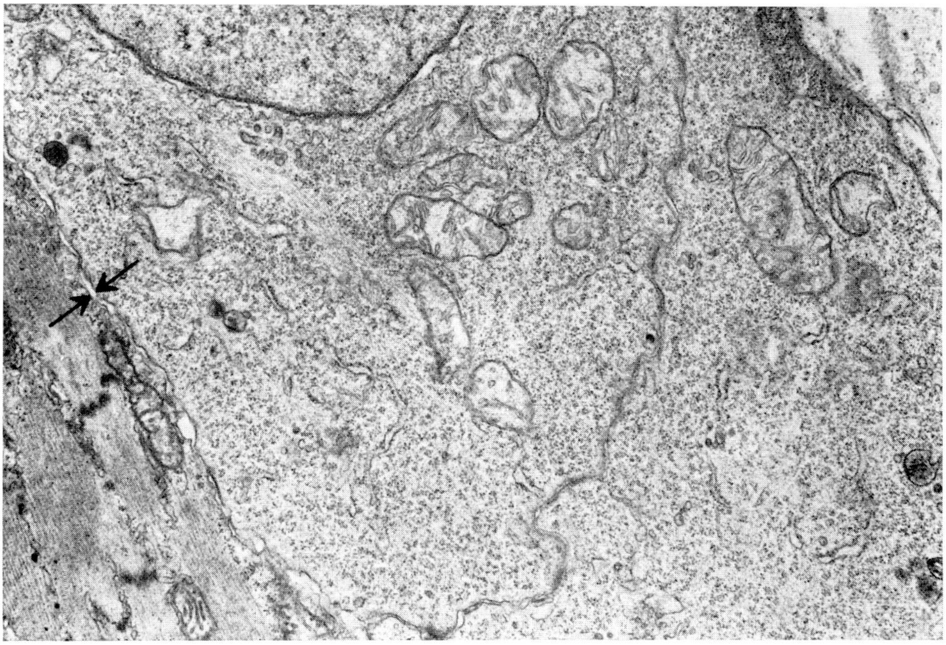

Fig. 109. Parts of 2 'activated' satellite cells containing numerous ribosomes, 3 days after injury. The apposed plasma membranes of a satellite cell and the muscle fiber are indicated by arrows. × 16,000.

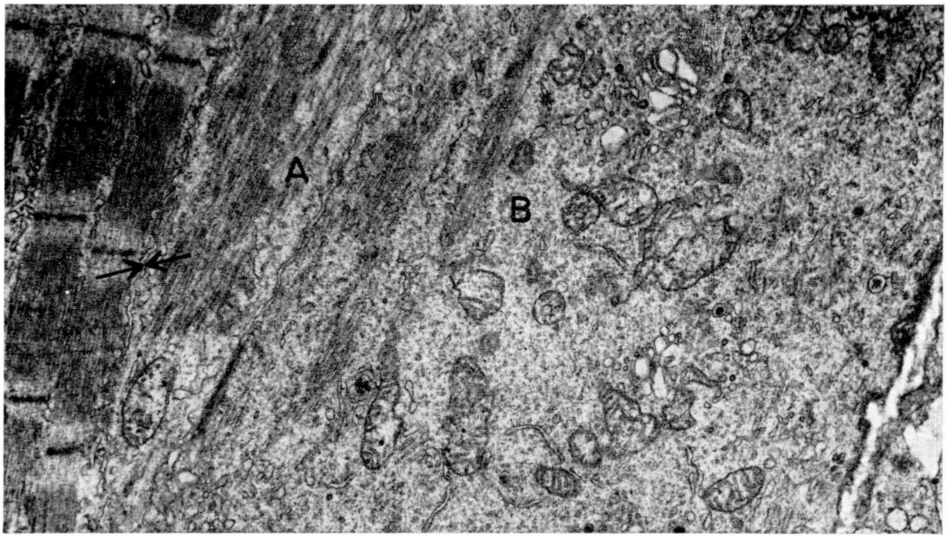

Fig. 110. Parts of 2 activated satellite cells (A and B) containing rudimentary myofilaments, 4 days after injury. The apposed plasma membranes of a satellite cell and the muscle fiber are indicated by arrows. × 14,400.

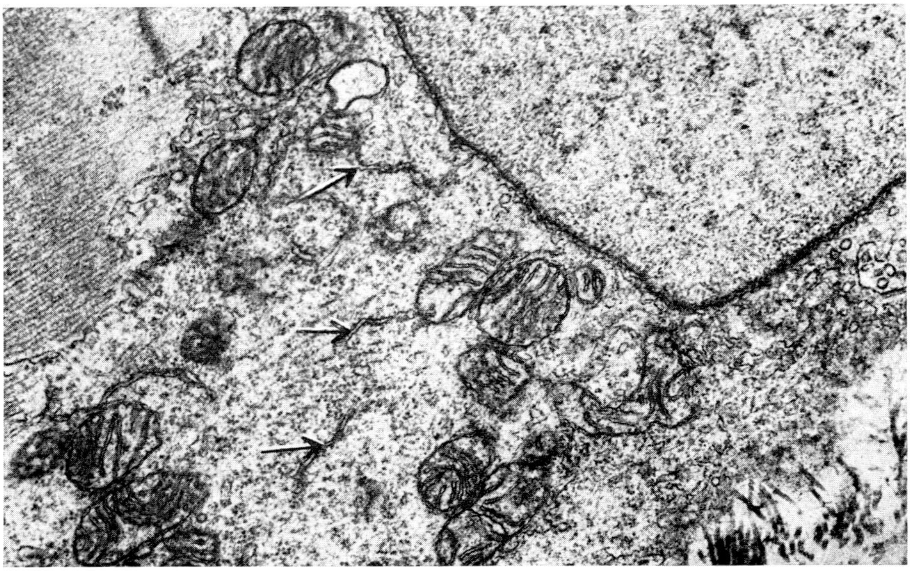

Fig. 111. Elements of rough-surfaced endoplasmic reticulum (indicated by arrows) and free ribosomes in the perinuclear area in healthy portion of a fiber adjacent to the crush lesion, 1 day after injury. × 22,400.

usual peripheral nuclei of normal muscle but are quite comparable to those seen in the sarcoplasmic buds and it looks as if they are the same nuclei which move into the buds.

UNIDENTIFIED SPEAKER: Dr. Shafiq, can you be sure that the nuclei in the sarcoplasmic buds (Fig. 114) are muscle nuclei as opposed to nuclei of the satellite cells?

S. A. SHAFIQ: Yes, I think so, because working with mice one can, to an extent, time the sequence of regenerative changes. Two days after the crush lesion small sarcoplasmic outgrowths start to develop from the injured fibers and in these little buds vesicular nuclei are present. At this time one also sees internalized vesicular nuclei in the healthy portions of the fibers and it seems probable that it is the internalized nuclei which have moved into the bud. The satellite cells at this time seem to be in the mitotic phase. Later on, when they have produced the cell clusters (Fig. 107), perhaps some of them do fuse with the growing sarcoplasmic buds and contribute their nuclei to them.

A. MAURO: You might emphasize that in the nuclei which are seen in Figures 114 and 115 there is no evidence of mitotic activity. That is a very important point.

S. A. SHAFIQ: Yes, we have seen mitotic activity only in the satellite cells, never in the nuclei of the muscle fibers proper or in the growing sarcoplasmic buds.

J. C. SLOPER: Everybody, I think, or at least the majority at any rate, have failed to see muscle buds after 48 hours of injury. If I am wrong here, I would like to know. Now are you maintaining that the nuclei in the muscle buds are derived from satellite cells that wandered into the muscle or am I jumping two or three steps in your argument?

S. A. SHAFIQ: In the present study we used a thin pointed forceps to make a localized crush injury (only about 2 millimeters in width). Fibers so injured do not degenerate completely but undergo only focal necrosis and the healthy portions of fibers become sealed off from the

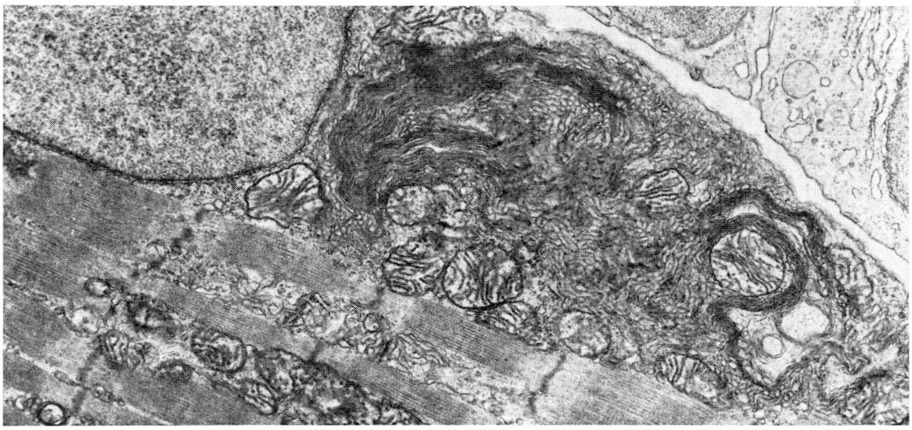

Fig. 112. Myelin figures (probably derived from degenerating mitochondria) in the perinuclear area of a muscle fiber, 2 days after injury. × 12,000.

injured portions by contraction clots. Two days after injury the clots begin to clear and sarcoplasmic buds start to grow from the healthy portions of the fibers. I am proposing that the nuclei seen in these young buds are the fiber nuclei which have migrated into the buds. At later stages, perhaps, the satellite cells or the myoblasts derived from them also fuse with the growing bud.

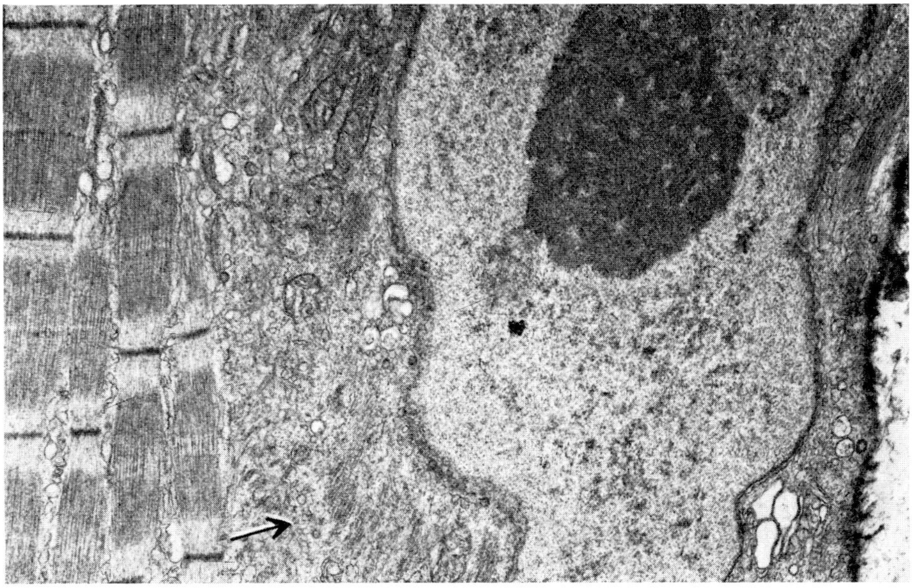

Fig. 113. Nucleus of a muscle fiber with prominent nucleolus and a filamentous material (probably myofilaments) (indicated by arrows) around it, 3 days after injury. × 14,400.

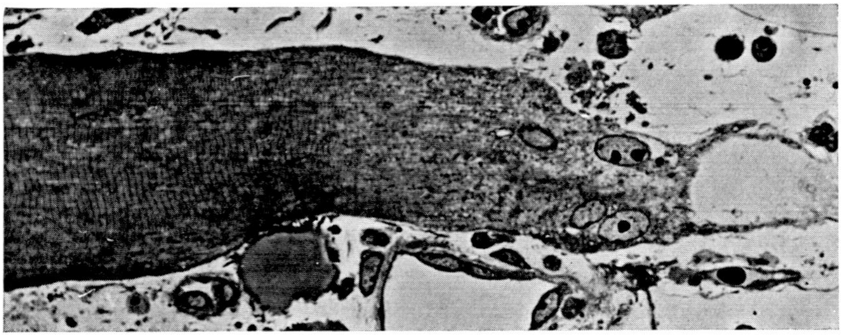

Fig. 114. Photograph showing a young sarcoplasmic outgrowth in the injured zone of the muscle, 2 days after injury. The outgrowth shows 4 nuclei; the healthy portion of the muscle fiber with which the sarcoplasmic outgrowth is continuous is on the left side. × 520.

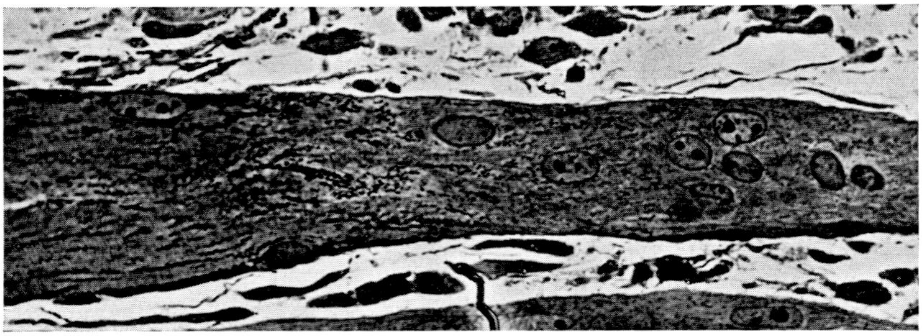

Fig. 115. Sarcoplasmic outgrowth (bud) at an advanced stage, 4 days after injury. Several nuclei and rudimentary filamentous material are present in it. The uninjured portion of the fiber is on the left side. × 520.

J. C. Sloper: Is there not a paradox though; you say that the muscle nuclei do not have mitotic figures and, therefore, do not multiply, but there are more nuclei in the regenerating area so where do they come from? Is there not an actual increase in the number of nuclei? If they migrated in from elsewhere in the fiber, then there should be some areas that have become depleted of the nuclei.

A. Mauro: I want to remind you, gentlemen, that it is well established in tissue culture – as was pointed out yesterday – that there is fantastic movement of nuclei, they just waltz up and down, they have a grand time.

I. R. Konigsberg: As was pointed out yesterday, you see this in early 'muscle' formation, where you do not have fibers packed with myofibrils. In the sarcoplasmic buds, however, there remains the question where the nuclei could have come from. Are there a sufficient number of nuclei in the vicinity which could migrate in?

A. R. Muir: There are, I think, about 80 to 90 nuclei to a millimeter, and about a tenth of these belong to satellite cells.

I. R. KONIGSBERG: Dr. Muir indicates that there would be enough nuclei and you would simply have to invoke a small movement.

H. HOLTZER: Do we not have the same conditions in various kinds of muscle disorders? Without either amitoses or mitoses, nuclei line up within narrow myotubes. I thought the simplest explanation for this was that the breakdown of myofilaments and their resorption as the muscle wastes simply brings separated nuclei closer together within the old sarcolemma. This gives the picture of a high nuclear/cytoplasmic ratio, whereas there is no true increase in nuclear number. Reports of amitosis . . .

A. MAURO: You mean it is interpreted as amitosis. Surely no one in Philadelphia believes in amitotic activity.

H. HOLTZER: Oh, I would like to believe in amitosis.

A. MAURO: Do you really, Howard? Seriously, we are finally getting to some biological principles here. Do you really believe in amitosis?

H. HOLTZER: Alex, my curious sense of humor is always getting me into trouble. I am under the impression that it was primarily the work of my colleagues and myself that laid low once and for all the notion of amitosis (Holtzer, 1959). Which incidentally, like so much of the data on regeneration, relied exclusively on conventional microscopic evidence. The histological evidence for amitosis, it is worth stressing, was most convincing. However, as soon as we tried an experiment like measuring the DNA or using labeled thymidine, amitosis as an issue faded.

But to come back to large numbers of aligned nuclei within myotubes: these can accumulate by reduction in cross-sectional area of muscle fibers and their movement within repairing myotubes.

A. MAURO: Your interpretation, then, would be that simply a migration is involved.

E. D. HAY: In the Figures that you showed, could not the so-called satellite cells be new cells? The issue here is whether what you are calling satellite cells in this regenerative phenomenon are the rather inactive appearing cells that Dr. Muir described yesterday, or whether or not they are actually new cells cleaving off the fibers?

S. A. SHAFIQ: Mitotic activity and myofibril synthesis which I described (Figs. 109 and 110) are in cells wedged between the fiber plasma and basement membranes and as such, I think, qualify for the anatomical definition of satellite cells. Changes, however, also occur in the nuclei and perinuclear sarcoplasm of the injured fibers. The nuclei and their nucleoli enlarge and the endoplasmic reticulum and Golgi apparatus in the perinuclear area become prominent. Sometimes one may see vesicles or tubules of endoplasmic reticulum in contact with fiber plasma membrane, but such profiles are perhaps fortuitous and seem unsatisfactory as evidence for formation of new cells from fiber nuclei.

E. D. HAY: That is the image that we see when these big muscle fragments are separating in the salamander limb regeneration (Fig. 5). During redifferentiation of muscles, of course, the reverse occurs. This is the kind of image you see where the cell membrane is disappearing, and also when cells fuse: rows of vesicles.

P. E. CONEN: But, there is a difference between what looks to be vaguely as though it might go on, and something that, in fact, does, is there not?

E. D. HAY: Well, it seems to me that very few of you are considering the possibility that what you are calling a satellite cell in this mammalian regenerative process is not, in fact, derived from Dr. Mauro's satellite cell or from the satellite cell that Dr. Muir defined for us yesterday, but arises from muscle *per se*.

P. E. Conen: In the identification of the satellite cell, I think we should stress one of the negative features that I mentioned yesterday. Typical satellite cells almost never have a 'nucleolus'. Is that the experience of most of us? When you look by light microscopy at cells with nucleoli like the one you showed, you know that these are unlikely to be satellite cells on electron microscopy of an adjacent ultrathin section. I think that once the cell develops myosin in its cytoplasm, as Dr. Muir suggested yesterday, that we should then term it myoblast and not satellite cell. I wonder whether there may be an analogy between the small lymphocyte and the satellite cell, since both cells have little cytoplasm and organelles. The small lymphocyte can be stimulated by phytohemagglutinin and other substances to transform into a large lymphocyte with nucleoli and considerable cytoplasm, and eventually it enters mitosis. Perhaps the satellite cell can be stimulated by some unknown substance to enter mitosis or transform by 'switching on' muscle protein synthesis and become an obvious myoblast with myofilaments.

D. Allbrook: It is true, I think, that groups of nuclei at the end of muscle buds stay around for a very long time, you see them weeks or months after injury without any apparent activity going on (Fig. 48). May be they are also waiting to be stimulated?

J. C. T. Church: Two short questions on the satellite cells: How far from the actual injury can you see any activation of these cells, and what is the condition of the muscle fibers against which they lie?

S. A. Shafiq: We have examined healthy parts of muscle fibers up to a distance of about 1 millimeter from the injured region. The satellite cells in this 1-millimeter zone appeared activated and the general architecture of muscle fibers under the satellite cells was well preserved.

J. C. T. Church: Have you ever seen cells which were activated beyond a millimeter from the lesion?

S. A. Shafiq: No, we did not look at muscle too far away from the lesion.

J. C. T. Church: I was trying to draw an analogy between this and other systems. In wound-healing experiments, for instance, cells are activated and go into mitosis some distance away from the site of injury. I presume this would also apply to muscle. One more point. As I said yesterday, we must think in terms of several satellite cell populations, and consider the myoblast as different from the satellite cell.

S. A. Shafiq: I think it is premature to subdivide the satellite cells in several categories at the present time. It is probably simplest to believe that in the embryo there is a homogeneous population of myoblasts which go on to different stages of myogenesis and the satellite cells represent the stragglers, the myoblasts which were left back in the process.

J. C. T. Church: Well, now you showed us pictures of cells called 'satellite' cells with myofilaments in them. Were those mononuclear cells, or had they already fused to form myotubes?

S. A. Shafiq: I have seen binuclear cells but those ones were probably mononuclear.

J. C. T. Church: Using a strict definition of the satellite cell, should we not exclude cells containing myofilaments?

A. R. Muir: Yes, I think if they have got myofilaments in them, then they should be called myoblasts.

J. C. T. Church: At least in this context then, there are two cell types, satellite cells proper, and other mononuclear cells containing myofilaments.

J. R. Coleman: One comes from the other.

E. D. HAY: I would like to second Dr. Church's remark. You do not know where the mononucleated cells come from. I think you prejudice your whole outlook if you call all these mononuclear cells under the muscle basement membrane satellite cells no matter what they look like. Dr. Muir has made very clear the definition of what we might call a satellite cell; it has a dense nucleus and scanty cytoplasm. To use that same term for what is really a different morphological entity during regeneration is simply asking for confusion.

S. A. SHAFIQ: If, as it seems, satellite cells become activated during regeneration and start dividing and proceed to different stages of myogenesis, is it not rather arbitrary at what stage you are going to stop calling them satellite cells?

E. D. HAY: You have not yet proved the transformation of satellite cell to myoblast and even if you do prove the transformation you should have a different series of names, like in the development of blood, we have promyocytes, myocytes and so on. If you do have a transformation, then you should develop a better nomenclature to describe it. But if you call all these seemingly mononucleated cells associated with muscle satellite cells, I do not see how you can avoid confusion.

S. A. SHAFIQ: Well, alright.

P. E. CONEN: There are further difficulties of satellite cell definition related to demonstration of myofilaments. If myofilaments are present in one section we will call the single cell a myoblast, but without many further sections through the same cell, above and below the plane of the original section, we cannot state that myofilaments are absent.

A. MAURO: It may be useful here to pose a question, namely, has it been shown in tissue culture that a presumptive myoblast can give rise, without fusion, to myofilaments? Has this ever been shown?

P. E. CONEN: Of course, it is true in early fetuses.

A. MAURO: Okay, so then there is no need to indulge in any metaphysical speculations about the fact that the so-called satellite cell has myofilaments and is mononucleated. We will call the satellite cell a presumptive myoblast.

E. D. HAY: That remains to be proven.

M. REZNIK: I would like to ask a question on the point of the actual discussion. Did I understand that the people here present believe that myofilaments are present in mononucleated cells?

A. MAURO: Yes, that was asserted and I think that is true; there is ample evidence.

M. REZNIK: What kind of myofilaments? I do not believe it and I would like someone to show me a picture of it.

H. HOLTZER: May I suggest you consult anyone of the following: Holtzer *et al.*, 1957; Okazaki and Holtzer, 1965, 1966; Bischoff and Holtzer, 1970.

M. REZNIK: Yes, but that is at the light microscopic level.

H. HOLTZER: We first described them using fluorescein-labeled antibodies; subsequently Allen and Pepe (1965), Przybylski and Blumberg (1966) and Heuson-Stiennon (1965) confirmed our findings in all details using electron microscopy. In the normal development of mouse, chick and salamander every single somite yields literally thousands of mononucleated myoblasts with typical striations including Z-, M- and H-bands. These cells are present days before bi- and trinucleated myotubes appear. On the other hand, in the anlagen of breast muscle there is no stage that we have observed when mononucleated myoblasts with cross striations dominate. Indeed in this muscle they are rare. In summary there is no obligatory sequential relationship between fibril formation and fusion. For many types of muscle *in vivo* and *in vitro* fusion precedes the formation of filaments.

M. NAMEROFF: Is there any evidence that a mononucleated cell that has cross striations can fuse as well as one without myofibrils?

M. ALOISI: A big question.

H. HOLTZER: I wonder if the real issue is why do multinucleated cells *cease* to fuse? Why do multinucleated cells not stuff themselves like little pigs with all those wonderful satellite cells out there just for the grasping? Why do satellite cells not initiate fusion with their neighboring muscle fibers? Dr. Bischoff has some evidence along these lines (Bischoff and Holtzer, 1969).

D. ALLBROOK: One final point on nomenclature. Can we agree that the *satellite cell* has no myofilaments, a *myoblast* is a mononucleated cell with a lattice of myofilaments in it, and a *myotube* is a series of fused myoblasts? It would be a great advance to agree on this nomenclature.

A. MAURO: Perhaps we should even drop the term satellite cell and just call it a subsarcolemmal cell or the subbasement membrane cell and let it go at that. The important thing is to agree on the histological observables.

We shall now call upon Dr. Reznik to give his view of the origin of satellite cells.

SATELLITE CELLS, MYOBLASTS, AND SKELETAL MUSCLE REGENERATION

MICHEL REZNIK

During regeneration of adult skeletal muscle, the origin of myoblasts and the mode of multinucleation of the regenerating myofibers are still controversial subjects (Betz *et al.*, 1966). On the first point, the major problem lies in the uncertainty of identifying myoblasts morphologically at an early stage of their existence. On the second point, light- or electron-microscopic investigations of fixed tissue do not allow demonstration of the dynamic modifications occurring in injured muscles.

Despite these difficulties, various experimental studies have produced specific sequential information upon which one may propose an unifying theory to correlate the structural changes observed during degeneration and regeneration of striated muscle. The results herein reported support the concepts that, during adult mammalian muscle regeneration: (1) myoblasts originate from pre-existing multinucleated muscle syncytium as mononucleated cells that pass through a satellite cell stage (Reznik, 1969b); (2) multinucleation of regenerating myofibers occurs by cytoplasmic fusion of myoblasts that have divided at least once before fusing (Reznik, 1968*b*); (3) myofibril formation is evident only in multinucleated cells (Betz *et al.*, 1966).

Investigations were made on the tibialis anterior of normal adult rabbits after acute ischemia, and the gastrocnemius muscle of normal adult mice (NIH, GP and R3a) after cold injury.

CYTOLOGIC ASPECTS

Injured muscles of both species follow a similar pattern of morphologic changes that have been extensively described at the level of either the light microscope (Field, 1960; Adams *et al.*, 1962*a*) or the electron microscope (Allbrook, 1962; Price *et al.*, 1964; Reznik, 1967*a*).

Either type of injury induces local acute muscle degeneration with inflammatory reaction. Two or 3 days after the injury, the damaged muscle can be divided into 3 irregular zones (Fig. 116): (1) the uninjured part; (2) the necrotic area; and (3) an intermediate region which is the most interesting because it is in this field that the process of regeneration is the most prominent.

During normal muscle regeneration *in vivo*, as well as during embryonic myogenesis, or when embryonic muscle develops *in vitro*, the appearance of multinucleated muscle fibers is always preceded by an active proliferation of mononucleated cells among which various types of connective tissue cells can be identified. Among these single cells are mononucleated myoblasts that are capable of forming new myofibers. Such myoblasts cannot be readily recognized in sections by the usual histologic techniques, although these cells are more easily identified when they are dissociated and growing *in vitro* (Konigsberg, 1963).

Some of the cellular modifications that occur during muscle regeneration can be illustrated by studying the variation of cell number (or nuclei) in the 3 zones of the damaged muscle, at various times after the local injury. In the normal gastrocnemius muscle of the adult

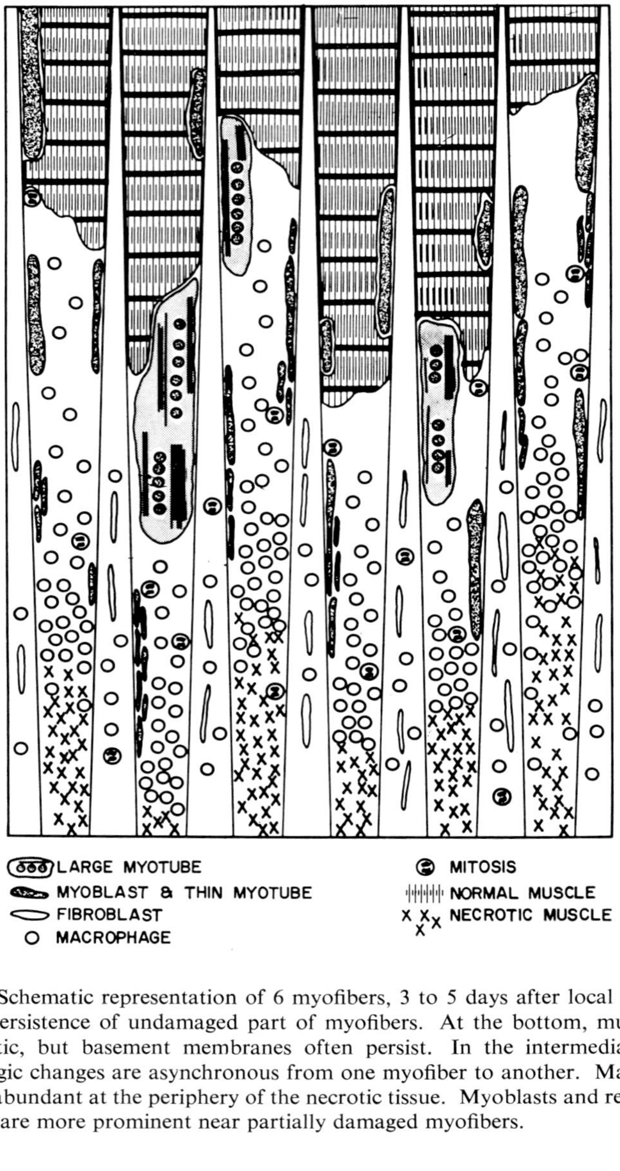

Fig. 116. Schematic representation of 6 myofibers, 3 to 5 days after local injury. At the top, persistence of undamaged part of myofibers. At the bottom, muscle fibers are necrotic, but basement membranes often persist. In the intermediate region, morphologic changes are asynchronous from one myofiber to another. Macrophages are more abundant at the periphery of the necrotic tissue. Myoblasts and regenerating myotubes are more prominent near partially damaged myofibers.

mouse, there are very few interstitial cells and almost all nuclei are inside myofibers. In 4- to 5-micron-thick histologic sections stained with hematoxylin and eosin, one can count an average of 25 nuclei per oil-immersion microscopic field (Fig. 117). Soon after the injury (12 to 24 hours), the number of nuclei drops sharply in the necrotic area, but all nuclei do not disappear. A few nuclei remain scattered in the necrotic fibers. There is no significant change in most of the undamaged part of the myofibers. In the intermediate region of the muscle, which includes the distal part of the undamaged myofibers and the proximal zone of the necrotic tissue, nuclei increase rapidly in number because of the presence of different cells. The mononucleated cells present in that area are leukocytes, histiocytes, macrophages,

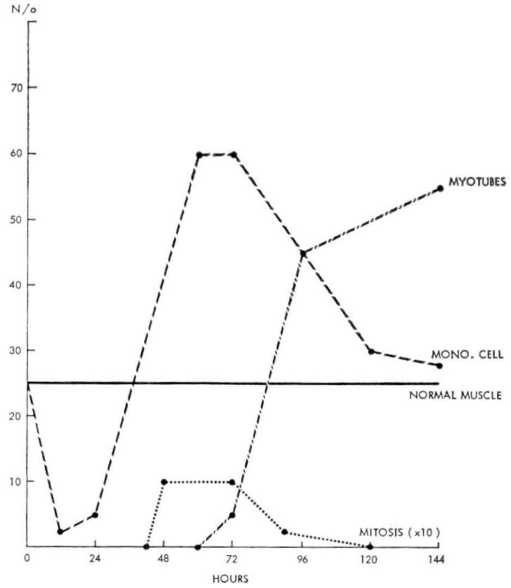

Fig. 117. Graph showing numbers of nuclei (per oil-immersion field, vertical line) in myofibers, myotubes, and mononucleated cells in gastrocnemius muscle of normal adult mouse after local cold injury (at time 0, horizontal line). See text for explanation.

fibroblasts, and fibroblast-like basophilic cells, among which are the presumptive myoblasts. When counting mainly the fusiform mononucleated cells with a basophilic cytoplasm (thus disregarding the leukocytes or the recognizable macrophages), the increase in cell number starts 24 to 30 hours after the injury, reaches its peak 48 to 72 hours after the injury, and then decreases progressively. These mononucleated cells are always the only ones dividing mitotically. The highest mitotic index is about 1.3% 48 to 72 hours after the injury, and then it decreases (Fig. 117).

Between 66 and 72 hours after the injury, long multinucleated cells become present in the intermediate region; they have been variously named myotube, early myofiber, or sarcoblast (Godman, 1958; Betz *et al.*, 1966). The number of nuclei in these developing structures increases abruptly from approximately 72 to 90-96 hours after the injury (Fig. 117). Later, the increase in number of nuclei in myotubes seems to be less rapid. There is never any sign of nuclear division inside the multinucleated myotubes or sarcoblasts (see Betz *et al.* (1966) for review of the previous literature).

If there is any relationship between the increase in number of mononucleated cells in the injured muscle, and the further development of regenerating multinucleated myotubes, then whatever reduces the number of these mononucleated cells should also inhibit muscle regeneration. For example, a mitotic inhibitor acting at a precise moment could alter the formation of myotubes. In order to test this hypothesis, the effect of local X-ray irradiation was investigated during muscle regeneration.

EFFECTS OF LOCAL IRRADIATION

Local X-ray irradiation of muscle has different effects on the process of regeneration. These effects depend on 2 major factors: (1) the dose of X-ray, and (2) the lapse of time between the moment of irradiation and the moment of injury (Reznik, 1967*b*; Reznik and Betz, 1967).

When a single local irradiation (1500 r Rx) is given 1 or 2 days after the injury, the muscle regeneration is completely inhibited. In these muscles (rabbit, mice) examined 1 to 15 days after irradiation, there is an inflammatory reaction but no, or very few, basophilic mononucleated cells are present; no mitotic activity can be detected. No multinucleated myotubes become apparent, even 2 weeks after the injury.

When a similar single local irradiation is made 4 to 5 days after the injury, muscle regeneration appears identical in the irradiated and the contralateral non-irradiated muscles of the same animal.

If irradiation is applied 3 days after the injury, the muscle regeneration is completely inhibited in some animals, or partially reduced in others. In the later group, it is possible to see nuclei of varying size in some of the regenerated myotubes; such alterations of size are described also in nuclei derived from a mitotic division occurring in the presence of nitrogen mustard or Myleran (Bassleer *et al.*, 1963).

From these experiments, one can conclude that mononucleated cells, dividing mitotically in the damaged muscle, 1 to 3 days after the injury, are necessary for the formation of multinucleated myotubes. When multinucleated myotubes are already present 4 to 5 days after injury, the further development of these myotubes is apparently not impaired by the local irradiation; therefore, it would seem that continued development of the myotubes does not require the presence of dividing mononucleated cells.

The effects of a similar irradiation applied before muscle injury bring additional information on the possible origin of myoblasts (Reznik and Betz, 1967). Although the normal adult skeletal muscle does not exhibit histologic alteration after such an irradiation, there is a latent lesion demonstrated when the muscle is induced to regenerate. If the muscle is irradiated locally (1500 r Rx), 1, 5, or 10 days before injury, the regeneration is almost completely inhibited. No myotubes are discovered in these muscles until 2 weeks after the injury. There is, however, a regular inflammatory reaction with the presence of mononucleated cells such as leukocytes or macrophages, but very few basophilic mononucleated cells; occasional mitosis is observed.

The effect of previous local X-ray irradiation is also dose- and time-dependent. For instance, 750 r Rx applied 1 day before cold injury strongly inhibits muscle regeneration in mice. A similar irradiation applied 10 days before cold injury has fewer inhibitory effects on the process of regeneration that occurs later and more slowly than in the control muscle from the other leg in the same animal.

From these results, one can hypothesize that the mononucleated cells needed for muscle regeneration are formed locally in the damaged muscle and further, that these cells originate from intramuscular structures that can be altered by previous irradiation. Therefore, the myoblasts, as opposed to the inflammatory cells, do not appear to be a part of the cell population brought to the damaged muscle from the blood stream. Most of these conclusions are confirmed by the following autoradiographic investigations and electron-microscopic observations.

LIGHT-MICROSCOPIC AUTORADIOGRAPHY

A sequential autoradiographic study of the incorporation of tritiated thymidine was investigated during regeneration of the cold-injured mouse gastrocnemius muscle (Reznik, 1968*b*). That research provides qualitative and quantitative data regarding the kinetic transformation of mononucleated cells into developing multinucleated myotubes. The results depend on the interval between the time of muscle injury and the time at which the labeled precursor of DNA is injected into the animal (Fig. 118).

When tritiated thymidine (4 to 5 microcuries per gram) is given by a single intraperitoneal

PERCENTAGE OF LABELLED NUCLEI FOLLOWING A SINGLE INJECTION OF H_3-THYMIDINE (100μc) AFTER MUSCLE INJURY.			
TIME OF INJECTION (HOURS)	TIME OF BIOPSY (HOURS)	MONO NUCLEATED CELLS	MYOTUBES
54	55	24	-
	57	32	-
	62	34	-
	71	34	0
	79	30	20
66	67	22	0
	69	34	0
	72	33	19
	78	36	31
	90	25	49
80	81	23	0
	83	22	0
	86	26	5
	98	26	9
	104	24	15

PERCENTAGE OF LABELLED NUCLEI FOLLOWING ONE TO FIVE INJECTIONS OF H_3-TYMIDINE (40μc) AFTER MUSCLE INJURY				
TIME OF INJECTION (HOURS)	NUMBER OF INJECTION	TIME OF BIOPSY	MONO NUCLEATED CELLS	MYOTUBES
48	1	54	27	-
54	2	60	33	-
60	3	66	45	0
66	4	72	72	85
72	5	78	71	92
		96	72	92
		120	72	89
		14 d	[60]	69

PRESENCE OF MYOTUBES (HOURS)	
62	-
66	±
72	+
84	+ +
96	+ + +
120	+ + + + +

Fig. 118. Graph showing quantitative results of light-microscopic, autoradiographic study in regenerating muscle after local cold injury and injection of tritiated thymidine. See text for explanation.

injection 54 hours after the injury, 24 to 35 per cent of the mononucleated cells are found labeled 1 to 24 hours later. No labeled nuclei are discovered in the normal myofibers; some labeled nuclei are very close to these muscle fibers, but their exact nature is often difficult to determine. In this series of experiments, 20 per cent of nuclei in the myotubes are found labeled 24 hours later.

If the same amount of tritiated thymidine is injected 66 hours after the injury, 22 to 33 per cent of mononucleated cells are labeled 1 to 24 hours later. At no time are there labeled nuclei in the normal myofibers. Some nuclei, however, which are surrounded by a dense basophilic cytoplasm, and situated at the periphery of partially damaged myofibers, are labeled. It is only 6 to 8 hours after the injection of tritiated thymidine that labeled nuclei are observed in myotubes; at that time, about 20 per cent of the nuclei inside the myotubes are labeled. Later, when the total number of nuclei increases in the myotubes, the percentage of labeled nuclei also increases; 24 hours after the injection of the tritiated thymidine, almost 50 per cent of the nuclei inside the developing myotubes are labeled.

The results are different if the precursor of DNA is injected 80 hours after the injury, at a time when numerous myotubes are already present in the regenerating muscle. As previously observed, only mononucleated cells (23 to 26 per cent) are found to be labeled 1 to 3 hours after the single injection of tritiated thymidine; no labeled nuclei are found immediately in the muscle fibers or in the myotubes. It is also 6 to 8 hours after the administration of the tritiated thymidine that labeled nuclei are discovered in myotubes. The rate of increase of labeled nuclei in myotubes is low, and comes to only 15 per cent, 24 hours after the injection. In this series, the labeled nuclei are found mainly in the smallest myotubes that are near the area of muscle necrosis.

From these sequential autoradiographic studies one may conclude that nuclei inside the myotubes are derived from the incorporation of mononucleated cells which have divided mitotically, at least once, before fusing (Reznik, 1968*b*). These findings also underline the abrupt formation of multinucleated myotubes between 66 to 72 hours and 90 to 96 hours after muscle injury. After 96 hours post injury, myotubes continue to form but at a slower rate, and incorporate fewer mononucleated cells. Such results are in agreement with recent findings on myotube formation *in vitro* (Bischoff and Holtzer, 1969).

Moreover, it can be demonstrated that almost all the nuclei inside the myotubes forming *in vivo* are derived from the fusion of recent mitotically divided mononucleated cells. For instance, if tritiated thymidine is given to the animal by several injections between 48 and 72 hours after the injury (Fig. 118), developing myotubes contain more than 90 per cent of labeled nuclei.

It should also be emphasized that the mononucleated myoblasts involved in the formation of multinucleated myotubes are very likely formed locally in the damaged muscle and are not derived from circulating cells. That assumption, already supported by the investigations of Walker (1963), is based on the present results herein reported and by various experiments that would exclude the concept that myoblasts could be circulating cells formed outside the injured muscle. Without going into detail, the following example is one from the latter type of experiments. In a group of mice, the gastrocnemius muscle of only one leg is injured. Two to 3 days later, each animal is given a load of several injections of tritiated thymidine in order to label the stem cells of a hypothetical center of myoblast production. Two to 3 days after these injections, the gastrocnemius of the other leg is injured. Both muscles are examined at varying times after injury. Regeneration appears comparable in both muscles of the same animal. However, only the first injured muscle contains myofibers with labeled nuclei. The muscle injured after the injection of tritiated thymidine contains labeled inflammatory cells but no labeled nuclei in the myotubes.

ULTRASTRUCTURAL OBSERVATIONS

Ultramicroscopic research provides more information on: (1) the possible origin of myoblasts during skeletal muscle regeneration; (2) the mode of multinucleation of sarcoblasts; and (3) some aspects of myofibrillogenesis.

Since the advent of electron-microscopic investigations of experimental muscle regeneration, 2 main hypotheses have been suggested for the origin of myoblasts. The first hypothesis is based on studies of the regenerating salamander limb where nucleated sarcoplasm arising from the fragmentation of sectioned muscle fibers is considered capable of becoming myoblasts (Hay, 1959; Lentz, 1969). Other authors (Church *et al.*, 1966; Shafiq *et al.*, 1967) support the second view that the muscle satellite cells described by Mauro (1961) are the precursors of myoblasts. Although there is good evidence that these satellite cells are presumptive myoblasts, the origin of such cells is still uncertain (Mauro, 1961; Ishikawa, 1966). Furthermore, satellite cells are rare in normal adult mammalian skeletal muscle (Muir *et al.*, 1965) although they seem to be more numerous in most pathological muscles, mainly in human neuromuscular disorders (Lee, 1965; Klinkerfuss, 1967; Shafiq *et al.*, 1967; Reger and Craig, 1968). Despite an extensive search, no satellite cells were observed by the author in normal muscle fibers of the adult rabbit tibialis anterior and the mouse gastrocnemius. On the contrary, satellite cells appear very numerous in regenerating areas of the muscle, mainly in partially damaged myofibers that are near the site of necrosis. In this particular region, although histopathologic and ultrastructural changes are asynchronous from one myofiber to another, it is possible to observe ultramicroscopic modifications which can be sequentially assembled to suggest that the so-called satellite cells are only a temporary morphological stage in the development of myoblasts (Reznik, 1969*b*).

The basic aspect and critical point regarding the origin of myoblasts is illustrated in Figure 119. This Figure shows a normal myofiber with a normal myonucleus, and a partially altered myofiber with an elongated mononucleated cell lying beneath its basement membrane. That mononucleated cell, which possesses a moderate amount of cytoplasm with ribosomes, few mitochondria, poorly developed endoplasmic reticulum and a centrosome, can be considered either as a satellite cell or a presumptive myoblast. It is now interesting to determine the possible relationship between: (1) the normal myonucleus and that type of mononucleated cell illustrated in Figure 119; (2) that same mononucleated cell and the myoblast involved in the formation of multinucleated sarcoblasts.

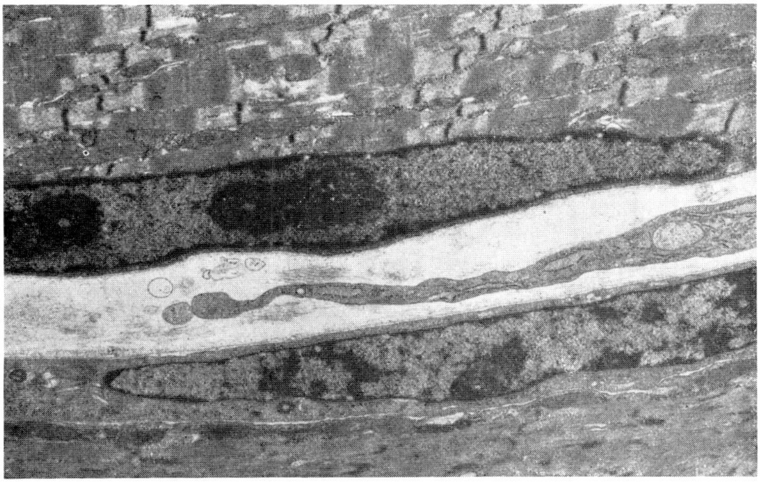

Fig. 119. Undamaged and partially degenerated myofibers in gastrocnemius muscle of adult mouse, 78 hours after cold injury. In the undamaged myofiber (top) the normal myonucleus is in contact with the myofibrils. In the partially degenerated myofiber (bottom) the nucleus, surrounded by a rim of cytoplasm and lying beneath the basement membrane of the fiber, is separated from the altered myofibrils. This mononucleated cell contains ribosomes, a few small mitochondria and a centrosome (near the center). Uranyl lead; × 7,400.

When one looks at most of the myofibers that appear essentially normal at low magnification in the intermediate region of injured muscle, it is obvious that many of the myonuclei and their surrounding cytoplasm have a different ultrastructural aspect from that observed in the undamaged region of the same muscle. Some of these myonuclei are larger (Figs. 121 and 123) than usual (Fig. 119). They are surrounded by a part of sarcoplasm devoid of myofibrils but richer in ribosomes. Furthermore, in many fibers, clefts and small vesicles are aligned beneath the basement membrane and also between the bundles of myofibrils and the part of cytoplasm rich in ribosomes (Figs. 120 and 122). The appearance of a new mononucleated cell is more impressive when such a myonucleus and its surrounding cytoplasm are almost separated from the muscle sarcoplasm (Fig. 123). A persistent area of cytoplasmic attachment of such a developing cell and the normal muscle sarcoplasm is illustrated at higher magnification in Figure 124. The origin of the vesicles and clefts that appear around some myonuclei and their surrounding rim of cytoplasm is unknown. Since developing mononucleated cells become separated from the muscle sarcoplasm, it is likely that vesicles and clefts forming the planes of cleavage must be, or must come, in contact with the extracellular space; in fact, connections of these clefts with the transverse tubular system are strongly suggested in some myofibers (Fig. 125).

When completely separated, the new cell is wedged between the muscle sarcoplasm and the basement membrane of the myofiber, as a satellite cell (Reznik, 1969*b*). At that stage, the cells also present the ultrastructural characteristics of myoblasts as they are described

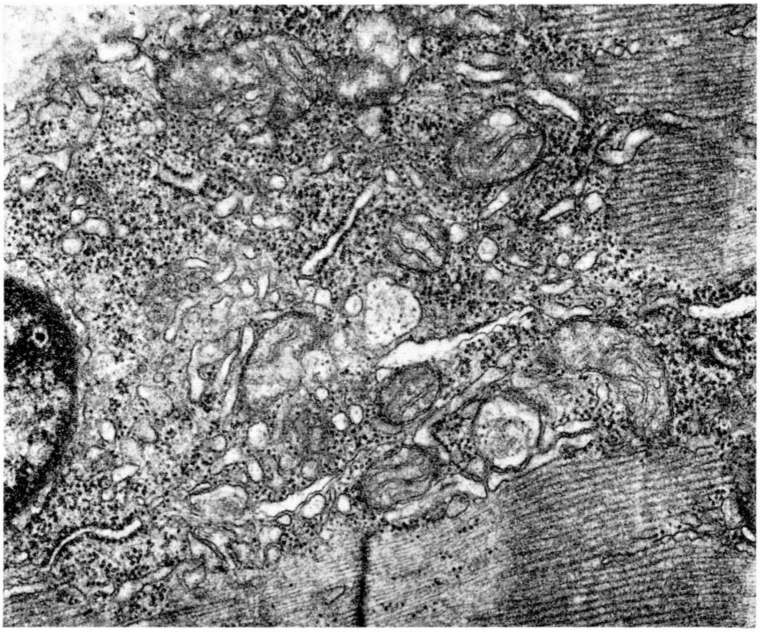

Fig. 120. Myofiber in the intermediate region of the muscle 68 hours after injury. A large number of ribosomes, endoplasmic reticulum and unusual clefts are visible between the normal myofibrils and the perinuclear sarcoplasm devoid of myofibrils. Uranyl lead; × 20,800.

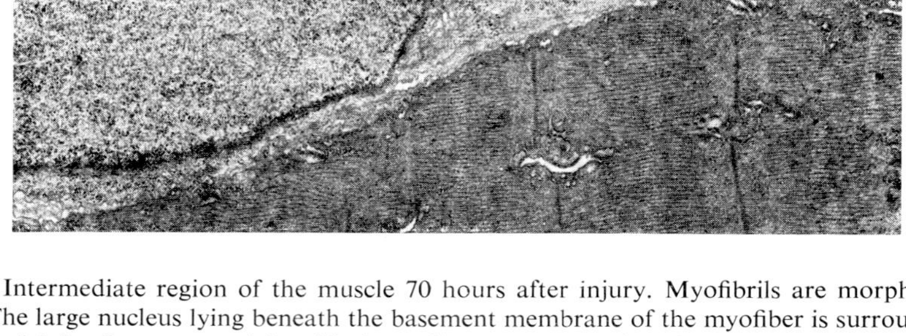

Fig. 121. Intermediate region of the muscle 70 hours after injury. Myofibrils are morphologically normal. The large nucleus lying beneath the basement membrane of the myofiber is surrounded by a cytoplasm rich in ribosomes and devoid of myofilaments. Uranyl lead; × 9,600.

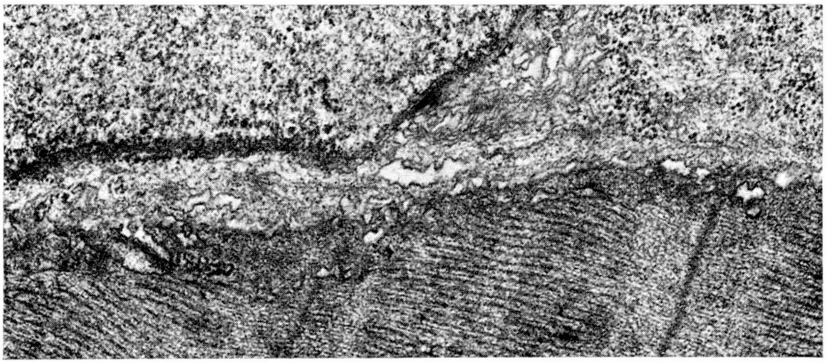

Fig. 122. Detail of Fig. 121. Although aligned vesicles are present, no clear separation is visible between the sarcoplasm with myofibrils and the perinuclear cytoplasm rich in ribosomes. Uranyl lead; × 24,800.

Fig. 123. Another undamaged myofiber near the site of necrosis 67 hours after injury. The cell, formed by a large nucleus and cytoplasm packed with ribosomes, is almost completely separated from the main sarcoplasm. This cell is in the same position as satellite cells, but exhibits ultrastructure of myoblasts (compare with Fig. 130). Uranyl lead; × 9,600.

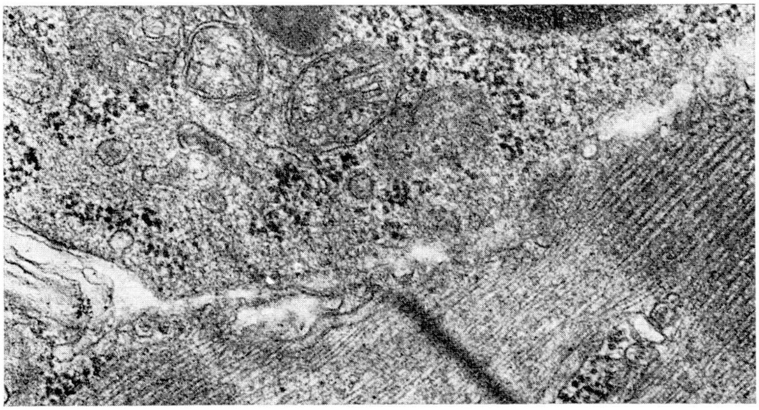

Fig. 124. Detail of Fig. 123 (far left), showing one of the cytoplasmic bridges (at right) between the myofibrillar sarcoplasm and the perinuclear cytoplasm of the developing cell. Uranyl lead; × 72,000.

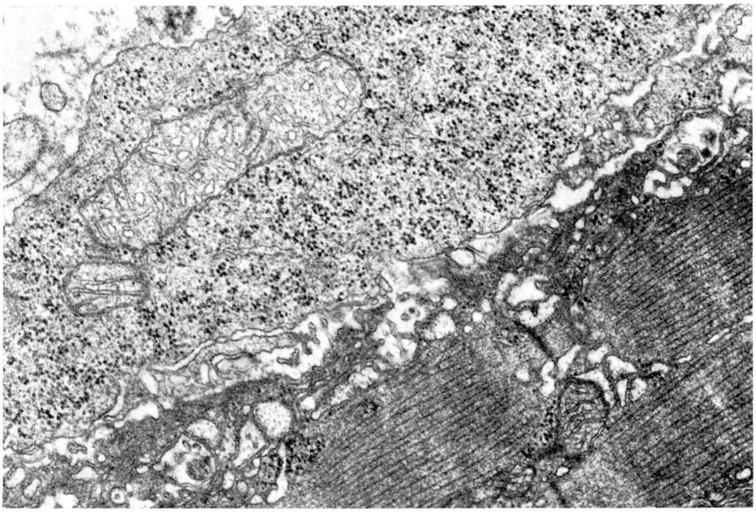

Fig. 125. Partially damaged myofiber in another muscle 67 hours after injury. Dilated segments of the sarcotubular system seen in connection with the clefts and vesicles that separate the peripheral cytoplasm rich in ribosomes from the morphologically normal myofibrillar sarcoplasm. Uranyl lead; × 21,600.

during embryonic development (Hay, 1963; Przybylski and Blumberg, 1966) or *in vitro* (Firket, 1967; Shimada *et al.*, 1967). Their cytoplasm is moderate in amount without prominent extensions; it is packed with ribosomes, and contains a poorly developed endoplasmic reticulum, a few scattered mitochondria, a small Golgi apparatus near the nucleus, a centriole, and some lipid droplets.

These mononucleated cells, which can be named either satellite cells or presumptive myoblasts at that stage, extend longitudinally beneath the basement membrane of the

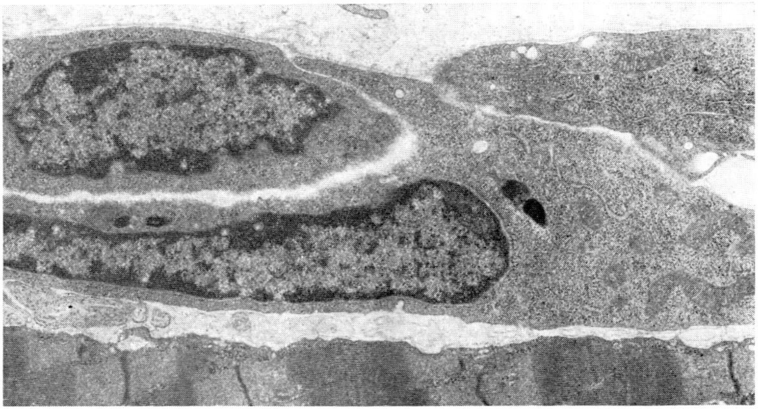

Fig. 126. Portion of 3 interdigitated mononucleated cells lying in a space between the normal myofibrils and the basement membrane (top) of the myofiber, near a site of necrosis, 72 hours after injury. These 3 cells exhibit similar ultrastructure. Uranyl lead; × 8,000.

pre-existing myofiber. In one sarcolemmal sheath, several similar cells can be in contact and overlap each other (Fig. 126). At that early stage of their development, all of these cells exhibit the same ultrastructural aspect (compare cells in Figs. 121, 123 and 126). Eventually,

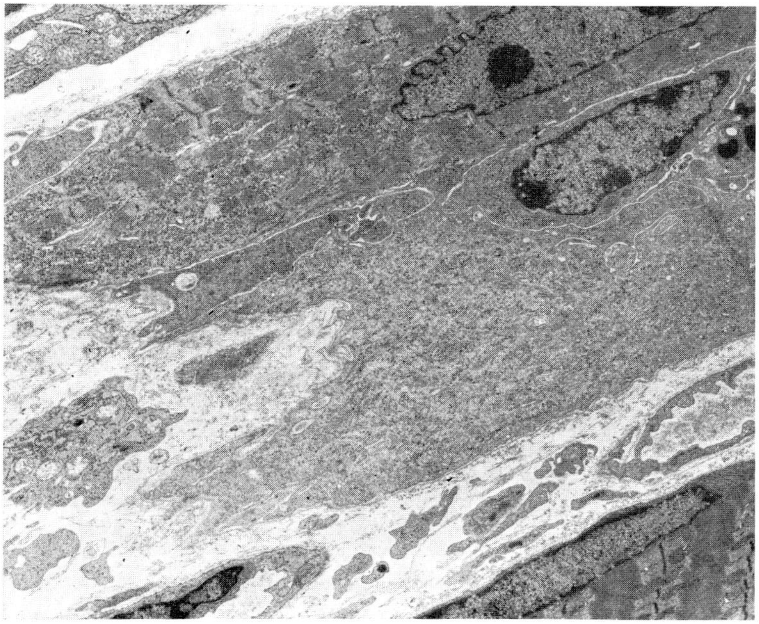

Fig. 127. Intermediate region of the muscle 78 hours after injury. A normal myofiber with a nucleus is seen at the bottom. Enclosed in the same basement membrane of a partially damaged myofiber (center) are: (*a*) part of the pre-existing myofiber with an enlarged myonucleus; (*b*) a mononucleated cell; and (*c*) a large portion of a cytoplasm packed with ribosomes but devoid of myofibrils. Uranyl lead; × 5,000.

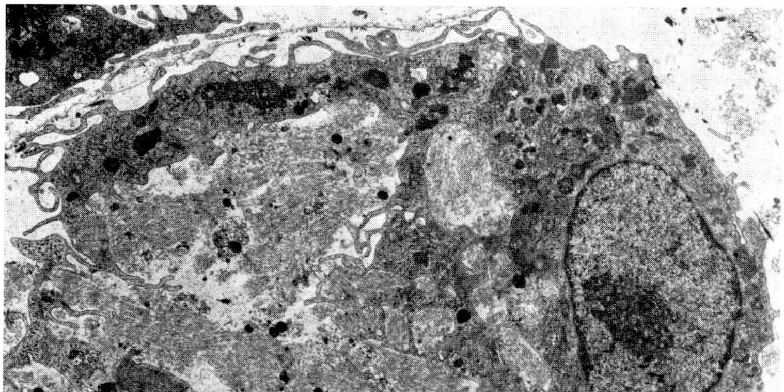

Fig. 128. Macrophage phagocytizing fragments of myofibrils 69 hours after injury. Note the numerous pseudopodic and cytoplasmic extensions encompassing necrotic material. Uranyl lead; × 4,560.

they form a large cytoplasmic mass which grows toward the extremity of the partially damaged myofiber, while remaining beneath its basement membrane (Fig. 127).

Where myofibers are damaged, the possible origin of myoblasts is more difficult to assess. During acute myofiber degeneration the great majority of the myonuclei become pyknotic and disappear. However, in the periphery of the necrotic areas, in proximity to the undamaged tissue, some elongated nuclei surrounded by a thin rim of cytoplasm are found between the degenerating myofibrils and the persisting basement membrane (Figs. 119 and 129). The cell illustrated in Figure 129 is very similar to the one observed in Figure 119, and appears as a possible surviving cell in the necrotic myofiber. Similar observations constitute for Church *et al.* (1966) proof of the reality of pre-existing satellite cells that escape necrosis because of their prior separation from the bulk of the degenerating myofibers. It is possible, however, that in this particular area where satellite cells could survive, some muscle nuclei

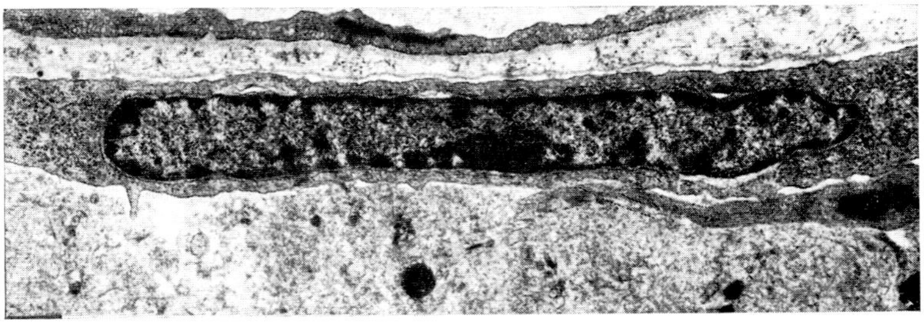

Fig. 129. Presumptive surviving mononucleated cell beneath the basement membrane of a necrotic myofiber (bottom) 73 hours after injury. Compare with the cell in Fig. 119. Uranyl lead; × 6,400.

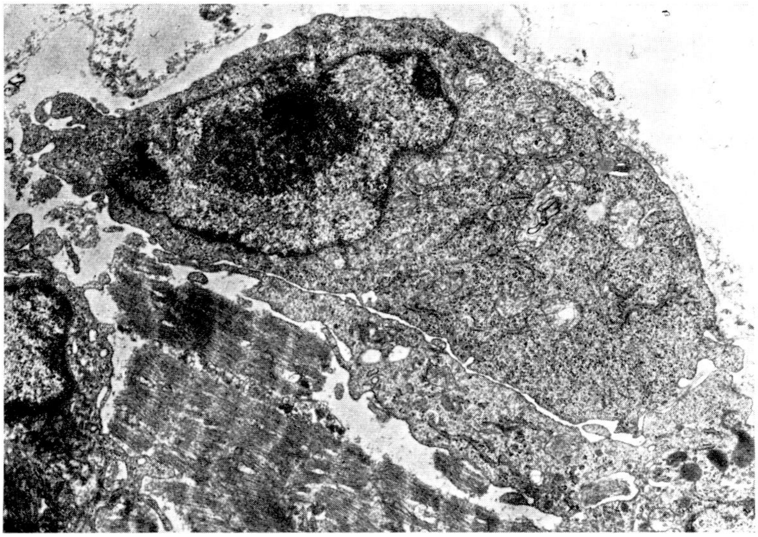

Fig. 130. Necrotic myofiber 73 hours after injury. Portion of 2 macrophages (left and bottom right) are observed. The mononucleated cell beneath the basement membrane (top) is recognized as a myoblast; it is packed with ribosomes, contains poorly developed endoplasmic reticulum, scattered mitochondria, and exhibits a few small pseudopodic extensions. Uranyl lead; × 8,800.

surrounded by a rim of cytoplasm can also survive an acute injury, whereas most of the sarcoplasm of the same syncytium becomes necrotic.

In the necrotic tissue, wandering cells are able to pass through the basement membrane and penetrate into the sheath of the necrotic myofibers. Most of these cells can readily be recognized as histiocytes and macrophages which phagocytize the necrotic material (Fig. 128). Such cells are found inside and outside the sarcolemmal sheath of necrotic myofibers; they contain a great number of mitochondria, a moderate amount of endoplasmic reticulum, a large Golgi apparatus, and a variable number of lysosomes and vesicles with necrotic fragments; these cells, which are very irregular in shape, also possess a great number of pseudopodic extensions (Fig. 128). The macrophages are in great number around necrotic material and eventually disappear altogether.

The cells recognized as presumptive myoblasts in the necrotic area are always closely associated with the internal side of the basement membrane of the myofibers (Figs. 129 and 130). Such cells are characterized by a cytoplasm packed with ribosomes, and contain scattered mitochondria, poorly developed endoplasmic reticulum, and small Golgi apparatus; they exhibit very few or no pseudopodic extensions. The myoblasts are not numerous in myofibers in which degenerating sarcoplasm is still present, but they become more abundant when the necrotic fragments are phagocytized and removed from the sarcolemmal sheath.

The mode of increase in number of myoblasts in that part of the muscle is not completely elucidated. Most of these myoblasts undergo mitotic division (Fig. 131), but that does not

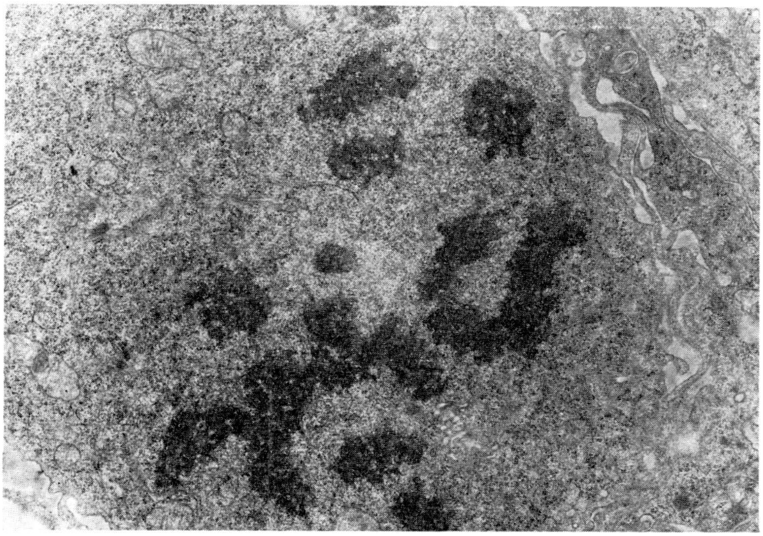

Fig. 131. Same area as in Fig. 130. Example of a mitosis in a cell with cytoplasmic ultrastructural characteristics of myoblast. Uranyl lead; × 8,800.

exclude the possibility of cells migrating from the partially damaged myofibers. Such a migration of cells is certainly suggested in the blastema of amphibian regenerating limbs (Hay, 1959; Lentz, 1969), or when muscle fibers developed *in vitro* from dissociated myoblasts (Konigsberg, 1963).

When they are more abundant, myoblasts demonstrate specific behavior characteristics such as a tendency to become aggregated among themselves and the ability to form a long thin strip of interdigitated mononucleated cells (Fig. 132). In such a disposition, myoblasts are in contact with each other on a significant portion of their membrane surface (Figs. 132 and 133). Sometimes, tight junctions are also observed between myoblasts (Firket, 1967; Fischman, 1967).

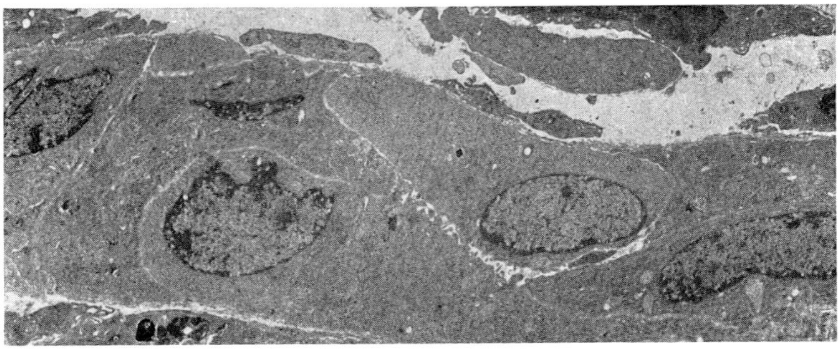

Fig. 132. Group of interdigitated myoblasts forming a thin chain parallel to the long axis of the muscle fibers 78 hours after injury. Myoblasts have extensive surface membrane contacts, one with another. Uranyl lead; × 3,120.

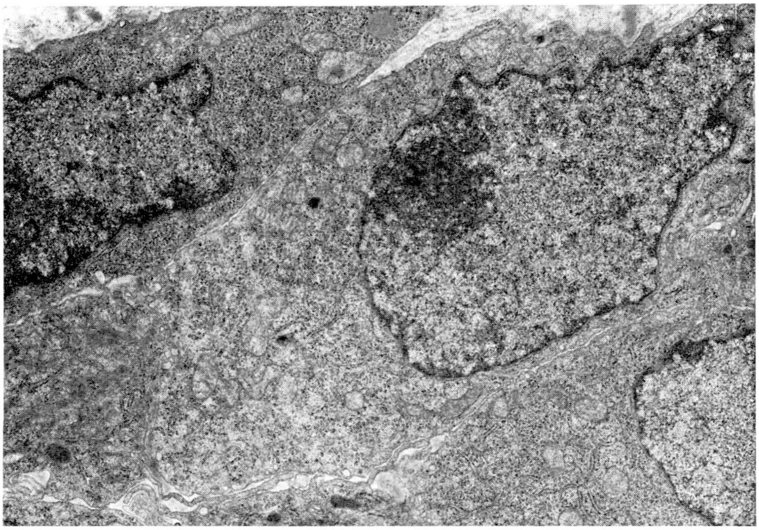

Fig. 133. Detail of 3 adjacent myoblasts, with similar ultrastructure and close cell surface contacts. Uranyl lead; × 7,040.

At a later stage, apposed elements of the unit plasma membrane of two adjacent myoblasts disappear and a cytoplasmic bridge between the two cells is visible (Fig. 134). Cell fusion can exist simultaneously between one myoblast and several adjacent mononucleated myo-

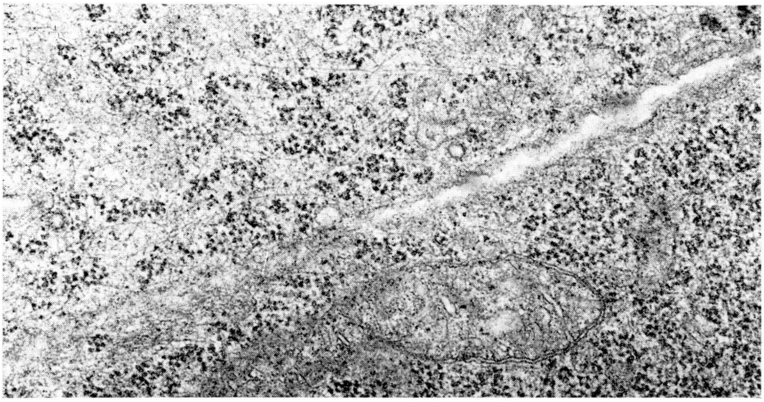

Fig. 134. Area of cytoplasmic fusion (at left) between 2 adjacent myoblasts. At right, the separation of the 2 cells is still visible, although the unit membrane is not apparent. Uranyl lead; × 46,400.

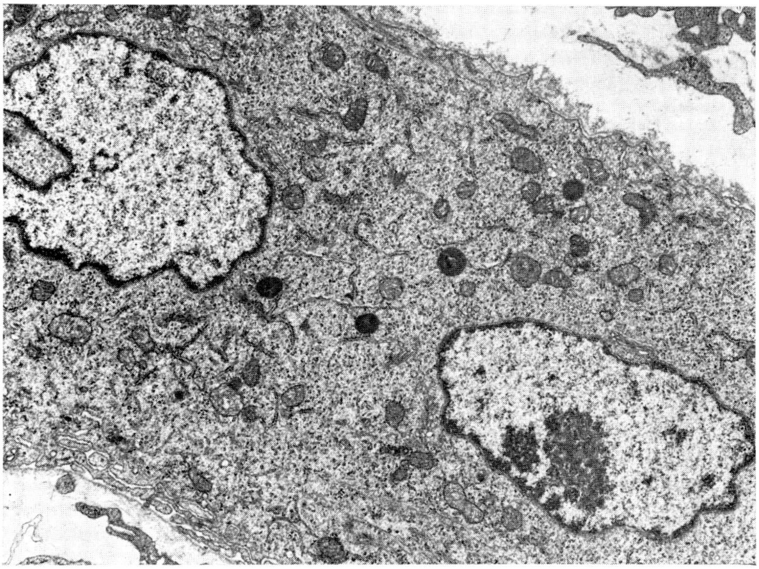

Fig. 135. Typical portion of a multinucleated myotube at an early stage of development. The large rounded nuclei are centrally located. The cytoplasm presents almost the same ultrastructure as mononucleated myoblasts. Scattered myofilaments are observed at the periphery of the cell. Uranyl lead; × 4,800.

blasts. Thereafter, multinucleated sarcoblasts are present in the tissue (Figs. 135 and 136). Their large nuclei occupy the central axis of the cytoplasm which, at the very beginning of their existence, is identical to the one of the mononucleated myoblasts (compare Figs. 126, 130 and 135).

It is in these myotubes that myofilaments and myofibrils become visible (Fig. 136). Ultra-microscopic aspects of myofibrillogenesis have been extensively studied in various conditions of muscle development (Betz *et al.*, 1966; Firket, 1967; Fischman, 1967; Kelly, 1969). Despite these investigations, there are still many questions about the time, the site, and the

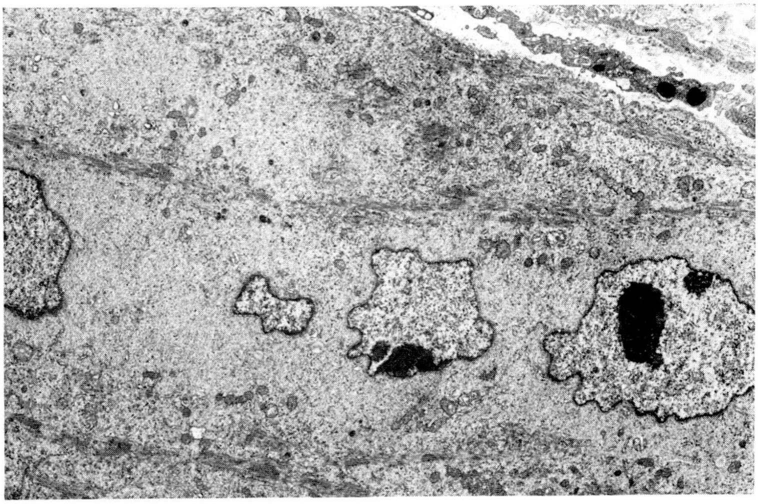

Fig. 136. Another multinucleated myotube in the same muscle as in Fig. 135, 90 hours after injury. Aggregation of myofilaments and development of a few longitudinal bundles of myofibrils are more readily recognized. Uranyl lead; × 4,160.

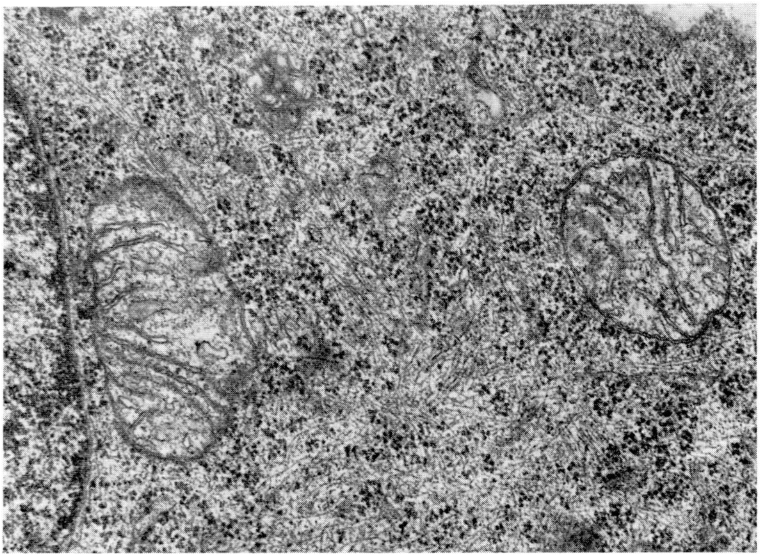

Fig. 137. Mononucleated myoblast 69 hours after injury. Thin filaments of various shape and orientation are irregularly dispersed in the cytoplasm. Uranyl lead; × 32,000.

mode of appearance and organization of the myofibrils in various conditions of muscle development. It is certainly hazardous to postulate that myofibrillogenesis follows a single schema in all species, and in different pathologic circumstances. Moreover, one should be cautious in naming 'thin' or 'thick' filaments as 'actin' or 'myosin'. For instance, Ishikawa *et al.* (1968) have suggested that many of the thin, or so-called actin filaments discovered in presumptive mononucleated myoblasts (Fig. 137) are not likely to be involved in myofibrillogenesis. This assumption is based on the fact that similar filaments are discovered in many cells which are completely independent of the muscle tissue (Ishikawa *et al.*, 1968).

In regenerating muscles studied in this work, the ultrastructural appearance of myofibrillogenesis exists only in multinucleated myotubes. In the experience of the author, it seems difficult to decide which type of myofilament is the first detectable. This uncertainty arises from the fact that: (1) all thin (50 to 60 Ångströms) filaments are not necessarily thin myofilaments (Ishikawa *et al.*, 1968); and (2) thick (100 to 130 Ångströms) myofilaments are not frequently found isolated but are very soon grouped in a bundle (Fig. 138). It is

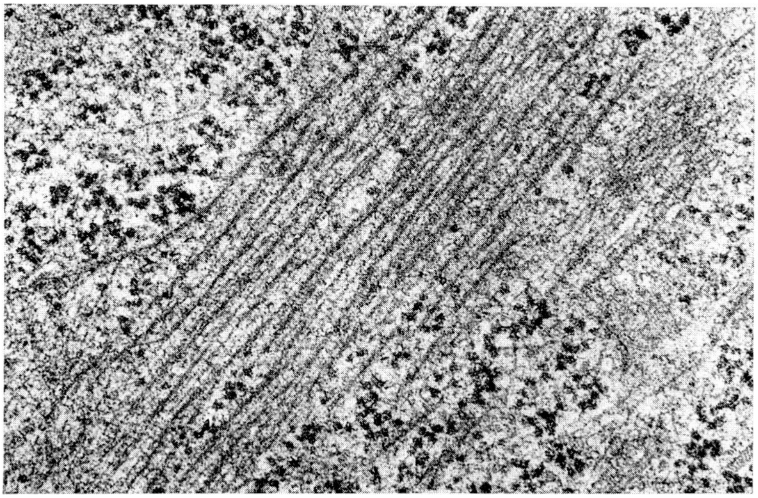

Fig. 138. Longitudinal section of early aggregating thin and thick myofilaments observed only in the cytoplasm of multinucleated myotubes. Uranyl lead; × 60,000.

also in such bundles that the thin myofilaments can be more readily identified morphologically. Therefore, it is the author's opinion that the two types of myofilaments appear simultaneously in the myotube, and that they can only be recognized as such with the electron microscope when they begin to lie associated in bundles (Fig. 138). These bundles of thin and thick myofilaments are the initiating structures of the developing myofibrils. Firstly they appear scattered in the cytoplasm of the myotubes (Fig. 136). The early myofibrils have a uniform aspect, without apparent banding. Later the A- and I-bands appear, followed by Z-line formation in the middle of the I-band (Kelly, 1969). The development of these myofibrils is asynchronous from one myotube to another, and even in the cytoplasm of the same multinucleated syncytium (Fig. 139). Progressively, the myofibrils become thicker, more numerous, and they occupy a larger portion of the cytoplasm (Fig. 140). Typical muscle cross striation occurs simultaneously (Kelly, 1969).

Many questions concerning myofibrillogenesis are still unanswered, but a few problems are becoming clarified. Polysomes are certainly involved in the process of muscle protein synthesis, although that is difficult to demonstrate in electron micrographs. The longitudinal

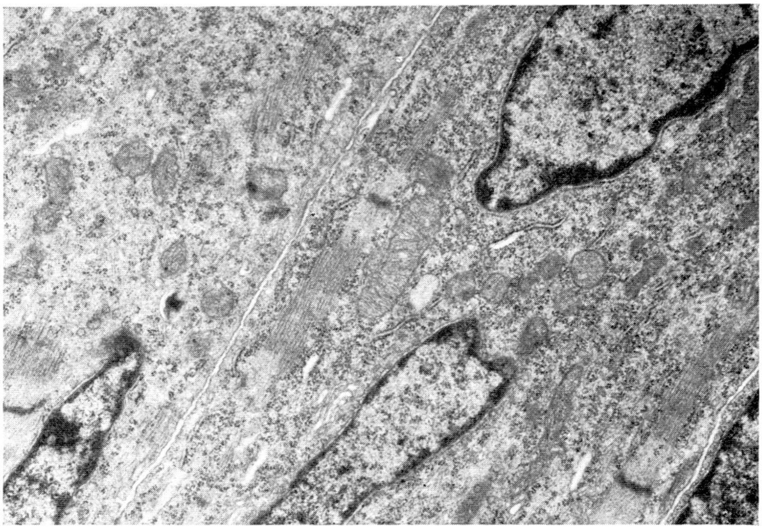

Fig. 139. Longitudinal section of myotubes containing myofibrils at a more advanced stage of development. A- and I-bands, with Z-bands, are present on some bundles of myofibrils. Uranyl lead; × 9,600.

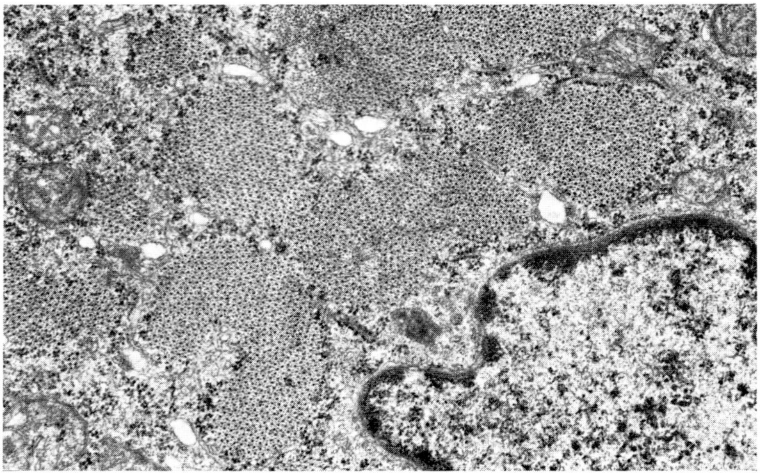

Fig. 140. Cross-section of a regenerating myofiber 5 days after injury. Myofibrils are larger and more abundant. Ribosomes are visible at the periphery of the myofibrils, but not inside the bundle of myofilaments. Scattered vesicles around the myofibrils are probably related to the formation of the sarcotubular system. Uranyl lead; × 19,200.

alignment of myofibrils probably depends on the presence of microtubules, although they do not appear to be always necessary (Warren, 1968).

During muscle regeneration, as well as during normal myogenesis, it is apparent that the transverse tubular system is developed simultaneously with the formation of myofibrils and the maturation of the new myofibers (Ezerman and Ishikawa, 1967; Schiaffino and Margreth, 1969). Ultimately, the future development and differentiation of the new myofibers will probably depend on the sort of physiologic activity they become involved in, and the type of their motor innervation (Adams *et al.*, 1962*a*; Betz *et al.*, 1966).

ELECTRON-MICROSCOPIC AUTORADIOGRAPHY

An electron-microscopic autoradiographic study to detect the cells incorporating tritiated thymidine in regenerating muscles of mice is now under investigation. Results are still preliminary (Reznik, 1969*a*), but generally confirm earlier observations. One of the most interesting findings is the discovery of labeled so-called satellite cells (Fig. 141) 2 to 6 hours

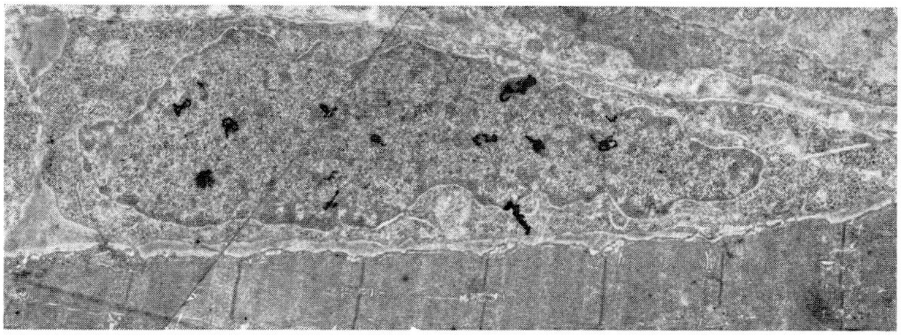

Fig. 141. Electron-microscopic autoradiography 68 hours after injury and 2 hours after a single injection of tritiated thymidine. So-called satellite cell (compare with the cells in Figs. 123 and 126) with a labeled nucleus. Uranyl lead; × 5,600.

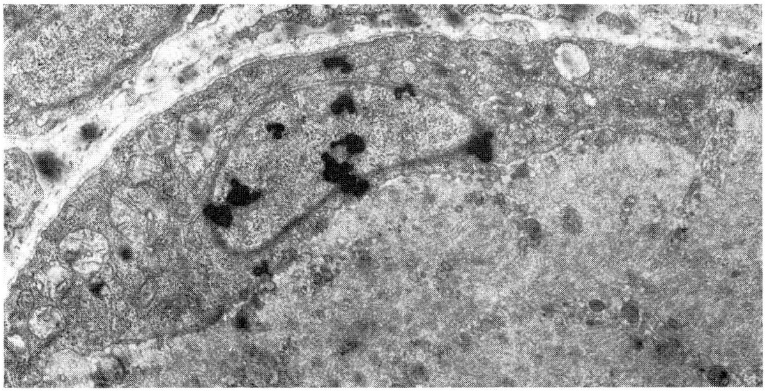

Fig. 142. Labeled myoblast in a necrotic fiber 90 hours after injury and 24 hours after a single injection of tritiated thymidine. Uranyl lead; × 12,000.

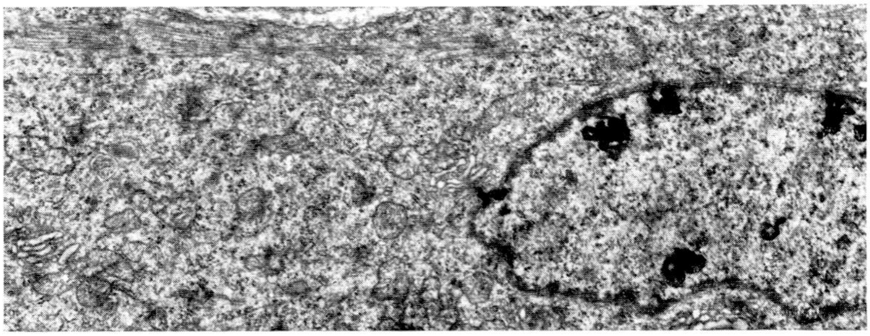

Fig. 143. Same muscle as in Fig. 142. Portion of a myotube with a labeled nucleus; a few bundles of myofilaments are present at the periphery of the cell (top left). Uranyl lead; × 8,000.

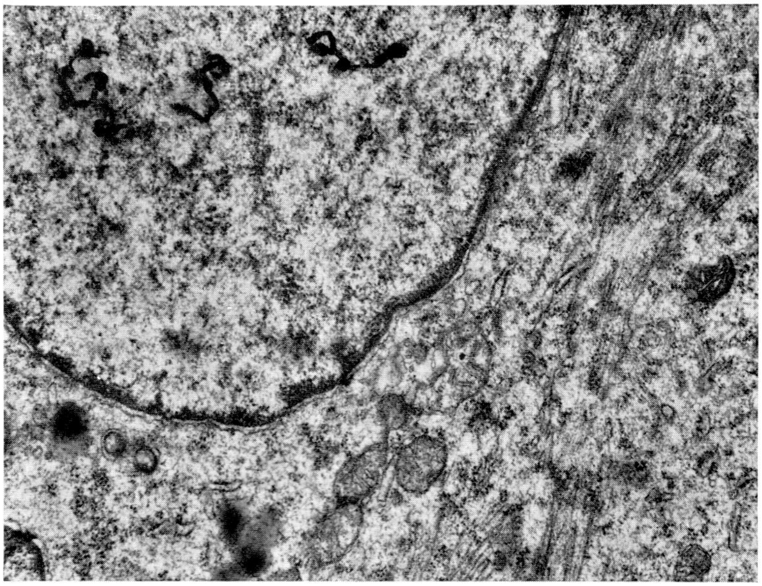

Fig. 144. Another myotube in the same muscle as in Fig. 143. Bundles of myofilaments are readily recognized near the labeled nucleus. Uranyl lead; × 16,000.

after a single injection of tritiated thymidine made 66 hours after cold injury. These observations undoubtedly indicate that these so-called satellite cells synthesize DNA during muscle regeneration, and confirm, indirectly, that they can divide mitotically. Labeled myoblasts are also found in necrotic myofibers (Fig. 142). It is not less than 8 to 12 hours after the injection of the tritiated thymidine that some nuclei in multinucleated myotubes are found to be labeled (Fig. 143). Developing myofibrils are observed in myotubes with labeled nuclei (Figs. 143 and 144).

CONCLUSIONS

The results reported herein can be interpreted to elaborate the schema of muscle regeneration in cold-injured gastrocnemius muscles of normal adult mice. Three successive yet coordinated stages are recognized:

1. During a first stage, between 24 hours and approximately 72 hours after injury, a pool of mononucleated myoblasts is formed mainly in the area between the undamaged myofibers and the necrotic tissue. This occurs in two sequences:

a. The initial step is the formation of a new generation of mononucleated myoblasts which are segregated from the pre-existing multinucleated myofibers. In the necrotic myofibers, some myonuclei surrounded by a thin rim of cytoplasm probably survive the acute injury while most of the sarcoplasm of the same syncytium becomes necrotic. In the partially damaged myofibers, the production of a myoblast results from the formation of new plasma membranes that isolate a previously normal myonucleus and its surrounding envelope of cytoplasm. During that process, the future myoblasts pass through a satellite-cell stage.

b. In the second step, this newly generated population of myoblasts becomes more abundant by mitotic division.

2. During a second stage, approximately between 72 hours and 96 hours after injury, there is an abrupt formation of multinucleated myotubes requiring two sequences:

a. Initially there is an organized aggregation of a great number of the myoblasts.

b. Then the adjacent myoblasts undergo cytoplasmic fusion to form multinucleated cells. The process of fusion involves only myoblasts that have divided at least once, and which have escaped the mitotic cycle before fusing. There is an interval of 6 hours minimum between the moment of DNA synthesis in a myoblast and its ability to fuse with another cell. No division of nuclei are observed in the multinucleated cells. It is in myotubes that myofibrillogenesis becomes detectable.

3. During the third stage, 96 hours and later after injury, myotubes develop into new myofibers. The process of maturation can be separated into two phases:

a. At first, multinucleated myotubes continue their development without necessity of further incorporation of mononucleated cells. Myofibrillogenesis is more prominent and typical cross-striation becomes visible.

b. Ultimately, the stability of regenerated myofibers will probably depend upon an efficient motor innervation.

J. C. T. CHURCH: What do you call that cell in the degenerating myofiber (Fig. 119)?

M. REZNIK: I would agree to name it a satellite cell. I see them only in an area of damaged muscle.

A. MAURO: May I just interject a question? This is a crucial point, involved with observables – not interpretation. Dr. Church, in your work, do you have any evidence or any suspicion of trauma or damage to the bats that you work with?

J. C. T. CHURCH: Do you mean in their normal habitat?

A. MAURO: Yes. As I recall, in your paper you said that you carried out a so-called satellite cell count and in some of the fibers you found that as high as 10 to 12 per cent of the nuclei were satellite cells. Am I not correct in that?

J. C. T. CHURCH: Yes.

A. MAURO: Now, did you have any suspicion in that work that the muscles were traumatized, injured, anoxic, that the bats had been beaten over the head the day before, or something like that? I mean, would you say that those fibers were in a normal state, that the muscles were in a normal state?

J. C. T. CHURCH: I have studied many normal muscles in the bat including well-protected muscles such as the external ocular muscles, and have always found satellite cells.

A. MAURO: Then Dr. Reznik's observations are unique in the literature.

J. C. T. CHURCH: I would be very surprised if, after a careful search, satellite cells were not found in normal rabbit muscles.

A. MAURO: Are you working with a special genetic strain of rabbit?

M. REZNIK: No.

A. MAURO: Where do you obtain your rabbits?

M. REZNIK: The rabbits are just ordinary rabbits and the mice I am working on are from the NIH, the general purpose strain.

S. A. SHAFIQ: I think in this picture (Fig. 119) you have the question of where the centrioles came from. We have been hearing that muscle nuclei do not have centrioles and, centrioles being generally considered 'self-replicating', here we will have to assume that, as this cell developed from the muscle nucleus, the centrioles originated *de novo* in it.

E. D. HAY: I think I can supply some information which comes from studies of ciliogenesis in our laboratory and in other laboratories. It is quite clear that centrioles do not have to arise by replication of existing centrioles (Steinman, 1968). It is quite clear that they can arise *de novo*, so I do not think that that is an objection.

A. R. MUIR: Can I ask you how long it takes?

E. D. HAY: In the *Xenopus* embryo when the neurula suddenly becomes ciliated, these centrioles appear within 2 hours from a precursor material that is very dense.

H. HOLTZER: Ciliates have numerous basal bodies derived from centrioles. Has it been shown in a tissue cell lacking a centriole, and centriole precursor regions, that in fact such a cell can *de novo* form a centriole?

A. MAURO: It is a good point, but I think we should come back to Dr. Reznik.

D. ALLBROOK: I would like to return to the question of nomenclature. I do not think we should call this (Fig. 131) a myoblast if myoblast is defined as a cell which we know for *certain* is going to form muscle, for the only way to be sure of this is to demonstrate actin or myosin in it. I suggest that you call this a presumptive myoblast.

M. REZNIK: I would agree.

D. ALLBROOK: Because this gets around the logical problem.

M. REZNIK: I think that the myoblasts are mononucleated cells which do not yet contain myofilaments. If you prefer to call it a presumptive myoblast, I would agree also. However, I believe that these cells will be involved in the formation of myotubes.

D. ALLBROOK: I agree with the logical argument, though I do not think that you can say 'that is a myoblast' with the evidence you have on that picture. I think you can say on your evidence that we *presume* it is going to be one. So let us call it a *presumptive myoblast* when the cytoplasm is devoid of filaments.

J. C. T. CHURCH: Dr. Reznik, now could you name the cells you have just shown us in the last 3 slides (Figs. 119, 121, and 123)?

M. REZNIK: No, do not ask me to give you a name now. We can discuss that afterwards.

J. R. FLORINI: Dr. Reznik, how long were your radioautographs exposed?

M. Reznik: About 3 to 4 months.

J. R. Florini: Well the difference between 3 disintegrations and no disintegrations over 3 to 4 months is not really such a significant difference, is it?

M. Reznik: But you do not see signs of disintegration elsewhere than in the nuclei.

J. R. Florini: I know, but if there are 1 or 2 disintegrations in one nucleus and not in another with a half-life of 11 or 12 years, does that really show conclusively that there is a difference? If you had 180 grains in that picture (Fig. 141), then I would be a lot more impressed.

M. Reznik: Yes, but before you can obtain an electron-microscopic autoradiograph with 180 grains, I think you would have to wait a long time before finding it. (Laughter).

H. Holtzer: Yes, but this is not altogether a 'made' point. The electron microscope shows us a great deal and, if we have some errors on our slide, the electron microscope magnifies the error equally well. I certainly do not think that this type of argument is to be put under the rug.

E. D. Hay: What is the argument? I do not understand it!

J. R. Florini: It is a question of radioactivity. Given a population in which only 3 disintegrations might occur in the course of 3 months, the probability that some disintegrations will be missed over that period is high.

E. D. Hay: Well, that is a very real possibility for light microscopy. The fact that he sees 3 grains in a section that is probably 800 Ångströms thick is very positive evidence.

J. R. Florini: I am not disputing the positive argument; it is the negative evidence I am concerned with.

E. D. Hay: Yes, but you would probably miss it in the light microscopy too.

D. Allbrook: I was convinced about this because you have many more disintegrations on some of the other slides.

J. R. Coleman: I think there is no argument about whether or not that nucleus is labeled. The point is there may be a number of other nuclei around that do not have grains, but nonetheless have tritiated thymidine in them.

M. Reznik: I would like to put a question to the audience: How many of you have seen satellite cells in the skeletal muscle of adult, normal, mammalian tissue? Perhaps I should make a distinction between man and other mammals, since most of the people who have seen satellite cells in man were looking at tissue obtained by surgical biopsy, which is not necessarily normal muscle tissue. Thus I ask the question: How many of you have seen satellite cells in muscles of normal, adult mammals other than man, in which species and at what age?

J. H. Venable: Mouse – anywhere from 40 days to 180 days. Tibialis anterior, biceps femoris and longissimus dorsi.

S. A. Shafiq: I saw them in tibialis anterior of mice and I would say around 5 per cent or less.

A. R. Muir: Healthy rats of 250 grams. There is no logic in restricting this to mammals because I and others have seen them in birds and amphibia.

M. Reznik: I have seen satellite cells only in humans, but I believe it was always in a pathological muscle.

S. A. Shafiq: But they were first described in non-humans.

M. Reznik: What I am trying to emphasize is that, since in normal adult muscle, the number of satellite cells is very low, it is difficult to explain the rapid increase of mononucleated cells (presumptive myoblasts) on the basis of pre-existing 5 per cent of mononucleated cells, even with a mitotic index of 13 per cent and a mitotic cycle of 14 hours (*cf.* Bischoff, this volume, pp. 218-231).

S. A. SHAFIQ: But in regard to your pictures showing continuity of mononucleated cells with injured fibers, which you interpret as stages in the origin of these cells from fiber nuclei, it must be pointed out that in the injured fibers satellite cells are dividing and are at different stages of rounding up. Their membranes are out of alignment with that of the muscle fiber so that one may easily get a false impression of continuity where plasma membranes have been cut obliquely.

A. MAURO: Dr. Reznik, I am afraid we must terminate this discussion since other participants would like to present their data.

MUSCLE REGENERATION IN MAN AND THE MOUSE; EVIDENCE DERIVED FROM TISSUE CULTURE AND FROM THE EVOLUTION OF EXPERIMENTAL AND SURGICAL INJURIES IN THE IRRADIATED AND NON-IRRADIATED SUBJECT

JOHN C. SLOPER

and

R. B. BATESON, D. HINDLE, J. WARREN

In analysing the mechanisms by which mammalian skeletal muscle is repaired after injury, our ultimate aim has been the elucidation of the abnormality in regeneration which has been postulated in human muscular dystrophy (Adams *et al.*, 1962*b*). The exact nature of this abnormality is still obscure. It clearly requires further study, too, in the light of the novel, but now widely accepted, concept of muscle regeneration recently advanced by a number of workers, and in particular Holtzer, Konigsberg and Walker. Moreover, the evidence for this new concept is based on animal work: it is essential that it should be shown also to apply to muscle regeneration in man.

In our human work we have relied largely on tissue culture, and on the rather speculative premise that the mechanisms of myogenesis are similar *in vitro* and *in vivo*. Our *in vitro* observations have confirmed that there is within human adult muscle a cell which participates in myogenesis. Our experimental observations in the mouse, which are based on the use of tritiated thymidine and the preirradiation of injured limbs, do, however, suggest that cells, probably derived from the circulation, may in fact contribute to this process.

In a Symposium devoid of a definitive programme, it seems justifiable to present our findings in the form of a series of questions, with the answers to which our researches have been primarily concerned.

DOES SKELETAL MUSCLE REGENERATE?

In Britain at least, there still appear to be many who are sceptical of the contribution made by new muscle fibres to the repair of injury in skeletal muscle (see, for example, Illingworth and Dick, 1963). Indeed, direct evidence in mammals seems to be lacking; but indirect evidence, based upon the examination of muscle lesions in experimental animals killed at various stages after injury, has long appeared conclusive (Weber, 1867; Le Gros Clark, 1946).

In man, however, because comparable data are so difficult to obtain, we have tended to base our interpretation of the changes seen in diseased muscle on the assumption that the cytological characteristics of experimental lesions observed in lower mammals, which have a life-span of some 2 years, are applicable even to the adult and ageing human subject. This probably is the case, but it is worth noting that in man we are considering the reformation of a multinucleate cell, thought, at least until recently, to persist without replacement for 70 years or longer.

For these reasons it seems justifiable to quote our own observations on the results of surgical injury to muscle in 6 subjects aged 12 to 65, suspected of suffering from muscle disease, thus amplifying the definitive analysis of this problem made by Walton and Adams in 1956, that is, at a time when the interpretation of muscle changes was based on muscle budding rather than the current myoblast multiplication theory. Sequential biopsies were made at intervals of 4 days (1 subject), 7 days (4 subjects) and 8 days (1 subject). It was hoped that these biopsies would aid in the differential diagnosis between neuropathy and myopathy, in that regeneration would tend to be diminished in the latter.

Regeneration seemed in fact to be slight in 2 patients. One was aged 45 and suffered from dystrophia myotonica; the other was a boy aged 12 suffering from muscular dystrophy. In the latter, the second biopsy from the gastrocnemius showed only 3 myotubes in the many sections examined. Conversely, in each of the 4 other subjects the second biopsy contained many myotubes. It was concluded that 2 of these suffered from neurogenic atrophy of muscle. In the other 2 the diagnosis remained in doubt, the first biopsy showing either slight simple atrophy of muscle or no recognizable abnormality.

These observations on surgical lesions confirm that muscle regeneration can occur in response to trauma even in ageing human subjects. The diminished response observed in a boy suffering from muscular dystrophy corroborates the similar finding of Walton and Adams (1956) in 3 severely affected subjects. Since the initial biopsy in our patient revealed abundant muscle fibres, it seems clear that a more fundamental disturbance than a mere diminution in the fibre content of the muscle is responsible for the poor regenerative response in muscular dystrophy.

CAN MUSCLE BE GROWN *IN VITRO* FROM NORMAL ADULT HUMAN SUBJECTS?

We have accordingly attempted to employ tissue culture as a further means of assessing the regenerative potential of adult muscle in the normal and in the diseased human subject; for, although conditions operating *in vitro* are grossly abnormal, it should not be overlooked that one is in fact studying a repair process in a tissue, injured at the time of biopsy, and thus in some measure comparable with muscle injured *in vivo*, save that the tissue is totally divorced from the circulation.

Particular attention has been paid to 4 points, namely:

1. the mechanisms of the formation of myotubes;
2. the identification of the multinuclear strap cells seen in culture as probable myotubes;
3. the absence of any evidence of pre-existing regeneration in the muscles taken for explant;
4. the time pattern of myogenesis in explants from *normal* muscle.

Bateson (Bateson *et al.*, 1967) began her tissue culture work on the chick embryo, employing, in general, Maximow cultures or Falcon flasks, and TC199 medium or 'conditioned' medium (medium which had already been used for the growth of skeletal muscle in culture). By the use of first trypsin, and later pronase, mononuclear subcultures were obtained.

In subsequent work on human embryonic muscle taken from 39 foetuses aged between 6 and 20 weeks, Bateson was able to confirm that myogenesis closely resembled that observed in biopsies made from the chick. In one experiment, Maximow cultures and subcultures were immersed in a medium which contained tritiated thymidine (0.3 to 0.5 microcuries per millilitre) for 20 to 30 minutes. These cultures were washed and then killed between 1 and 5 days later. From 1 to 13 hours after labelling, mononuclear cells were labelled but multinuclear cells were not.

It was thus shown that the nuclei of human myotubes cultured *in vitro* do not divide. This observation was first made by Stockdale and Holtzer (1961) on chick muscle, and, taken in conjunction with Konigsberg's (1963) observations on the growth of chick myotubes from single clones, first established that myogenesis *in vitro* probably involves the multiplication of mononuclear cells and their fusion to form myotubes.

With respect to the *identification of multinucleate strap cells* (Pearce, 1963) with myotubes, this presents little difficulty in the case of chick muscle cultured *in vitro*: for cross-striated myofibrils are easily recognizable in such cultures. They were, however, scanty in human cultures, but were found to be more easily identifiable if cultures were grown on Melinex (an I.C.I. plastic), which facilitated study by light and electron microscopy. This technique was developed by Firket (1966), and enabled us to identify occasional cross striations in myotubes cultured for adult human subjects.

In such subjects we were especially concerned to exclude the presence of *pre-existing areas of regeneration* in the biopsies taken for culture. This seemed essential if we wished to establish unequivocally that new muscle fibres could be derived, at least in part, from cells existing in the normal muscle of the adult human. For Pearce (1963) has recently stressed the difficulty of growing muscle from normal mice as opposed to injured muscle in mice; he tentatively suggests that, at any rate in the mouse, existing areas of myogenesis are responsible for muscle growth *in vitro*.

The data available in the pioneer observations made on human adult muscle by Popogeff and Murray (1946) and Geiger and Garvin (1957) lack in fact detailed information concerning the source of the muscle cultured. Bateson, Hindle and Warren have cultured myotubes from biopsies made from 41 human subjects, 36 of them adults (27 to 79 years), a half of the latter being females. Most biopsies were made from the lower limb, 4 from the upper and 17 from the trunk. Eight biopsies were from normal adults and 3 from normal subjects aged 20 months to 12 years. These subjects were undergoing operations unassociated with recent trauma or disease of muscle; several operations, for example, were arthrodeses of the hip for osteoarthritis. All biopsies explanted were established histologically as lacking in areas of regeneration. We have, therefore, satisfied ourselves that muscle will grow *in vitro* from a cell present in normal adult muscle.

In this way, too, we have established a *time pattern* for the first appearance of mononuclear cells and of multinucleate cells in cultures from normal subjects. This time pattern is wide and variable. It is essential that it should be firmly based, if abnormalities in the time pattern of growth *in vitro* are to be identified in muscle diseases.

NEW MYOTUBES ARE HELD TO ARISE EITHER FROM EXISTING MUSCLE FIBRES BY BUDDING OR FROM MONONUCLEAR CELLS OF UNCERTAIN ORIGIN. ARE THESE THEORIES NECESSARILY INCOMPATIBLE?

The myoblast multiplication theory is based on 3 main groups of *in vivo* experiments. The crucial experiment has yet to be repeated: it is based on the observation that myotube nuclei in lesions in the mouse are always diploid (Lash *et al.*, 1957), and therefore cannot have undergone amitotic division. Mitotic division seems equally never to have been seen in myotubes, even after the local apposition of colchicine (Pietsch, 1961*a*). On the other hand, the predecessors of many of the nuclei of myotubes must have undergone mitotic division: for many myotube nuclei are labelled if tritiated thymidine is given to mice 24 or so hours after muscle injury (Bintliff and Walker, 1960). These observations are best explained on the assumption that myotubes stem from the fusion of mononuclear precursors, mitotic division being confined to the latter.

It seems, however, to have been overlooked that this concept of regeneration is not incompatible with muscle budding, provided that the nuclei observed in muscle buds are solely derived from the nuclei surviving in the injured fibre by a process of translocation. Just how important this process of budding could be has yet to be assessed. Le Gros Clark (1946), the only worker to have, in recent years, studied the connection between bud and mother fibre by serial sections, clearly believed in its significance. He did not, however, exclude the possibility that the 'muscle buds' observed reflected the secondary fusion of an independent myotube with an injured adult fibre.

Whatever the case, it must be insisted that muscle budding is not incompatible with the formation of myotubes from newly formed myoblasts. It is unlikely, however, that muscle budding is particularly important.

WHERE DOES THE MONONUCLEAR MUSCLE CELL PRECURSOR ARISE?

1. Possible sources of muscle cell precursors

Muscle cell precursors could originate either:

a. from the nuclei of injured or uninjured muscle fibres by dedifferentiation;

b. from 'satellite cells', lying within the sarcolemma of injured or uninjured muscle fibres;

c. from interstitial cells within the area of muscle injury. These cells might be undifferentiated cells persisting into adult life, in this respect fulfilling a role comparable with that attributed to the 'satellite cell': or again, they might conceivably be differentiated cells such as fibroblasts; these would presumably require to dedifferentiate before becoming myoblasts;

d. from circulating cells, immigrating into muscle lesions from the blood-stream.

These possibilities are all feasible, although it must be admitted that the persistence of undifferentiated cells into adult life is not an easily acceptable hypothesis. Equally, the concept of the circulating cell (Bateson *et al.*, 1967) may seem strange. Yet it is worth noting that work with tritiated thymidine indicates that many of the mononuclear cells observed in muscle lesions appear to reach such lesions from the circulation. Is there then any reason why some of these cells should not participate in myogenesis, rather than in phagocytosis, immune mechanisms and repair?

2. Investigations with tritiated thymidine on the assumption that none is reutilized

The experimental investigation of myogenesis *in vivo* at present centres on 2 techniques: the first, the use of tritiated thymidine as a means of identifying those cells preparing for mitosis at the time of radioisotope injection; the second, the irradiation of uninjured muscle with the aim of preventing, or largely preventing, the subsequent mitosis of cells capable of dividing in the repair of injured muscle. With respect to the use of tritiated thymidine, Walker's pioneer work was based on the premise that labelled thymidine, or some derivative, once taken up by a dividing cell could not subsequently be reutilized. In the light of the work of Hughes *et al.* (1958), it was assumed that uptake by the nuclei preparing for mitosis was complete within 1 or 2 hours of injection of the radioisotope.

Bintliff and Walker (1960) claimed that radioisotope, given before injury, labelled many inflammatory cells, which they therefore assumed had immigrated into the lesions from the blood-stream. The nuclei of myotubes, they claimed, were not labelled; and they thus inferred that these nuclei stemmed from precursors which did not immigrate into lesions like inflammatory cells. When, however, Bateson *et al.* (1967) repeated certain of Walker's experiments, we found that when we gave tritiated thymidine *before* injury, we could label myotube nuclei. This finding, in terms of Walker's premises, strongly suggested that circulating cells could contribute to myogenesis.

Our experiments involved 8 adult mice (each weighing 30 to 40 grams), which we injected intraperitoneally with tritiated thymidine (50 microcuries per animal) 4 times successively at 6-hour intervals. In 2 control cases the last injection was given 48 hours before injury to the right leg. In the other 6 cases, the left leg was crushed 6 hours before the first injection of tritiated thymidine, the last injection thus preceding injury to the right leg by 12 hours (1 animal), 24 hours (2 animals), and 48 hours (3 animals). The pathological effects of crush injury appeared to be very similar to those following the incision employed by Bintliff and Walker. Labelling in autoradiographs we assessed by making granule counts on 100 nuclei in myotubes.

There is a methodological problem here: namely, the measurement of labelling intensity. In the light of current interest in the reutilization of derivatives of tritiated thymidine, this merits perhaps rather more attention than it has received in earlier work on muscle. We have arbitrarily classified our findings in terms of:

a. low labelling (1 and 2 grains; these levels probably representing 'background' labelling);
b. moderate labelling (3 to 10 grains);
c. conspicuous labelling (above 10 grains).

Mean percentages for 'moderate' and 'conspicuous' labelling in legs injured *before* the injection of tritiated thymidine were 54 and 23: and injured *after* injection, 47 and 31. Medians varied between 5 and 8 in the first group; between 5 and 12 in the second.

Most interesting of all, medians in 2 animals in this experiment were lower (5 and 6) in legs injured before thymidine injection than in the contralateral legs injured after injection (6 and 8, respectively). Moreover, in the animals in which only the right leg was injured (48 hours after injection of thymidine), there was conspicuous labelling of myotube nuclei.

In terms of Bintliff and Walker's (1960) premises, these findings would indicate that a proportion of myoblasts had, like macrophages, immigrated into the lesion from the circulation. It should be added that this immigration must be assumed to occur at some undefined time between thymidine injection and the death of the animal: an interval in our experiments of some 6 days and longer.

3. The reutilization of thymidine or a derivative

There is, however, the likelihood that tritiated thymidine (or a derivative) is released from labelled cells, and reincorporated in further cells when they divide. It seems probable, too, that the amount thus made available by, for example, successive waves of cell death in marrow and gut is very considerably less than that available to cells preparing for mitosis within 1 hour of isotope injection (see, for example, Galassi, 1967). If this is so, then the 'conspicuous' labelling noted in legs injured *after* the injection of tritiated thymidine, whilst it might reflect thymidine reutilization, might nevertheless be better explained in terms of the participation in myogenesis of a heavily labelled precursor cell. One can visualize the immigration of a highly labelled myoblast precursor into the lesion from, for example, the marrow, lymphoid tissue or thymus: or again, the transference of labelled thymidine to locally proliferating myoblasts from circulating cells akin to 'trephocytes'. Whether these cells merely act as vehicles for labelled thymidine, or again are capable of participating more directly in the formation of myoblasts, remains to be determined. The problem is one of semantics, and possibly analogous to that which exists concerning the relationship between circulating lymphocytes and monocytes and, for example, fixed tissue histiocytes.

Irrespective of our interpretation of these findings, the question arises as to why, if reutilization is as cogent a factor as has been suggested, myotube labelling has not previously been noted when tritiated thymidine has been given before injury. The explanation may lie in different interpretations of what constitutes 'labelling'; unless granule counts are made, the significance of this term is difficult to define. Another possibility is that the method of administering labelled thymidine misses a hypothetical mitotic peak. By spacing our injections evenly through 24 hours we were less likely to do so than those who have given single injections, or again 4 injections in such a way as to miss periods of 12 hours, presumably through the night. Indeed, in recent experiments where we have given single injections, we have been struck by the low percentage of 'conspicuously' labelled myotube nuclei.

For example, when tritiated thymidine (170 microcuries) was given intraperitoneally at 5 p.m., respectively 48 hours before (3 mice) and 48 hours after (3 mice) crush injury, grain counts on myotube nuclei revealed 'conspicuous' labelling (over 10 grains) in 7, 4, and 1 per cent of the first group, and 8, 10, and 17 per cent of the second group. The fact, nevertheless, that there were 'conspicuously' labelled nuclei in the first group is further evidence either of reutilization or, as seems more likely, of the participation in myogenesis of a heavily labelled circulating precursor.

THE PREIRRADIATION OF A SITE OF MUSCLE INJURY CAN INHIBIT MUSCLE REPAIR. IS THIS CONSISTENT WITH THE PARTICIPATION OF A CIRCULATING CELL IN MYOGENESIS?

Mechanisms which inhibit muscle repair act in 2 ways: namely, first, directly upon the formation of new muscle fibres; and second, indirectly by modifying the inflammatory reaction. The latter, in essence, serves to remove dead tissue and to provide a scaffold for the newly formed myotubes.

In recent years increasing stress has been laid on the circulatory as opposed to local component of the inflammatory response. In mammals this is characterized primarily by the entry of white cells into the tissues from the blood-stream. Of the latter many are phagocytes or participate in immune mechanisms; but the function of many, among them lymphocytes and probably a proportion of monocytes, is unknown. Local tissue cells also contribute to the inflammatory reaction, primarily by furnishing the fibroblasts and new blood vessels seen in granulation tissue. With regard to *myogenesis*, it is clear that local cells, perhaps after dedifferentiation, do yield myoblasts, which must then multiply before they fuse to yield myotubes: but, as we have noted, a haematogenous element may also contribute. The interplay of these 2 aspects of muscle repair is probably reflected in the extent to which cortisone (0.3 milligram daily, given before and after crush injury) retards both the removal of debris and the formation of myotubes in crush injuries in mice (Sloper and Pegrum, 1967). Since many macrophages are haematogenous, the apparent reduction in their number could well be due to the inhibition of cell division in precursor cells in the marrow and lymphoid system.

In the case of myogenesis, myoblast proliferation undoubtedly occurs in marked degree at the site of injury. It is this aspect of muscle regeneration which is probably inhibited by local irradiation, especially if this is given 48 hours and longer after injury, that is, at the time of the maximal proliferation of myoblasts (Williams and Pietsch, 1965). A similar explanation could well account for the local inhibition of myotube formation seen in the region of microabscesses (Sloper and Pegrum, 1967).

Such observations do not, however, clarify the source of myoblasts. One way of investigating this lies in the irradiation of the uninjured leg; for preirradiation of this kind, given over varying periods, up to weeks before injury, appears to prevent cell division in local cells capable of dividing in response to subsequent injury. In the case of hepatic cells or glia (Leong *et al.*, 1959; Cavanagh, 1968), this inhibition is marked with doses of the order of 1000 rad. Reznik and Betz (1967) performed similar experiments in muscle, giving 750 and 1500 rad to the hind limbs of mice 10, 5, and 1 days before opening the skin of the leg over the gastrocnemius and soleus muscles, and causing muscle necrosis by the apposition of CO_2 snow for a minute to an area 0.5 centimetre in diameter. In an unspecified number of animals the lower dose led to a considerable diminution in the number of myotubes ('sarcoblasts'), although some were observed. At the higher dose, inhibition was very marked, only rare rudimentary myotubes being found. No difference was noted in animals irradiated at different times before injury. These observations suggest that a local and radiosensitive precursor cell plays a key part in myogenesis. Although it does not exclude the participation of (unirradiated) circulating cells in this process, evidence in favour of their participation would be stronger if, at a dose sufficient to inhibit the multiplication of the majority of local cells capable of dividing, abundant myotubes were, nevertheless, formed.

Unaware of Betz and Reznik's work, we tested this hypothesis, choosing, in the light of published work, 1000 rad as the critical dose. That a very high dose of preirradiation would inhibit myogenesis we established by the irradiation of a single hind leg in each of 6 anaesthetized adult mice with 5000 rad (230 kilovolts; 2 millimetres Cu half-value layer; 15 milliamperes; 50 rad per minute). These experiments were performed in the Department of Radiotherapy, dosimetry being calculated by the radiophysicist, Mr. Ramsey. Four days later the calf muscles of both hind legs were crushed by Spencer Wells forceps, 2 millimetres wide, and the surviving 5 animals were killed 5 days later. Phagocytosis of debris was incomplete in the irradiated legs, and myotubes were many fewer in number than in the unirradiated legs; varying numbers of myotubes were, however, found, some of them normal in appearance, many with poorly formed cytoplasm and unusually large nuclei. In 4 areas from each section where there were myotubes (in all 224,000 square microns), the figures for myotubes in irradiated limbs were 7, 7, 5, 6, and 3, and in unirradiated limbs 16, 19, 36, 49, and 19. These animals were given tritiated thymidine (170 microcuries intraperitoneally in a single 5 p.m. injection) 24 hours after crush injury. Autoradiographs have yet to be fully counted: but in 1 animal the percentages of conspicuously and moderately labelled myotube nuclei were respectively 53 and 19 (unirradiated) and 49 and 17 (irradiated).

These findings indicated a severe but incomplete inhibition of myogenesis by 5000 rad given 4 days before crush injury. The thymidine studies confirmed the contribution of myoblast proliferation to the formation of at least some myotubes: budding may, of course, have been a factor. The inhibition obtained seems to be intermediate between the degrees of inhibition described by Reznik and Betz at 750 rad and 1500 rad (*e.g.* 10 days before cold injury). Our findings thus differ considerably, a discrepancy which must reflect the different procedures used: it will be best resolved when the 2 experiments are repeated in parallel. The greatest differences lie in the nature of the injuries made; in our use of tritiated thymidine and in the fact that we injured 4 as opposed to 10, 5 and 1 days before injury. At these 3 periods Reznik and Betz noted no difference, a point they emphasized in an earlier paper concerned with ischaemic injury (1965). Moreover, Cavanagh (1968) and others have performed similar experiments on other tissues, preirradiating with effect many weeks before injury. In short, it may be that in our experiments the time gap (15 as opposed to 10 days) allowed some repair of radiation-induced damage: but this is unlikely.

Our second and critical experiment involved 11 adult mice, the left legs receiving 1000 rad (200 kilovolts; 2 millimetres Cu half-value layer; 15 milliamperes; 40 rad per minute). Particular care was taken to avoid the introduction of infection at the time when crush injuries were made, 5 days later: for in irradiated tissues infection was likely markedly to inhibit myogenesis. Similar precautions were more difficult in rats given 5000 rad, for this led to severe epilation, and at times ulceration. Tritiated thymidine (170 microcuries) was given at 17.00 hours either 48 hours after injury or before injury, and animals were killed 8 days after injury. Myotube counts (see above) for the irradiated legs lay between 11 and 24 (mean 18.5) and for the unirradiated between 10 and 27 (mean 17). When thymidine was given before injury, in the irradiated legs 42.3 per cent (mean) of myotube nuclei were moderately labelled (3 to 10 grains) and 4.8 per cent (mean) conspicuously: the figures for the unirradiated legs were 45 and 3 per cent. When thymidine was given after injury the figures were 42.4 and 6.8 per cent, and 34.8 and 6.8 per cent.

In short, we were unable to establish any difference in muscle repair in legs given 1000 rad and those not so treated. It was apparent, moreover, that myoblast proliferation had been active in both. These experiments suggest that myoblast precursors, under certain conditions, are relatively radio-resistant: or that they recover quickly from radiation damage, or, perhaps more plausibly, that exogenous unirradiated cells actively participate with unaffected local myoblasts in forming new muscle.

CONCLUSIONS

It is confirmed that new-formed muscle fibres contribute to the repair of surgical wounds in muscle in man as in other mammals.

Myogenesis in man is simplest studied *in vitro*: observations on human foetal muscle establish that the mechanisms of myogenesis are similar to those observed in chick explants.

Muscle repair *in vivo* involves a similar mechanism, namely, the multiplication and fusion of myoblasts, but it is noted that this theory is not incompatible with the budding of myotubes from injured fibres and the translocation of nuclei into these muscle buds.

The origin of myoblasts remains obscure. It is shown, however, with the aid of tritiated thymidine, that muscle repair in mice is not totally explained by the origin of myoblasts from tissue at the site of injury. Even when thymidine reutilization is taken into account, there is a strong possibility that circulating cells participate in myoblast proliferation, just as they participate too in the inflammatory reaction to injury.

Moreover, under certain experimental conditions, the preirradiation of sites of injury with a dose of 5000 rad causes incomplete inhibition of myogenesis: at a dose of 1000 rad, under these same conditions, myogenesis was not inhibited. These observations are consistent with the participation of circulating (unirradiated) cells in myogenesis.

NATURE OF DIVIDING NUCLEI IN SKELETAL MUSCLE OF GROWING RATS

FRANCIS P. MOSS and CHARLES P. LEBLOND

The purpose of our work was to reconcile some apparently conflicting observations concerning the proliferation of nuclei within the muscle fibers of growing animals.

The classical view has been that, although in the embryo myoblasts divide and the daughter cells may fuse with multinucleated myotubes, the nuclei of the myotubes themselves do not divide nor do those of postembryonic muscle fibers. This view is supported by numerous *in vitro* studies (*e.g.* Stockdale and Holtzer, 1961).

On the other hand, Enesco and Puddy (1964) and Moss (1968), using DNA determination, observed a considerable increase in the total number of nuclei in the muscles of growing rats and chickens. Moreover, MacConnachie *et al.* (1964) observed mitoses within the basement membrane of muscle fibers of growing rats.

MacConnachie and associates proposed a hypothesis to reconcile these various observations. They suggested that the mitoses, which they had seen, were not of true muscle nuclei but belonged to muscle satellite cells. These are mononucleated cells, having scanty cytoplasm and are located between the basement membrane and the plasmalemma of the fibers. They were first observed in the electron microscope by Mauro (1961) and are not distinguishable from true muscle nuclei in the light microscope. It was suggested that after division one or both daughter satellite cells may be incorporated into the associated fiber, *i.e.*, that satellite cells function as myoblasts.

Subsequently, Shafiq *et al.* (1968) using colchicine observed mitotic figures in satellite cells but not within the fiber proper.

The technique employed was radioautography in the electron microscope using ^{3}H-thymidine. Six male rats, 17 days old and weighing 25 to 30 grams were given a single injection of 25 to 60 microcuries ^{3}H-thymidine per gram body weight, and sacrificed at intervals of from 1 to 72 hours later. The tibialis anterior was fixed in glutaraldehyde paraformaldehyde (Karnovsky, 1965), dehydrated in acetone and embedded in Epon. Sections were subjected to electron-microscope radioautography using Ilford L4 emulsion and poststained with lead citrate (Reynolds, 1963).

Approximately 15 per cent of the nuclei within the basement membrane belonged to satellite cells. In radioautographs at 1 hour after injection of ^{3}H-thymidine approximately 3 per cent of all nuclei within the basement membrane were labeled. These invariably belonged to satellite cells; not a single true muscle nucleus was labeled at that time. The results were similar at 6 and 10 hours after injection. However, at 24, 48, and 72 hours after injection, labeled satellite cells and labeled true muscle nuclei were observed.

Since within the first 10 hours after injection 55 labeled satellite cells but no labeled true muscle nuclei were observed, it is concluded that satellite cells divide but true muscle nuclei do not. Moreover, since a dose of ^{3}H-thymidine is metabolized within a few hours after injection, it follows that the true muscle nuclei labeled at 24 hours or later must have come from nuclei labeled immediately after injection, *i.e.* from satellite cells, which had subsequently been incorporated into the fiber.

The results therefore show that, although true muscle nuclei do not divide, their number increases as a result of division of satellite cells and the subsequent incorporation of one or both of the daughter nuclei into the associated fiber. Satellite cells thus function as myoblasts.

Since no labeled true muscle nuclei were observed during the first 10 hours following injection of ^{3}H-thymidine, it appears unlikely that incorporation of satellite cells occurs during or immediately after mitosis. The duration of the interval between division of the satellite cells and their incorporation into the fiber is not yet known but is currently being investigated.

The existence of labeled satellite cells 72 hours after injection suggests that some daughter cells may not be incorporated but may undergo further divisions. On theoretical grounds this would appear to be necessary to prevent the satellite cell population becoming exhausted. Moreover, labeled mitoses were observed at 72 hours using the light microscope with 1-micron Epon sections; this is additional evidence that satellite cells may divide more than once prior to incorporation into the fiber.

We have also developed a technique for radioautography of whole muscle fibers. Although satellite cells cannot be distinguished from true muscle nuclei with this technique, we observed an approximate doubling of the total number of labeled nuclei between 1 hour and 24 hours after ^{3}H-thymidine injection; there was no further increase between 24 and 48 hours.

Using this technique, the distance along the fiber between successive labeled nuclei was measured, and a frequency distribution of these distances was plotted for the different time intervals. Between 1 and 10 hours there was a considerable increase in the number of distances less than 100 microns but little change in the number of longer distances; this is attributed to satellite cells which had divided but whose daughter cells were still close together. Between 24 and 48 hours after injection, there was a decrease in the number of distances less than 100 microns with a commensurate increase of the longer distances; this is attributable to the daughter satellite cells moving away from each other, either before or after incorporation into the fiber.

SATELLITE CELLS IN DEVELOPING MUSCLE AND TISSUE CULTURE

HARUNORI ISHIKAWA

I have worked on the satellite cell in embryonic muscles. The materials which I used were mostly human embryos. Perhaps human muscles are not very good for studying the satellite cell. However, I could confirm the results from the human (embryonic) muscle by observing cat muscles.

If we stick to Dr. Mauro's definition, we can see many satellite cells in embryonic muscles. Recent studies favor the idea that the satellite cell is indeed a myoblast, or functions as a myoblast. However, since the nature of the satellite cell has not been completely elucidated yet, I think we had better stick to the definition, otherwise we may lose communication on what cells we are talking about.

Figure 145 shows developing muscles from a 10-week human embryo. Several muscle cells (myotubes) are grouped into a bundle. As you see, a small mononucleated cell with a dark nucleus is found in close contact with the myotubes. The basement membrane is

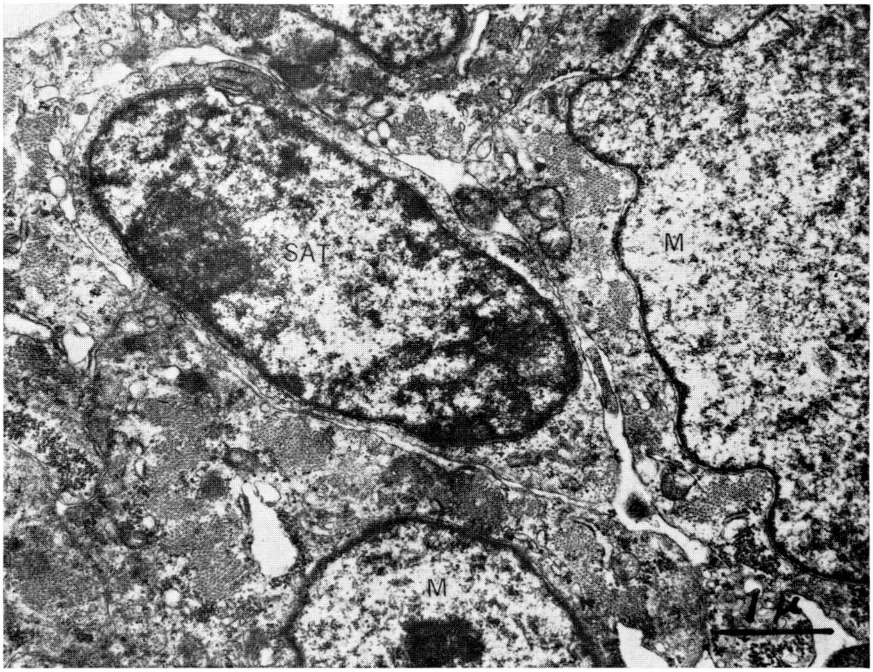

Fig. 145. Electron micrograph of a 'fibroblast-like' cell (SAT) seen deep inside the bundle of muscle cells (M); 10-week-old human fetus gastrocnemius muscle. × 14,400. (*from*: Ishikawa, 1966)

barely visible along the myotube at this stage. These small undifferentiated cells are not distinguishable from interstitial cells (fibroblasts), except that the former are in close association with myotubes. I once called these cells 'fibroblast-like'. This is a morphological term rather than a functional one. Thus, I did not mean this cell is fibroblastic in nature.

At the 15-week stage the basement membrane can be seen. The 'fibroblast-like cells' are found closely associated with the myotubes and are now enveloped by a common basement membrane. Now these 'fibroblast-like cells' may be regarded as 'satellite cells'. The situation becomes more complex when one considers, in addition to satellite cells, myoblasts and presumptive myoblasts which are present in developing muscles. The definition and morphology of the so-called myoblast varies in the literature. Since it is not possible to unequivocally identify a 'myoblast', all the mononucleated, undifferentiated cells within the basement membranes of myotubes are regarded as satellite cells. If such cells within the basement membrane are indeed myogenic cells, then the situation would become simple.

Figure 146 shows a low-powered electron micrograph of a transverse section of muscle fibers from a 25-week-old human fetus. Several satellite cells are seen here. Myotubes are

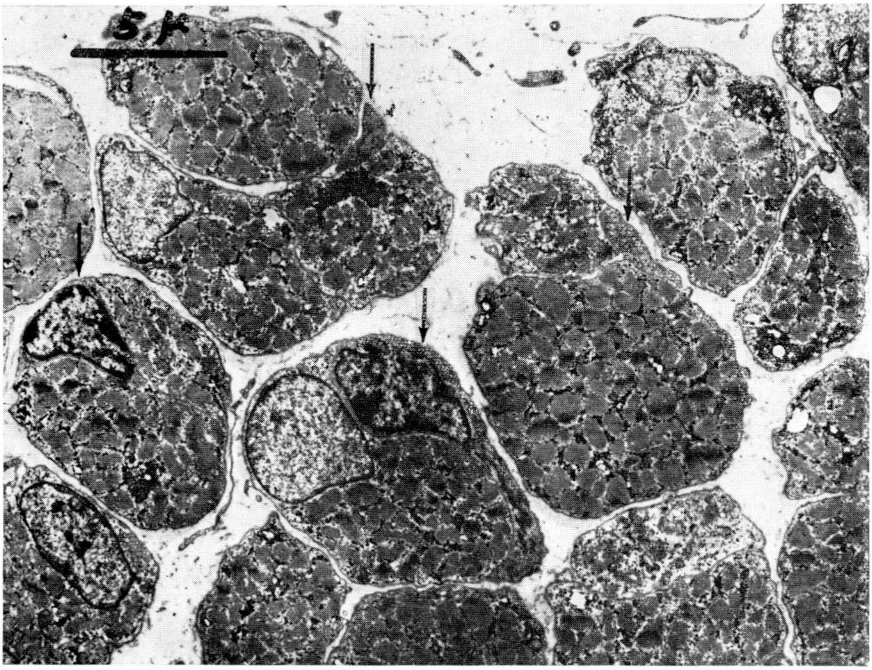

Fig. 146. Low-powered electron micrograph of a transverse section of muscle fibers; 25-week-old human fetus. Several satellite cells (arrows) are seen closely associated with the muscle fibers. × 4000. (*from*: Ishikawa, 1966)

closely associated with each other. I would like to point out here that the satellite cells in developing muscles do not exhibit a unique morphology. They are spindle-shaped, or very elongated: more than 50 microns in length. Sometimes, they are almost round in shape. In many cases, they have wing-like cytoplasmic projections, often extending between the muscle fibers. Also, they often show multilayered rough endoplasmic reticulum. Therefore, they do not seem to be latent or silent in the developing muscles.

J. H. VENABLE: What do you call this at the bottom (Fig. 146)? Is this a myoblast?

H. ISHIKAWA: This is a satellite cell.

J. H. VENABLE: There is not a glycocalyx intact, is there?

H. ISHIKAWA: Yes. This is surrounded by the basement membrane (glycocalyx).

J. H. VENABLE: It is.

UNIDENTIFIED SPEAKER: Dr. Ishikawa, it looks as though there are many more satellite cells in this slide (Fig. 146) than maybe we have seen so far. Is that true?

H. ISHIKAWA: Yes. We can see many such cells in developing muscles and the number of satellite cells decreases in an electron microscopic section as the muscle grows older.

A. MAURO: In a brief paper by Shafiq *et al.* (1968) – a sequel to the work of MacConnhiace *et al.* (1964) – the meaning of the mitotic activity seen in the post-neonatal period is analyzed. In this paper it is shown by means of electron microscopy that mitotic figures are always in the satellite cells and not in the nuclei of the multinucleated fiber.

J. H. VENABLE: A word on quantitation, in the mouse gracilis muscle, the number of satellite cells at birth is about 40 per cent of total nuclei that are associated in the myotube. By 10 days they have dropped to 16 to 18 per cent but this is a drop in percentage not in total population; actually the total cell population is continuously increasing.

D. ALLBROOK: Hellmuth and Han working in my laboratory have demonstrated a fall in percentage (Hellmuth *et al.*, 1970) of satellite cells (Table VII). The study indicated (1)

TABLE VII

The declining percentage of satellite cell nuclei with respect to myonuclei in relation to age after birth in the m. subclavius of the rat and the m. peroneus longus of the mouse

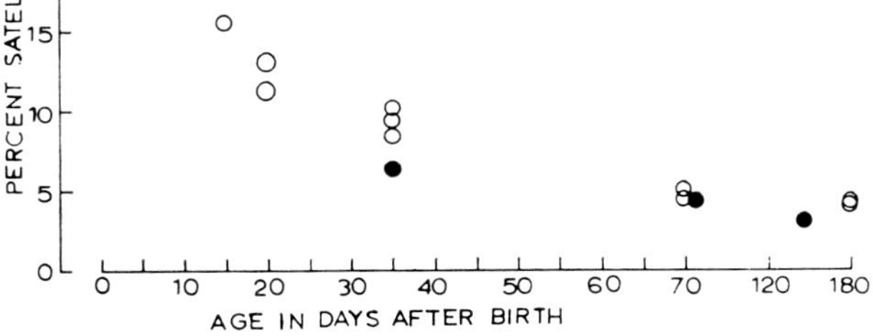

that there is a decline in the percentage of satellite cell nuclei by a factor of about 8 between birth and maturity, (2) evidence of myofibril formation in cells occupying the anatomical position of satellite cells, and (3) evidence of mitotic activity of the satellite cells (Figs. 147, 148, 149, 150). I have also noticed that under certain conditions satellite cell cytoplasm is irregular. For example, mouse neonatal satellite cells have been seen with cytoplasmic

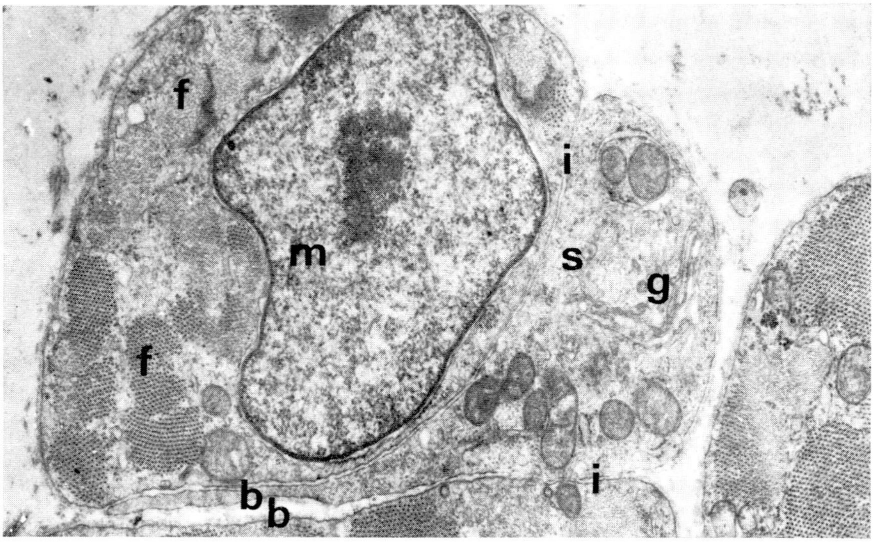

Fig. 147. The partial separation of the satellite cell-myocyte pair from the fiber by external lamina; 1-day-old rat *m. subclavius.* Uranyl acetate and Reynold's lead citrate; × 13,500.

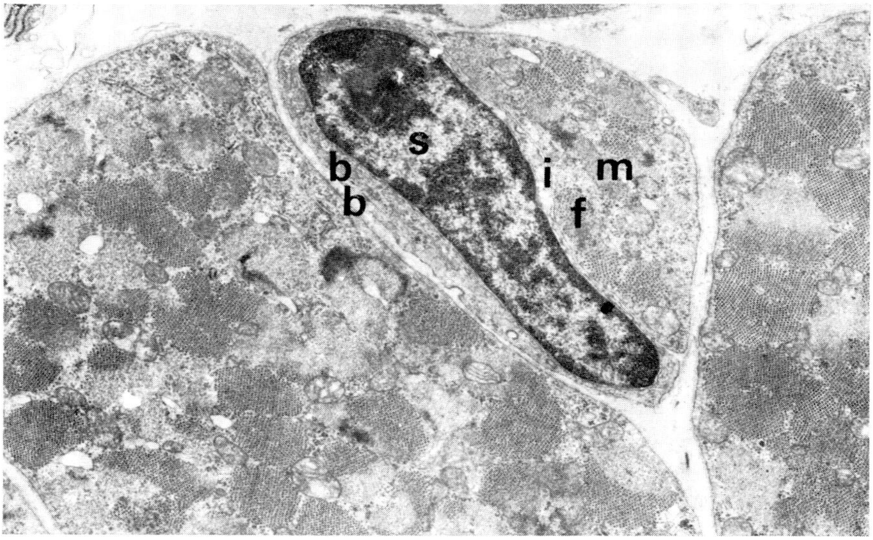

Fig. 148. A double layer of external lamina separates the satellite cell-myocyte pair from the adjacent fiber; 1-day-old rat *m. subclavius.* Uranyl acetate and Reynold's lead citrate; × 9,600.

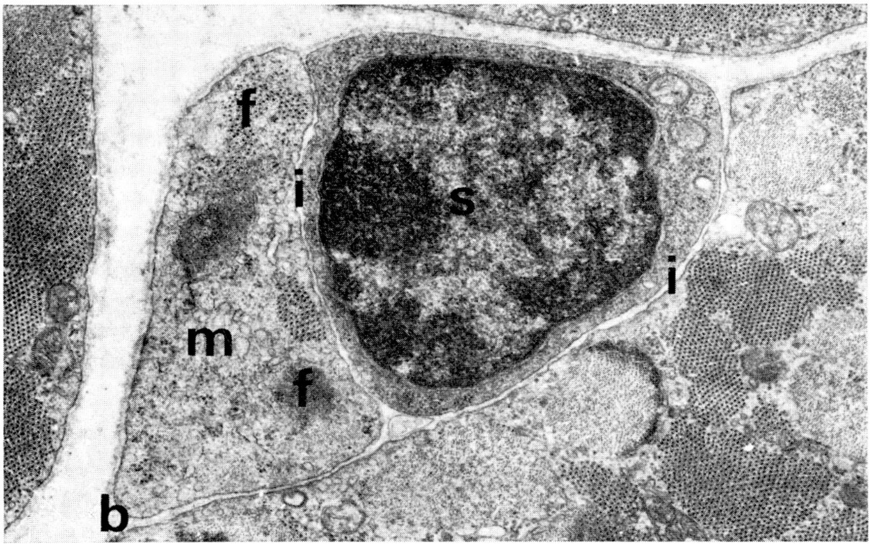

Fig. 149. A pairing of satellite cell and myocyte, and the association with the fiber brought about by a common external lamina; 1-day-old rat *m. subclavius.* Uranyl acetate and Reynold's lead citrate; × 17,600.

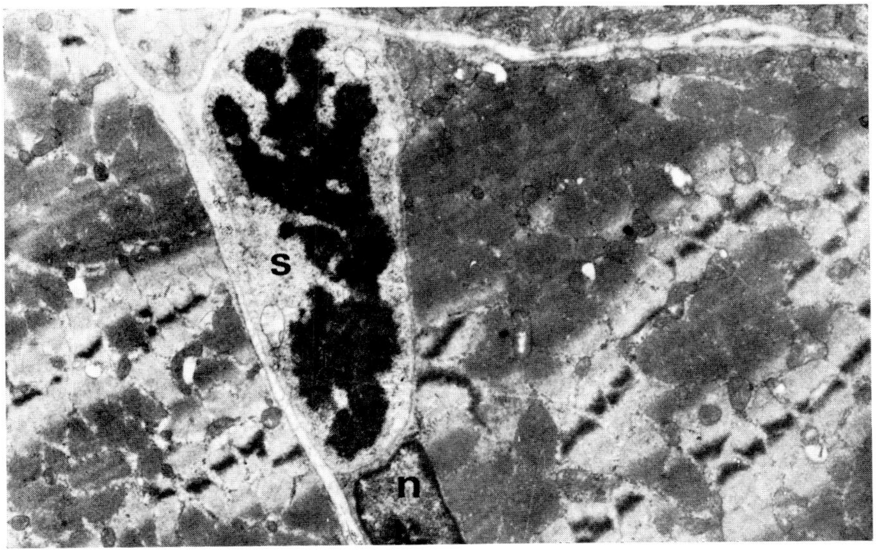

Fig. 150. Mitotic figure in a satellite cell of 20-day-old rat *m. subclavius.* Uranyl acetate and Reynold's lead citrate; × 5,600.

projections into the adjacent fiber sarcoplasm. I have noted the same in marsupial muscle (*Setonix brachyurus*) recovering from acute vitamin E deficiency (which causes acute myopathy in these wallabies, reversible by giving vitamin E).

H. ISHIKAWA: Figure 151 shows one of the cases in which the basement membrane can be clearly seen, and a satellite cell forms pseudopodial, cytoplasmic projections. Here, I would like to comment on where the satellite cells come from. We cannot definitely answer this question even from the work on embryonic muscles. However, I can say that before the basement membrane appears or becomes visible, we see the same type of cell in the same position as that of the definitive satellite cell. Very possibly such cells are present from very early stages of muscle development. In fact, there are some cells which seem to be transitional forms between the satellite cells and immature muscle cells containing a few nascent myofibrils. Some of the 'satellite cells' in developing muscles may differentiate into definitive myotubes.

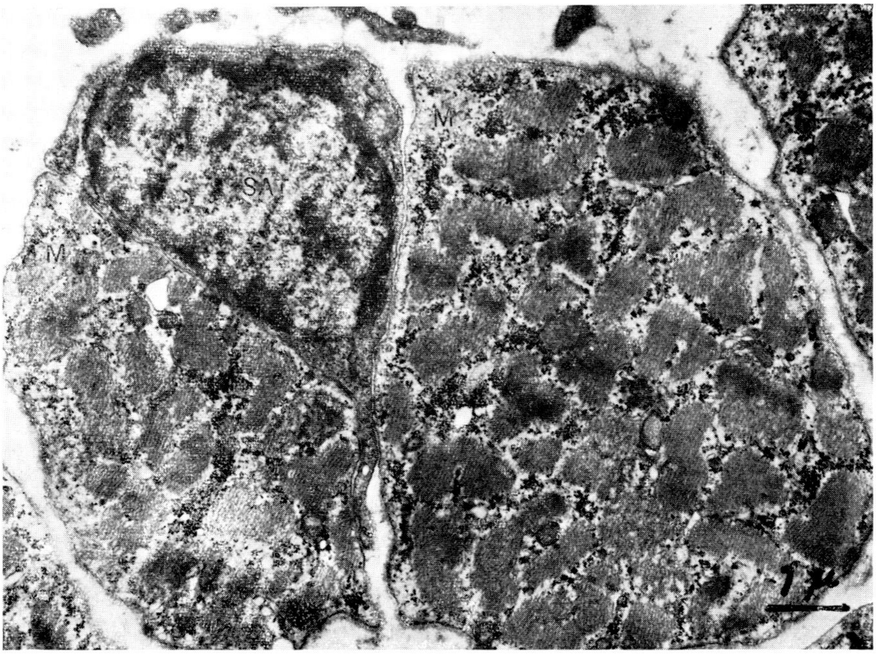

Fig. 151. Satellite cell (SAT) intervening between 2 adjacent muscle fibers (M); 25-week-old human fetus. × 11,200. (*from*: Ishikawa, 1966)

UNIDENTIFIED SPEAKER: Are those black dots ribosome granules or glycogen granules?

H. ISHIKAWA: These are glycogen granules.

J. H. VENABLE: But there are some ribosome granules in there, are there not?

H. ISHIKAWA: Yes. An interesting thing is that the satellite cells do not contain any detectable glycogen granules, whereas myotubes do.

In Figure 152 we see a typical satellite cell from a 25-week-old human embryo. My impression is that the satellite cell remains almost the same in size, whereas the muscle fiber is growing in size. Sometimes we can see very round cells which are surrounded by the

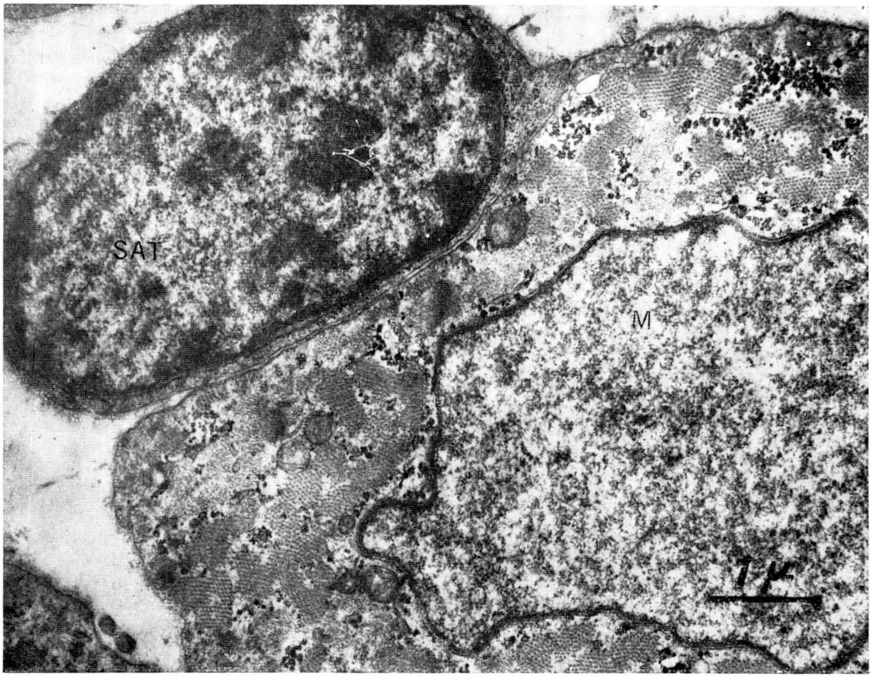

Fig. 152. Typical satellite cell (SAT) closely associated with a muscle fiber (M); 25-week-old human fetus. × 14,400.

basement membranes of the muscle fibers. Here I would like to ask you if you have seen peculiar forms of satellite cells in mature muscles. We cannot rule out the possibility that some of the satellite cells may be migrating out and/or in through the basement membrane. The basement membrane itself may not serve as a barrier to cell movement.

J. H. VENABLE: Would you repeat that statement again? Two different types of satellite cells?

H. ISHIKAWA: I mean the morphology of satellite cells in embryonic muscles is very variable. Some are round in shape, and others are elongated.

J. H. VENABLE: Yes. I have noted the same thing.

H. ISHIKAWA: I do not know whether this represents different cell types or different functional states. Now I would like to discuss the satellite cells in tissue culture. The situation of the satellite cells in cultured skeletal muscles is the same as that in embryonic skeletal muscles.

Figure 153 shows a cross-sectioned myotube from a 7-day muscle culture of chick embryo breast muscle. You can see that the myotube is closely attached to a fibroblast-like cell which is rich in rough endoplasmic reticulum. The basement membrane is hardly visible in this case. However, some well-developed myotubes in this 7-day culture are seen surrounded by basement membranes. Therefore we can identify so-called satellite cells, which show an undifferentiated cytoplasm and lie between the basement membrane and plasma membrane of the myotubes. Therefore, there is no doubt that satellite cells can be seen in skeletal muscles grown *in vitro*. De Rényi and Hogue (1934) described certain fibroblastic cells intimately attached to muscle fibers in tissue culture. These fibroblastic cells probably correspond to the satellite cells defined in electron microscopy. These cells again may repre-

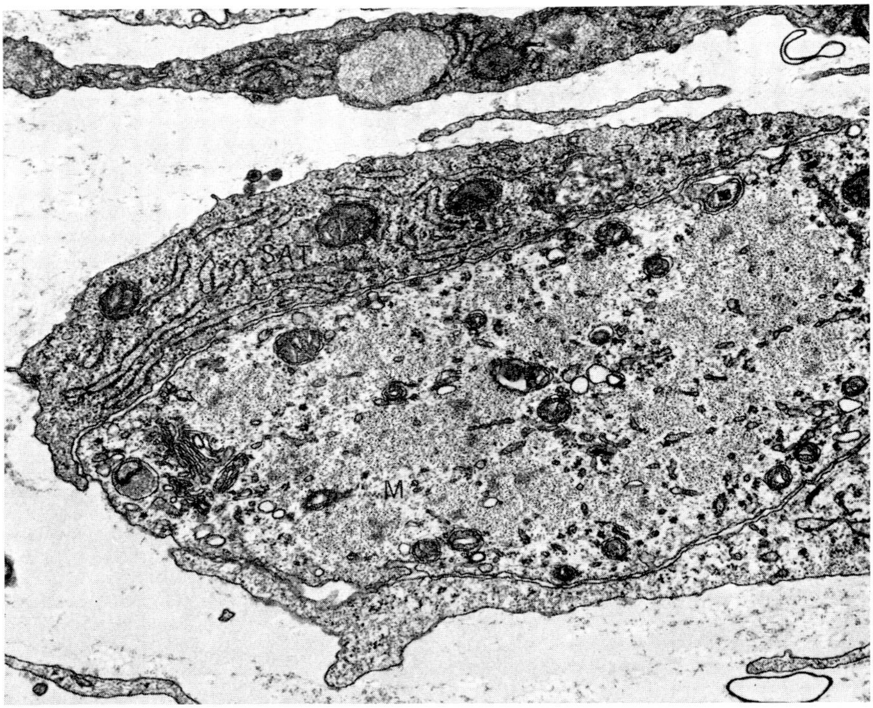

Fig. 153. Electron micrograph of a 'fibroblast-like' cell (SAT) closely attached to a myotube (M); 7-day muscle culture. The basement membrane is hardly visible. × 19,200.

sent presumptive myoblasts, but we cannot identify them as myoblasts on a morphological basis. Many of the satellite cells contain fine filaments of 50 to 100 Ångströms in diameter. I would like to stress here that the fine filaments do not necessarily serve to identify muscle cells. In other words, we cannot identify myoblasts or differentiating muscle cells on the basis of the presence of those fine filaments. We have to wait until we find thick (myosin) filaments, myofibrils and other clues for muscle differentiation.

Figure 154 is from a 13-day culture. We can see the basement membrane here. Here a mononucleated cell is seen wedged between the basement membrane and the plasma membrane of the myotubes, and I would call it a satellite cell. Some satellite cells contain well-developed rough endoplasmic reticulum (see Fig. 155).

J. H. Venable: What actually was put into the culture? Single cells?

H. Ishikawa: Yes. Breast muscle tissues from 11-day chick embryos were dissected and trypsinized to obtain single cell populations. They are plated on plastic culture dishes. In cross section of the muscle fibers from a 16-day culture small cells are seen wedged between the basement membrane and the muscle fiber. But if you look carefully, they also contain myofibrils.

Satellite cells are rarely observed in 16-day culture or later. I think this may be a very interesting fact. I think there will be a comment on this by Dr. Bischoff later, in terms of secondary culture or 'subculture'. When you trypsinize such older muscle cultures to obtain single cell populations for subculture, you cannot get any differentiated muscle cells in such subcultures.

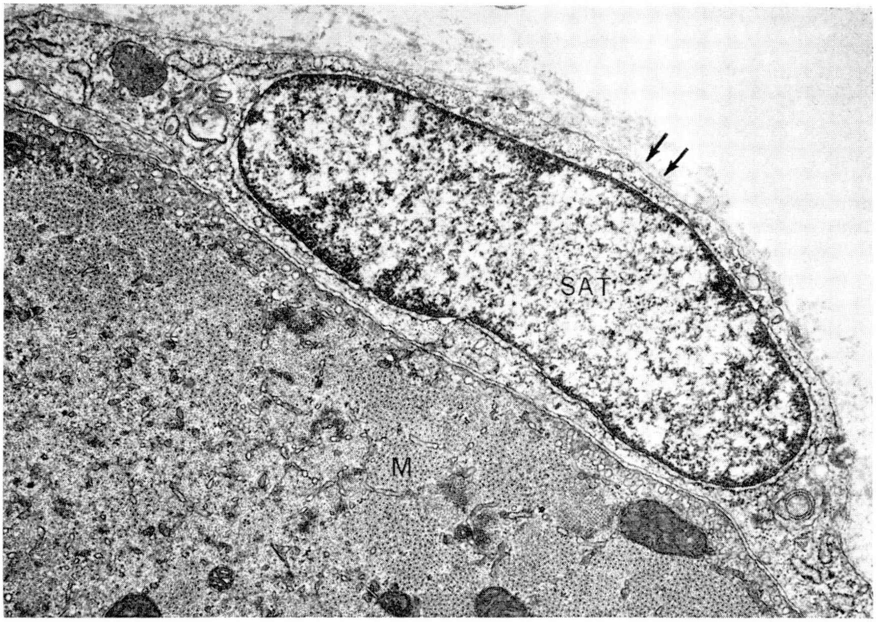

Fig. 154. Electron micrograph of a satellite cell (SAT) in a 13-day muscle culture. The basement membrane (arrows) is clearly visible along the surface of the satellite cell and the myotube (M). × 15,000.

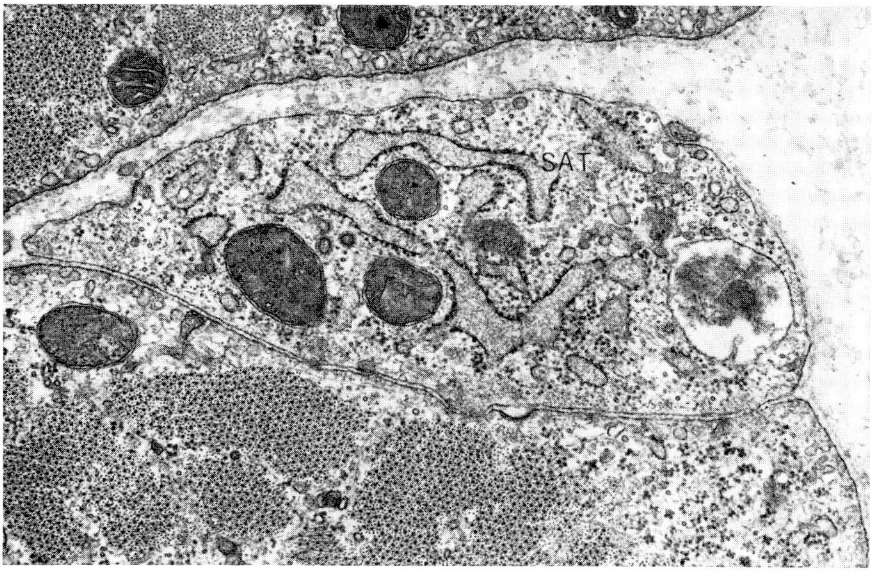

Fig. 155. Electron micrograph of a satellite cell (SAT) containing well-developed rough endoplasmic reticulum; 13-day muscle culture. × 25,200.

M. NAMEROFF: Dr. Ishikawa, *in vitro*, you do get splitting of muscle fibers. Is this definitely a mononucleated cell?

H. ISHIKAWA: The muscle culture is started from undifferentiated single cells.

M. NAMEROFF: No, what I am asking is that if you get serial sections back, would this be a mononucleated cell or might this join up and be a nucleus in a fiber that has a split in it?

H. ISHIKAWA: No, I cannot answer this. I believe this is another multinucleated fiber, but a very small one. I have never seen any mononucleated cells which contain myofibrils in this type of culture. I do not have any evidence for the idea that muscle fibers can split longitudinally. I often see branched muscle fibers in the cultured muscles, though.

P. E. CONEN: There is an interesting point here. In this slide (Fig. 155) you have collagen and you have basement membrane; some previous slides did not show it. I wonder at what stage did the basement membrane develop?

H. ISHIKAWA: In the tissue cultures we observed, the basement membrane appears on the seventh day of culture.

H. HOLTZER: I know Dr. Mauro is impatient to get on with the next speaker, but there is one point which I think is pertinent to what has gone on and to what will come. Though studying this material for over 2 years now Dr. Ishikawa is such a poor microscopist that he is still unwilling to describe fusion. Yet we know that fusion occurs at a fabulous rate in these cultures. Within 24 hours between 40 and 60 per cent of these mononucleated cells fuse. Several people here have invoked the presence of small vesicles as indicating cells are pinching off or fusing. I would underscore the fact that Dr. Ishikawa has not convinced himself, or anybody in my lab, that we know *anything* about fusion on the electron-microscopic level, or for that matter on any level. Accordingly I would suggest great caution in interpreting electron photomicrographs as revealing muscle nuclei budding off injured muscle fibers by the formation of new plasmalemmal material.

J. R. COLEMAN: Dr. Ishikawa is not alone in not being able to see fusion.

J. H. VENABLE: Following Dr. Reznik's technique I would like to take just a moment for everybody to hold up their hands who have seen a picture so that they can say "I know what fusion is". I, for one, do not know what fusion is; or we might ask the question: "How many here have seen pictures of a cell which apparently was away from the parent fiber but there was some sort of structure, for example, a cytoplasmic bridge* which could be interpreted as evidence for fusion?"

H. HOLTZER: That is the point. We all have preconceived notions of what should be happening, but this time there is no well-established dogma to guide us in recognizing fusion of cells from the postulated fragmentation and *de novo* formation of cell membranes advocated by Drs. Reznik and Hay.

A. R. MUIR: I would agree with Dr. Holtzer. I could show you cells resembling these satellite cells in their intercellular relationships which never fuse. I refer to cardiac muscle where it would be very easy to believe that cardiac muscle cells were in the process of fusing with each other at intercalated discs, as micrographs, just like Dr. Reznik's can be obtained when membranes are sectioned obliquely, and there you know that they are not fusing. All this is just a limitation of thin-section electron microscopy.

D. ALLBROOK: I think one cannot say from a given electron micrograph whether two cells are fusing or not. Does not the process happen fairly rapidly?

* Dr. Hay states that such bridges can be found in developing myotome (Hay, 1963) as illustrated in accompanying Figure (Fig. 156). Thymidine uptake in a cell containing myofibrils is shown in Fig. 157.

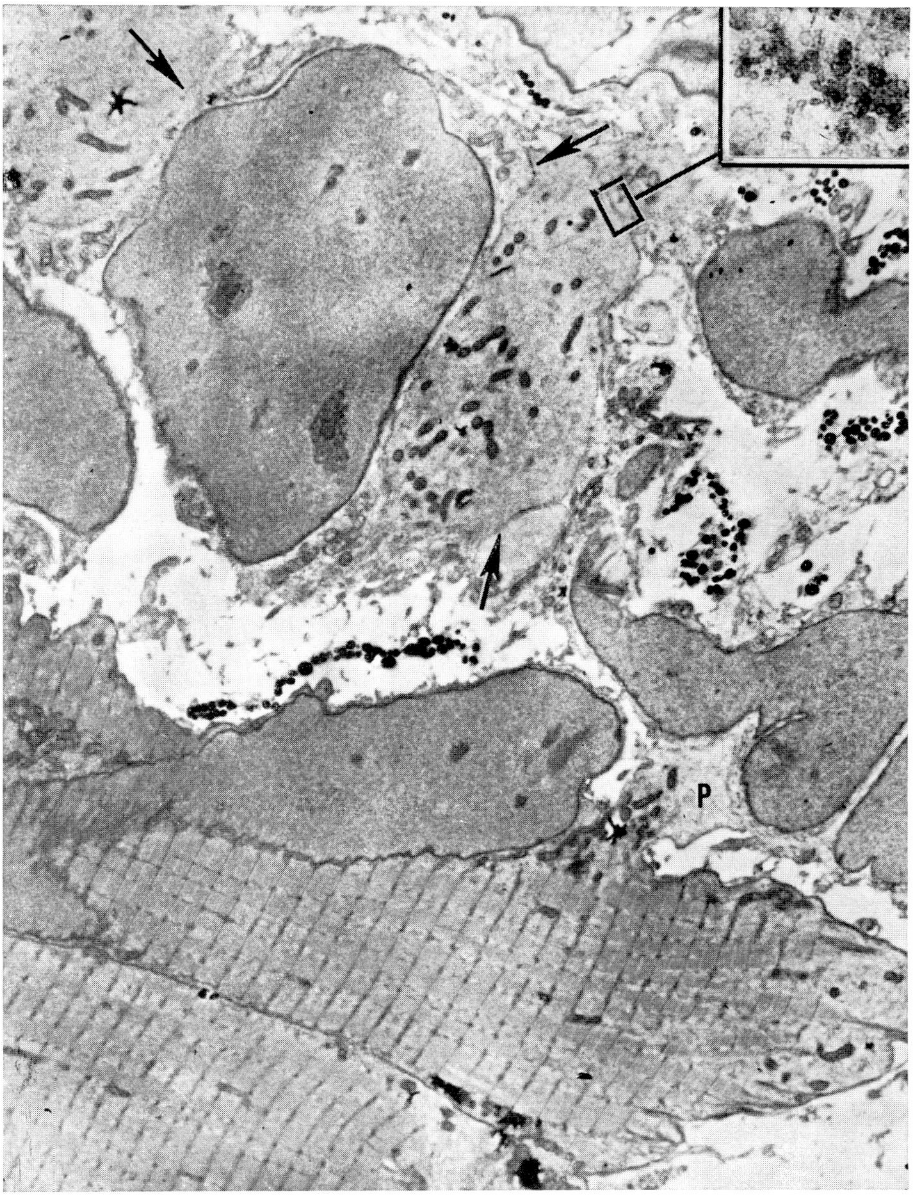

Fig. 156. Electron micrograph depicting myoblasts in the process of fusion during differentiation of tail muscle in an *Amblystoma punctatum* larva. The myoblasts are separated by rows of vesicles which, at this low magnification, appear as ill-defined linear densities across the cytoplasms (arrows). The vesicles are shown at higher magnification in the inset. Since these vesicles subsequently disappear, Hay (1963) suggested that they represent a step in the dissolution of apposed cell membranes during the fusion of myoblasts. A cell process from one of the myoblasts in this chain of cells has attached (P) to one of the formed myofibers. Fusion of myoblasts in this way to the lateral surfaces of myofibers has been described in the living tadpole tail by Speidel (1938). × 3,600, inset × 49,000. (*from*: Hay, 1963; copyright by Springer Verlag, Berlin)

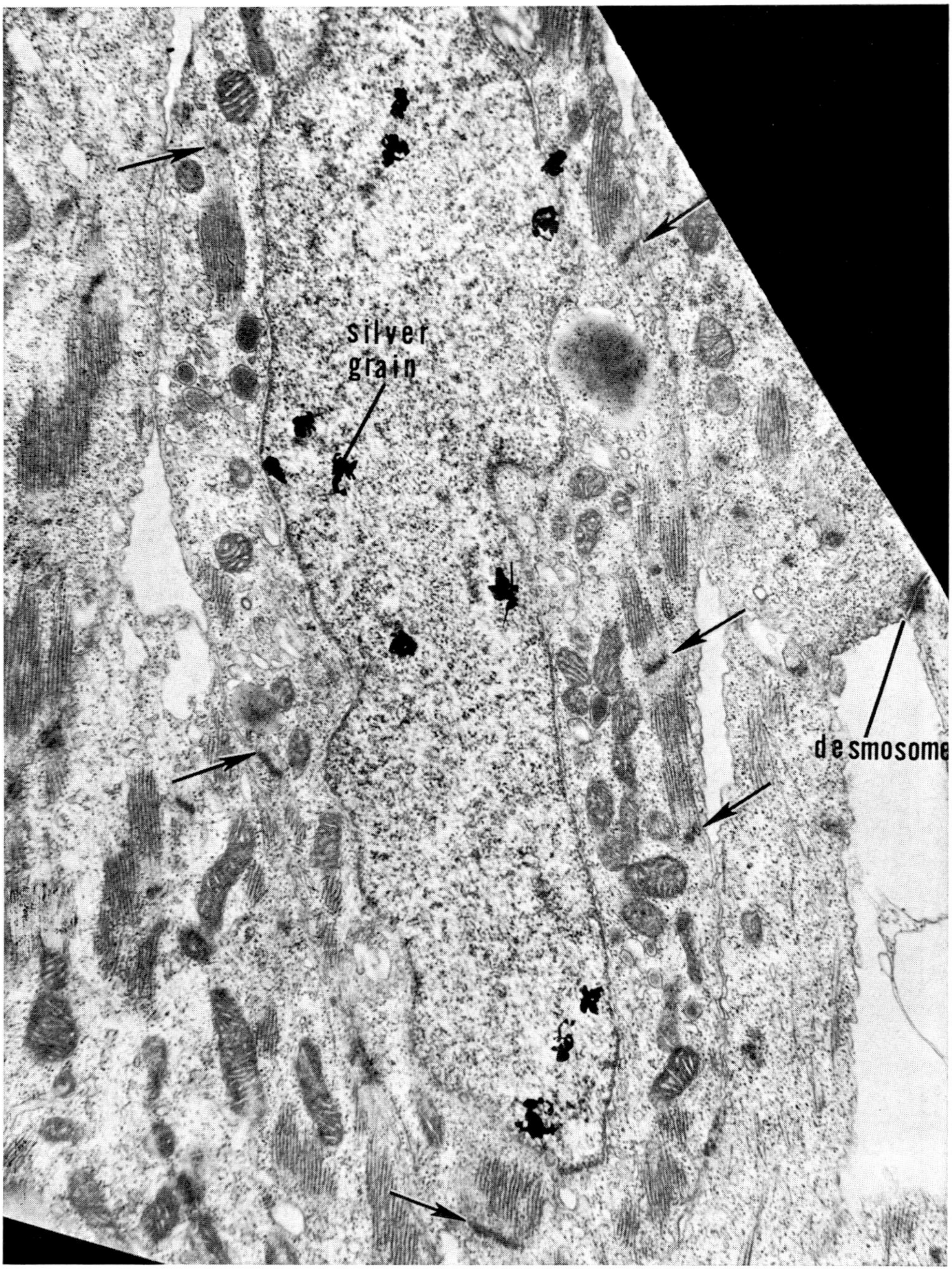

Fig. 157. Electron micrograph-autoradiograph of a radioactive cardiac muscle cell from a 5-day-old chick heart that was incubated for 1 hour in tritiated thymidine prior to fixation. Cardiac muscle consists of single cells joined by desmosomes and intercalated discs. There are no myoblasts present. All of the cells that are synthesizing DNA contain myofibrils and presumably are functional muscle units. There is no significant difference in the degree of differentiation of muscle cells which are radioactive, as compared with muscle cells that were not in the S-period at the time of fixation. Thick and thin myofilaments can be detected in the A-bands, and the Z-bands (arrows) are similar in appearance to those in adjacent cells. There are 12 silver grains in the emulsion overlying this nucleus. × 17,500. (*from*: Weinstein and Hay, 1970)

I. R. KONIGSBERG: Judging from time lapse photography of myoblasts in tissue culture, it must happen very rapidly.

H. HOLTZER: When myoblasts are prepared to fuse they do so in less than 30 minutes. On the other hand we have observed cells in intimate contact for much greater periods before fusing. And then there are always those contacting cells that we are actually photographing which the next day blithely slide away from one another. In the experiments with Bischoff we demonstrated that the cells in contact for many hours before fusing are probably cells that were not initially in the right mitotic cycle.

A. MAURO: Dr. Aloisi, would you like to make your comments?

PATTERNS OF MUSCLE REGENERATION

MASSIMO ALOISI

I should like to show you some pictures directly or indirectly concerned with the problem of muscle regeneration. However, relevant though they may be to the discussion here, they are unfortunately heterogeneous in character: namely, some are of tissues of experimental animals, like those performed by Studitsky and Carlson; the others are of tissue of patients with muscle diseases. Most of the electron micrographs I have to show belong to the latter group. For the regeneration experiments and for the experiments with tissue culture, I must limit myself to conventional microscopic observation.

POSSIBILITY OF A 'CONTINUOUS TYPE' OF MUSCLE REGENERATION

We realize, and this conference seems to confirm, that regeneration of muscle fibers *per continuum* from viable stumps of the fibers is now seriously questioned. In the past, several authors have shared the opinion that in all or in many instances regeneration *per continuum* is the only type of muscle regeneration (Millar, 1934; Le Gros Clark, 1946; Litver, 1955, 1959, 1962; Kitiyakara and Angevine, 1960; and others). The conflict in the Soviet Union between Studitsky, on one side, and Litver, on the other, about this problem is well known.

If we follow a partial regeneration of the fibers after cutting or some other kind of damage, we sometimes find pictures which seem to give evidence of a growth *per continuum* from stumps.

In 1965, and again at this conference, Shafiq has shown electron-microscopic evidence of multinucleated sarcoplasmic tips at the extremities of cut fibers, but while he considers this change related to the processes leading to regeneration, he still holds the opinion that final regeneration is possible only *per discontinuum.*

The observation of living muscle tissue cultures seems to add some details, such as the accumulation of nuclei in growing muscle fibers, to the general picture of terminal 'budding'. But this may well be an artifact due to sarcoplasmic movements and the occasional contraction of the new fiber. Speaking of ameboid movements, in experiments *in vivo* with the frog, we can see pseudopodial processes emerging from symplasts of myoblasts and very young myotubes and adapting their cytoplasm to the surrounding environment of connective tissue.

THE 'DISCONTINUOUS TYPE' OF MUSCLE REGENERATION

Let us for the moment suspend any decision as to the reality of, and the possible contribution to, regeneration of fibers from the terminal buds of fiber stumps and move on to a discussion of the discontinuous type of regeneration of muscle. It is now the largely accepted type of

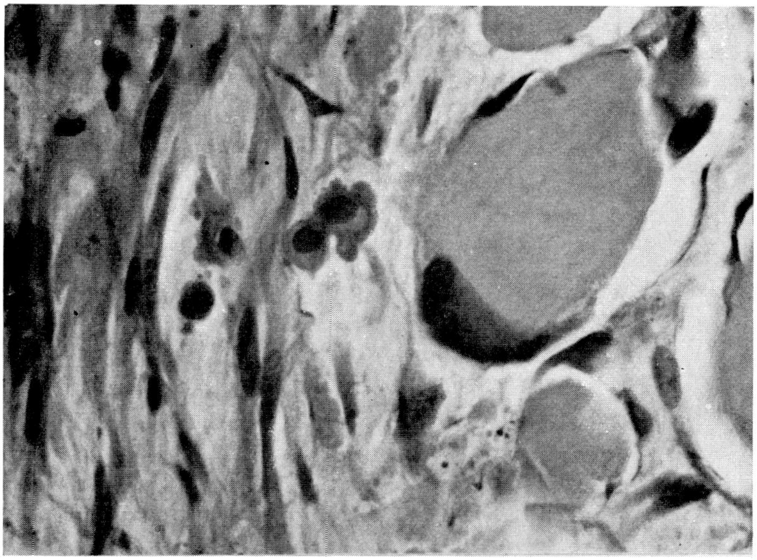

Fig. 158

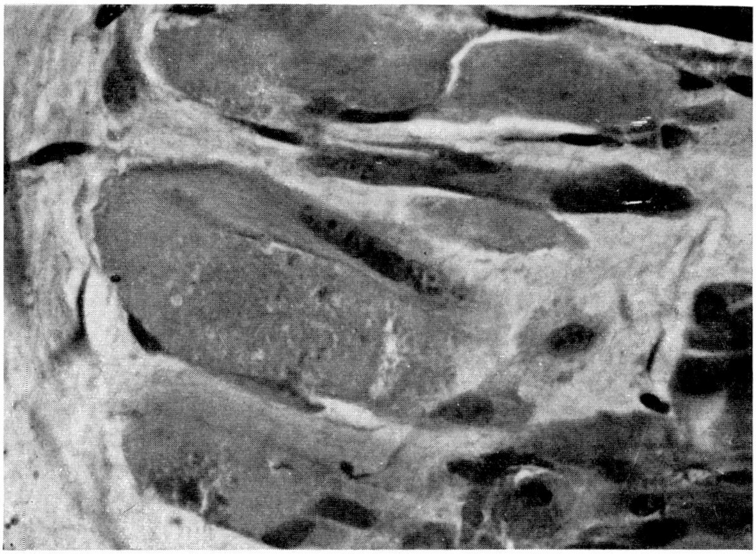

Fig. 159

Fig. 158. Rat muscle explant in a millipore (0.45 micron) diffusion chamber located in the peritoneum of the same animal; 7 days. A cross section of a hyalinized fiber presents a strongly peripheral basophilic and multinucleated semilunar cap which is considered a regenerating portion of sarcoplasm. H and E; × 450.

Fig. 159. Rat muscle explant as in Fig. 158; 7 days. Approximate longitudinal section of a degenerated muscle fiber. On one side of it there is a multinucleated sarcoplasmic layer which is not degenerated; on the contrary, it appears to be in a regenerative stage. H and E; × 450.

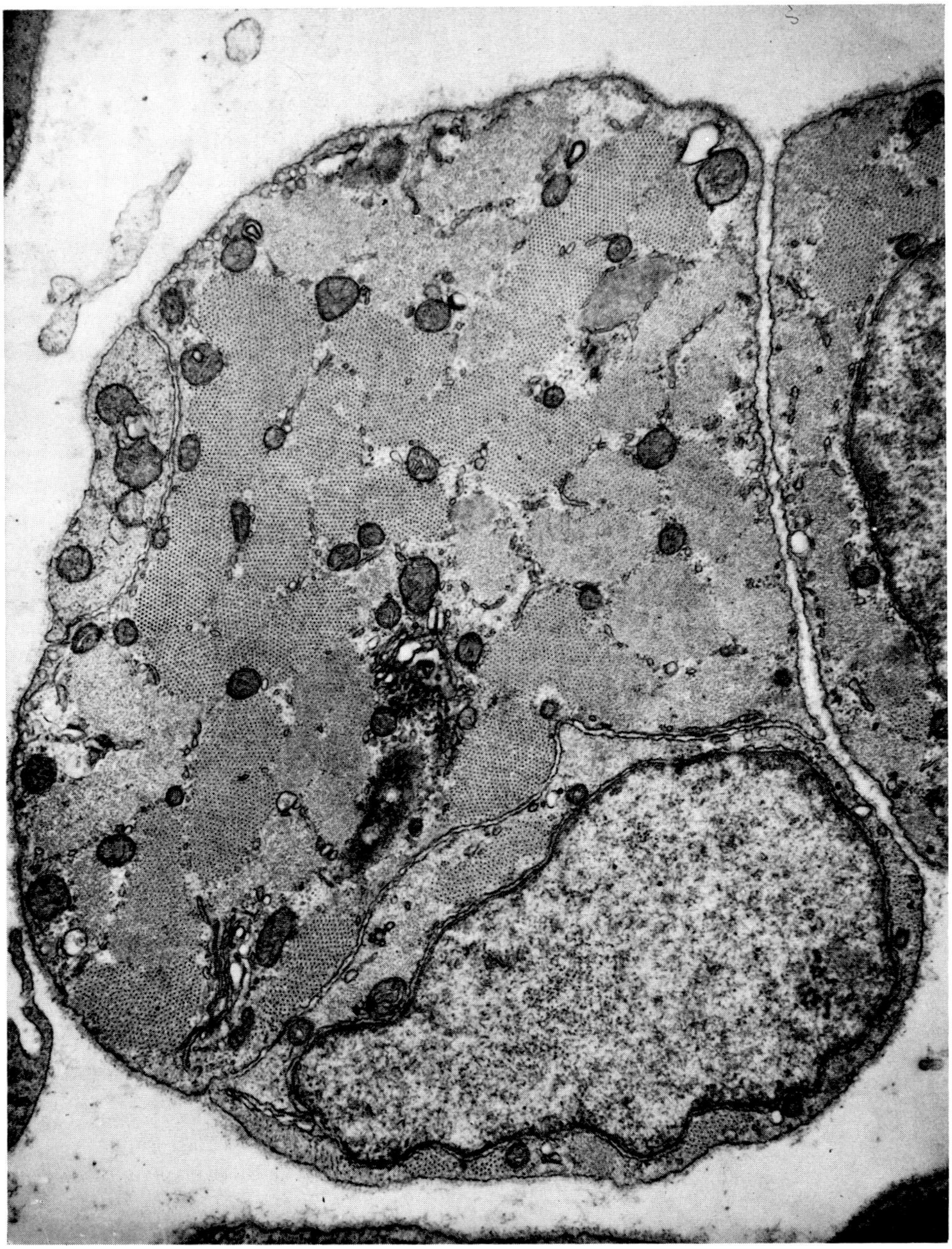

Fig. 160. Rat gastrocnemius at birth. A young muscle fiber with an adjacent satellite element (included under a common basal membrane) which shows several sets of myofilaments; it looks more like a young satellite fiber than a satellite cell. Electron micrograph, × 17,200.

regeneration and, moreover, represents the repetition of the initial ontogenetic process; in this respect it has a more general value.

Muscle regeneration *per discontinuum* is easily observable after massive destruction of muscle fibers (extensive degeneration like that occurring in hereditary dystrophies or in vitamin E deficient animals, or extensive irreversible denaturation of the muscle substance brought about by deep-freezing or by ischemic injury). One good example is the Studitsky procedure for the regeneration of large parts of muscle tissue – what he calls 'the regeneration' of the entire muscle (Studitsky, 1959).

This procedure, which we use on frogs and rats (autologous and homologous rats), has been mentioned by Dr. Carlson during the presentation of his paper.

We have also used the technique of explanting small portions of minced muscle tissue in a millipore diffusion chamber and implanting the latter in the peritoneum of the same animal or in others of different species (experiments in collaboration with Dr. Giacomini and Dr. Tessari). With this procedure, we can follow the sequence of events leading to the 'liberation' of nucleated portions of sarcoplasm or single myoblastic cells from the degenerating and reabsorbed muscle fibers (Figs. 158 and 159). So far, we have followed this process only by conventional microscopic investigations, and therefore we cannot solve the problem of the origin of the myoblastic cells.

In the next section, I shall advance some indirect arguments regarding this problem, but here I simply want to summarize the results we obtained with the diffusion chamber: (1) onset of regeneration is quicker if the pores of the chamber are large enough (0.80 micron) to permit the passage of leucocytes, which help greatly in clearing the explanted tissue from its dead parts; (2) there are great species differences in the regeneration potency of muscle tissue in interspecies experiments: adult rats are both good donors and good hosts, whereas adult guinea-pigs are bad donors and bad hosts and even in autoplastic explants the regeneration remains very poor.

SATELLITE CELLS – THEIR NUMBER AND STRUCTURE

Clearly the discovery of satellite cells (Mauro, 1961, in muscle of frog and rat) has offered a good histological and cytological basis for an explanation of the discontinuous type of regeneration (see, *i.a.*, the paper by Shafiq *et al.*, 1967). Notwithstanding, a theory which would limit the source of the cellular elements responsible for this type of regeneration to satellite cells is likely to be inadequate, particularly when we consider the adult muscle.

Satellite cells are abundant during embryonic and fetal life. We cannot give definite ratios of the number of ordinary fiber nuclei and that of satellite cells at different stages in ontogenesis, but, in the rat, there is a striking loss of satellite cells after birth. (Approximate ratios in the rat would be 1 satellite nucleus out of 3 or 4 at birth and only 1 in 100 in adult life). This situation appears to be different from that described by Muir *et al.* (1965) and by Church *et al.* (1966) in the fruit bat, where the ratio appears to be as high as 1 in 10.

When cross sections of muscle fibers (from young animals) are observed and the associated satellite cells contain sets of myofilaments, thick and thin, oriented according to the longitudinal axis of the fibers (regular cross sections of the myofilaments), it is likely that it is no longer a satellite cell, but rather a satellite fiber, a young ribbon-like fiber attached to a more mature one (Fig. 160). As a matter of fact, longitudinal sections may sometimes show elongated and multinucleated satellite elements. In these cases, we need to observe carefully the behavior of the basal membrane before deciding whether the associated element is 'satellite' in nature, or if it is a completely independent unit; perhaps, however, such distinctions have no relevance when considering the dynamic aspect of this morphology.

Fig. 161. Longitudinal section of a muscle fiber in a case of human dystrophia myotonica. The peripheral part of the fiber shows a thickening or extension of the subsarcolemmal sarcoplasm in which are visible 2 nuclei, mitochondria and other organelles. There is no membrane between this portion of the sarcoplasm and the rest of the fiber. Note the external basal membrane and the point of emergence (arrow) of this sarcoplasmic extension. Electron micrograph, × 6,700.

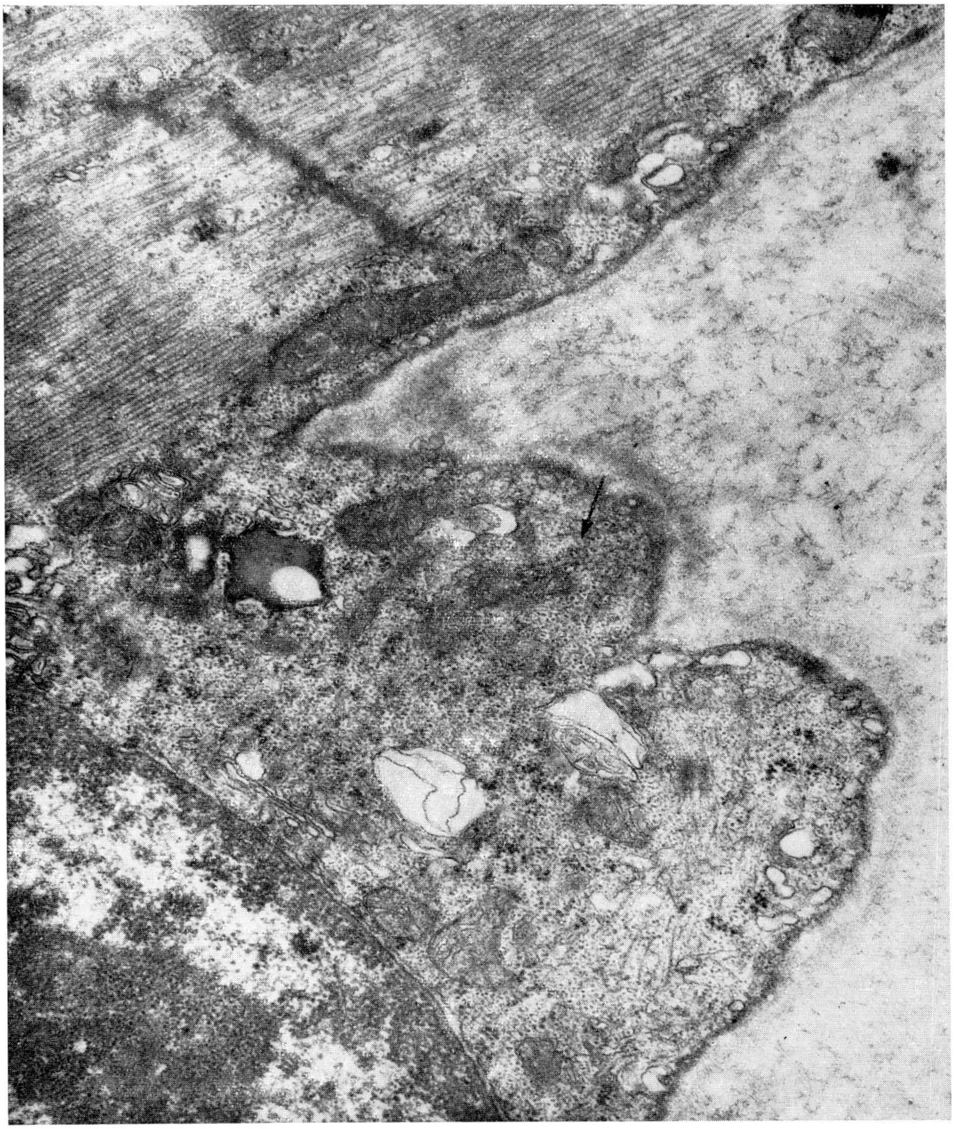

Fig. 162. Human dystrophia myotonica. Longitudinal section of a muscle fiber with a sarcoplasmic extension, one extremity of which is only visible in the picture. Note the bulging toward the external space and the permanence of the basal membrane. The sarcoplasmic extension shows mitochondria, glycogen particles, vesicles and myelinic figures, some myofilaments and portions of rough reticulum (arrow). Electron micrograph, × 25,000.

Fig. 163. A serial non-consecutive section of the same fiber as in Fig. 161. The sarcoplasmic extension here shows mitochondria, sarcoplasmic reticulum, T-systems with triadic structures (in the middle), ribosomes, polysomes and portions of rough reticulum. Myofilaments are also visible: the discrete thickenings among them are due to initial portions of Z-lines (Z-bodies). The fiber shows early signs of degeneration in the form of rarefaction of the thin myofilaments, dilations of the longitudinal reticulum and of the cisternae. Electron micrograph, × 18,000.

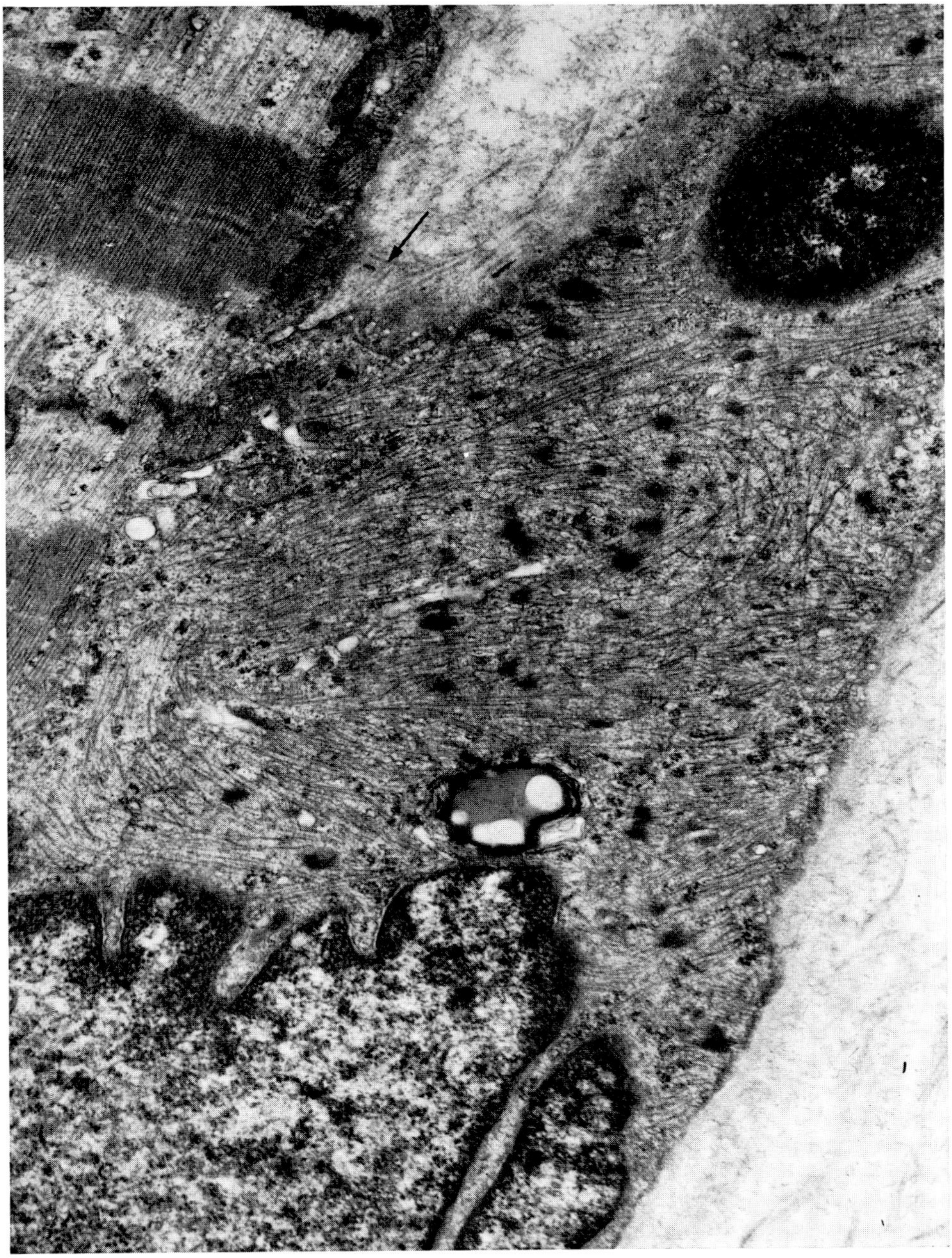

Fig. 164. Human dystrophia myotonica. A degenerated muscle fiber is partially shown on one side of the picture. Directly attached to it there is a sarcoplasmic extension which is bulging toward the external space (arrow), but it is still covered by the basal membrane. The sarcoplasmic extension contains nuclei, organelles, and partially and randomly oriented myofilaments. For other details, see Fig. 162. Electron micrograph, × 18,000.

POSSIBILITY OF A REGENERATION OF 'CONTINUOUS' OR 'DISCONTINUOUS TYPES' ORIGINATING FROM NUCLEATED PORTIONS OF UNDIFFERENTIATED SARCOPLASM

In all instances in which regeneration of discontinuous type takes place, the first stage is characterized by a rapid and abundant formation of delineation of semilunar or sickle-shaped cells, which with their basophilic cytoplasm are located around the degenerated and partially reabsorbed material of the damaged fiber (Figs. 158 and 159). Using the Studitsky procedure in experiments with rats, we find more than one of these cells along the sarcolemmal profile in a cross section of a degenerating or necrotic fiber; this is true also of frog muscle.

The incidence of these subsarcolemmal cells suggests that their number is far greater than it should be when one considers the distribution of satellite cells along the fibers of adult rats.

It is, of course, possible that some divisions of satellite cells take place early. Mitotic divisions certainly can be seen in the cells inside the empty sarcolemmal tubes; but this occurs after the appearance of the sickle-shaped cells. Church *et al.* (1966) also reported this.

A. MAURO: You say you do not see mitotic activity in this phase?

M. ALOISI: Yes, I have not been able to find it in this phase. In general, you can find mitosis after the absorption of the myoplasm. In the empty sarcolemmal tubes, in which the cells are less crowded, they are dividing mitotically.

S. A. SHAFIQ: I am surprised that there is no mitotic activity up to 4 days in your experiment. In injured mouse muscle we have seen mitoses on the second day after injury.

A. MAURO: Dr. Carlson, you should make a comment on this. While presenting your paper you said there was mitotic activity.

B. M. CARLSON: Although I have not yet done quantitative autoradiographic studies on these early stages, I have seen mitotic figures on ordinary histologic sections. On the simple basis of scanning, however, I am not sure that the mitotic activity could account for this entirely.

M. ALOISI: From about 3 days you can see a number of cells just inside the sarcolemmal tube; however, some of them are macrophages. Other slides from diffusion chamber experiments show analogous features. The sequence of events is relatively slower, and we can better appreciate important details. In most cases, even in longitudinal sections, you can see 2 areas in the fiber, one hyaline and degenerated and the other, peripheral, basophilic and nucleated with the appearance of a lateral portion of the cytoplasm; it looks like a muscle cytoplasm, which has escaped degeneration and is preparing for regeneration. Of course, it would be better if we had some knowledge of the ultrastructure of this material, but perhaps this can best be shown later with the help of the clinical-pathological material.

In denervated pigeon breast muscle, prepared for electron microscopy 15 days after denervation one can see numerous satellite cells. In my experience with denervated pigeon breast muscle, I have never observed mitosis, so this accumulation of satellite cells is difficult to understand.

By the way, the increased number of satellite cells in a denervated muscle may be the morphological aspect of the so-called 'plastic' state if we adopt the Studitsky terminology mentioned by Dr. Carlson this morning.

D. ALLBROOK: I do not really understand the 'plastic' terminology; I think that what the Russian school means is that, if you denervate, you get an increase in the number of free cells in muscle which then allows it to regenerate better than it otherwise would. Is that correct?

M. ALOISI: Well, as far as I know, the intention is to denote the status of a fiber which facilitates regeneration (or the implantation of rather large portions of muscle tissue). Denervation produces changes in the muscle fibers; both the velocity of the atrophic processes and the date of onset of true degenerative events vary. These changes are different in different species, but there are also differences in kind if you consider the myoplasm versus the sarcoplasm. For instance, in the frog, where the atrophy of myoplasm is extremely slow, you can follow the reactive changes in the sarcoplasm (Muscatello *et al.*, 1965) for a longer period of time. These changes consist of a proliferation of the sarcotubular system, with an increase in the turnover of its proteins, an increase in ribosomes and polyribosomes, and an increase in the number of mitochondria. Under these conditions the nuclei also appear to be activated and, as shown in the case of the pigeon breast muscle, satellite cells may increase in number. In any case, there are assemblages of mononucleated cells of a muscular nature at definite sites. This situation may favor a regenerative process, if such a muscle undergoes further damage, or is implanted according to Studitsky or explanted in tissue cultures.

M. REZNIK: How do you explain that denervation induces an atrophy of the muscle fibers and at the same time improves the ability of the muscle to regenerate?

Many years ago Denny-Brown described the relationship between denervation and regeneration. He showed that 1 month after denervation the process of regeneration is nearly the same as in the control. But after a few more months, 4 or 5 perhaps, the ability of the muscle to regenerate disappears.

M. ALOISI: Well, I am not defending the Studitsky terminology or the 'plastic state' as such. I am only interpreting it on the basis of facts arising from experimentation. In any case, if you discuss the fate of the denervated muscle and the possible reactive processes in its sarcoplasm, you have to distinguish carefully between the different behavior patterns shown by the different species with which you are dealing. The timing of events is extremely different. In pigeon breast muscle, for instance, atrophy and degeneration are already well advanced after 3 days, whereas in frog muscle an appreciable degenerative atrophy can be seen only after 3 months. This question of the time-course of the pathological processes in muscle is very important, particularly in the management of patients with muscle diseases. The different microscopic observations may be due largely to the different time-courses of the atrophic and degenerative changes in different muscles of the same patient. This has an impact also on the regenerative capacities.

D. ALLBROOK: In 1951 I showed that the peripheral stump of a nerve implanted into a muscle graft was necessary for maturation of the fibers to full diameter. I think *maturation* will be inhibited with denervation of the fiber, but I found no evidence for the notion that for adult muscle regeneration the presence of peripheral nerve endings was necessary. In the devascularized anterior tibial muscle of the rabbit, regeneration occurred to the same extent in the renervated or the denervated muscle.

M. ALOISI: Now let us move on to the slides which show mononucleated (or even multinucleated) cells in the isolated peripheral portions of a damaged fiber, the presence of which could explain the discrepancy between the number of pre-existing satellite cells, the number of mitoses and the speed with which the muscle can regenerate. The material I am presenting for this is derived chiefly from muscle of patients with myopathies which showed regenerative processes or at least preparative stages leading to a possible regeneration.

We have some electron micrographs of muscle of a patient with dystrophia myotonica, which may furnish some evidence for the above mechanism. In many places on the fibers,

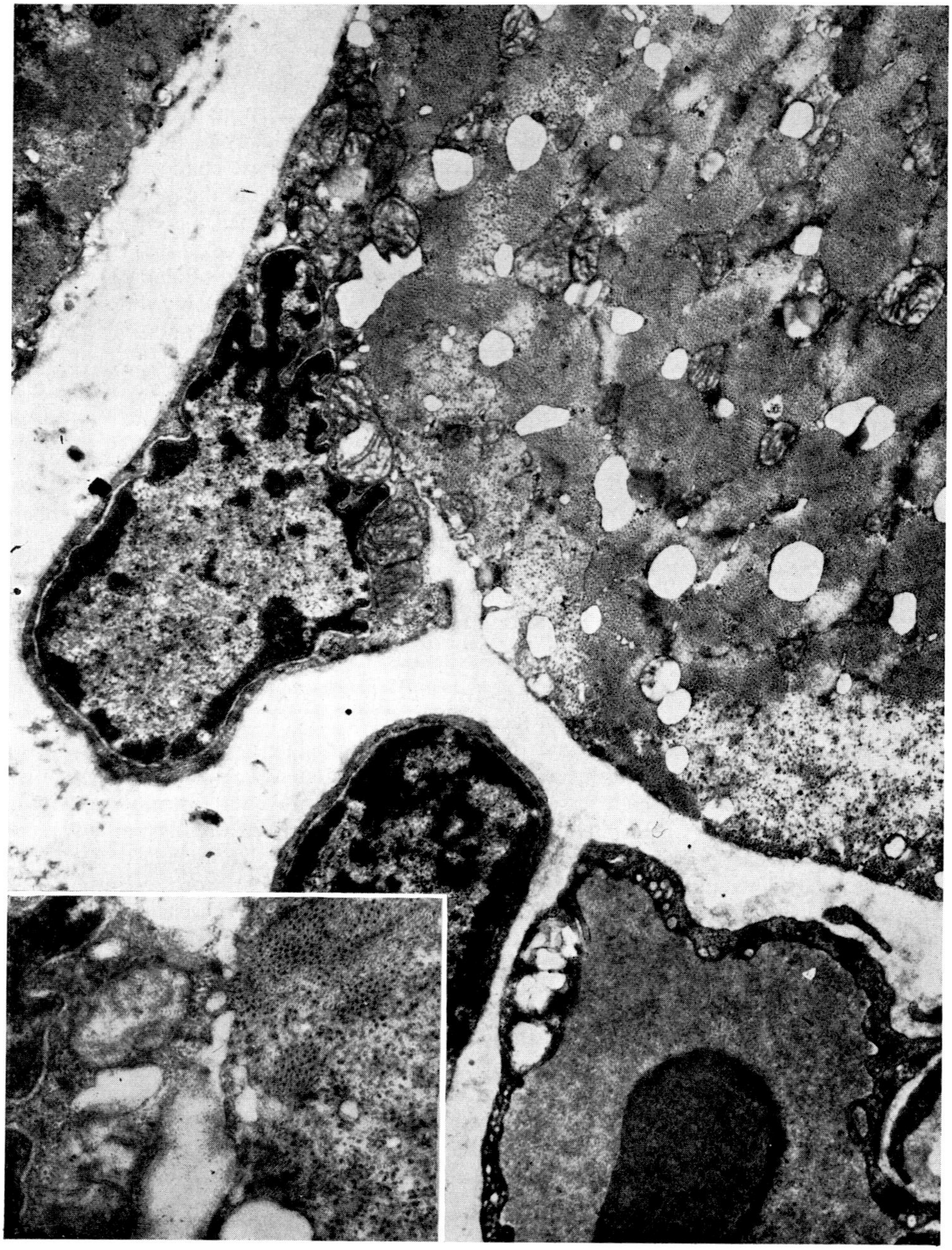

Fig. 165. Case of familial hypokalemic paralysis. Adjacent to a degenerated fiber there is a cell apparently similar to a satellite cell. In many points its cytoplasm seems to be in direct connection with the sarcoplasm of the fiber. The insert is a micrograph taken from a serial non-consecutive section and shows the details of the connection at a higher magnification. Bridges of sarcoplasm are clearly visible together with their content in glycogen and small mitochondria. It is supposed that the cell was formerly a nucleated portion of the fiber sarcoplasm. Electron micrographs, × 12,000 and × 24,000.

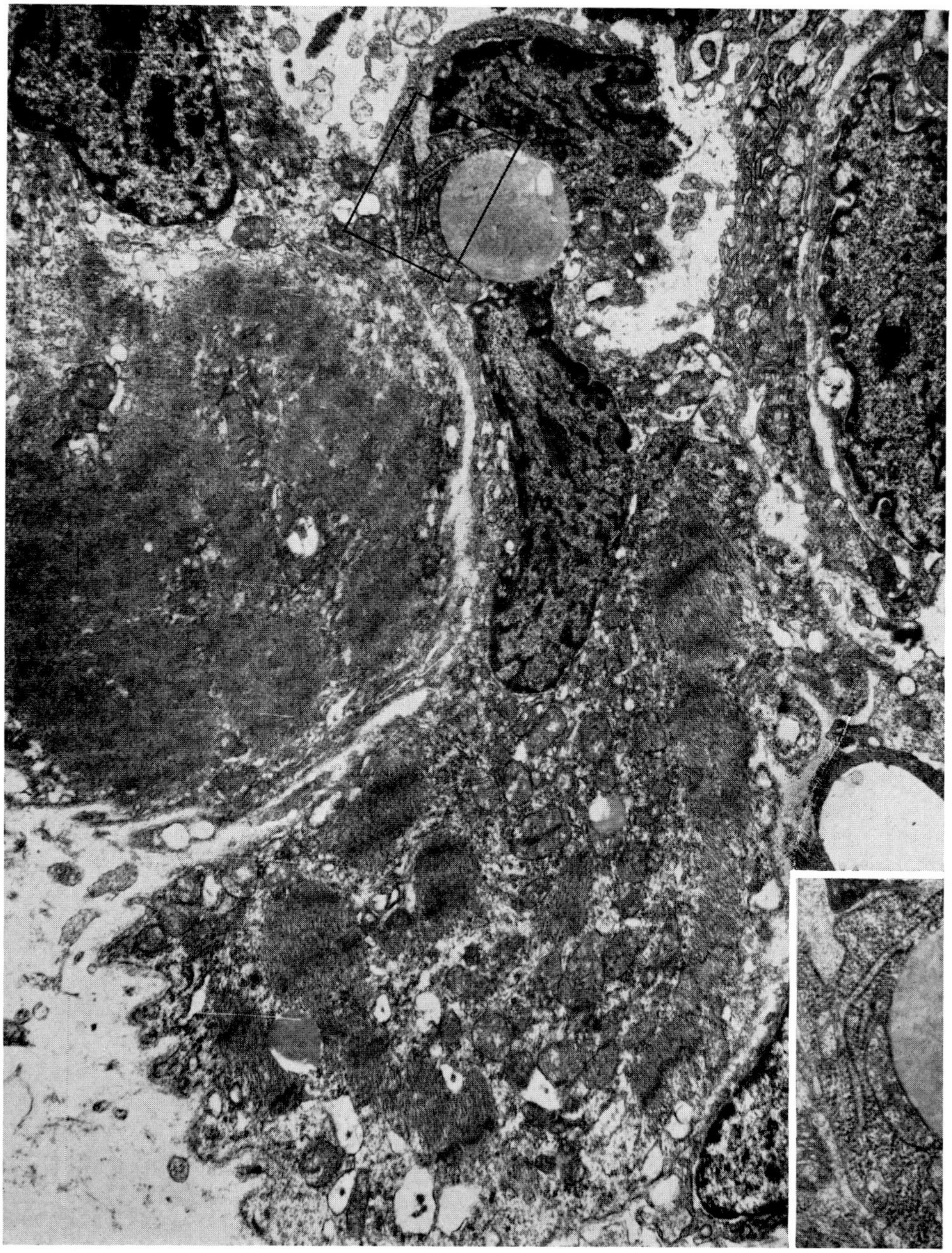

Fig. 166. Levator ani of a female rat of 5 days. The muscle fibers are physiologically atrophied and degenerated. From one of them a portion of nucleated sarcoplasm is bulging toward the external space and is likely to become a separate healthy cell. The insert shows, at higher magnification, details like mitochondria, glycogen, ribosomes, and clear portions of rough reticulum. Electron micrographs, × 12,000 and × 24,000.

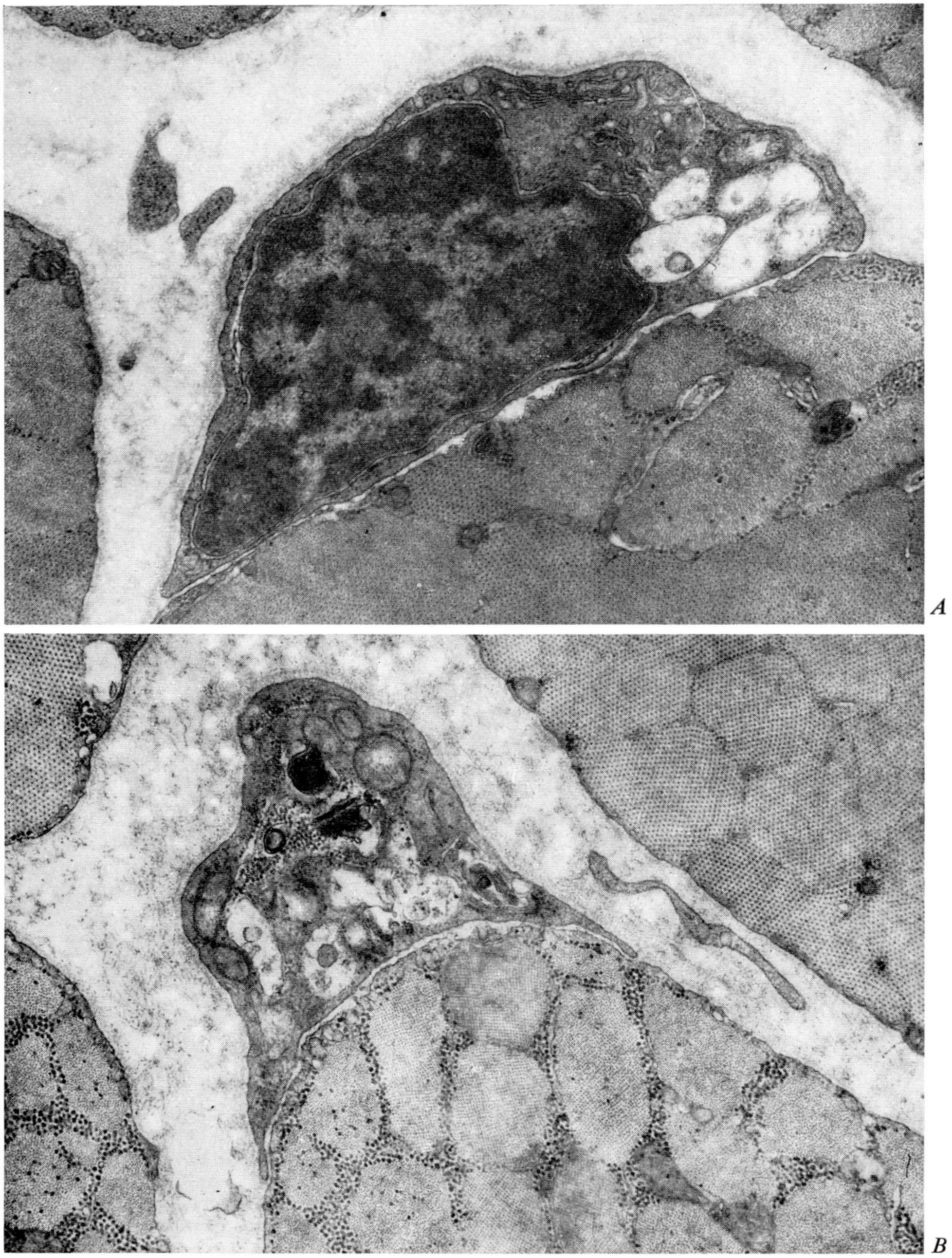

Fig. 167. Case of uncommon congenital hypotonia; boy, 15 months. Well-preserved muscle fibers (only thinner than in normal condition at the same age) have degenerated satellite cells. In both sections A and B of the figure the satellite cells present large vacuolizations due to large lysosomes and segresomes, some containing glycogen particles and remnants of organelles. Myelin figures are also present. Electron micrographsm. A, × 22,500; B, × 27,000..

the peripheral parts of the sarcoplasm are particularly thick, with numerous and often activated nuclei, ribosomes, and polyribosomes. They contain also endoplasmic reticulum, mitochondria, glycogen particles and, frequently, also myofilaments; these structures appear normal. This kind of fiber should correspond to those which, in the light microscope, appear basophilic or with an external basophilic layer. Whereas the remainder of the fiber tends to degenerate, these peripheral portions of the sarcoplasm become hypertrophic and protrude outside the main profile of the fiber (Figs. 161 to 164). You can clearly see that there is no membrane separating these nucleated portions from the rest of the fiber.

In these specimens we observe only a few, often randomly dispersed, myofilaments. In some places, initial formation of Z-bodies can be seen, but not true myofibrils. You can also find good examples of triadic systems. All this is cogent evidence that these sarcoplasmic portions are not degenerative in nature. In the last slides, you can see how they become hypertrophic and tend to detach from the rest of the fiber. The entire process occurs inside the basal membrane.

The next slides show a specimen of muscle from a patient with a familial hypokalemic periodic paralysis. A nucleated cell is connected loosely to a degenerated fiber. With sufficient magnification the connections appear to be true sarcoplasmic bridges. It is difficult to interpret this as a satellite cell fusing with a fiber, since the latter clearly is degenerated. It appears much more probable that it is a detaching cell (Fig. 165).

The next pictures are taken from an atrophying levator ani muscle of a female rat, 5 days old (taken from an investigation in our laboratory by Dr. V. Hanzlíková and Dr. S. Schiaffino). A nucleated cytoplasmic extension, although still directly connected, protrudes outside an atrophied and degenerated fiber. This cytoplasmic extension shows a well-preserved nucleus, mitochondria, glycogen, ribosomes, and clear portions of rough reticulum (Fig. 166).

Before ending my presentation relevant to the role of satellite cells, I want to show you some other electron micrographs of specimens from a patient with congenital hypotonia (Dr. I. Mussini). The boy was 15 months old at the time; the muscle showed a small number of satellite cells, but almost all of them were degenerated. Large vacuoles and cisternae, many lysosomes, and segresomes, fatty bodies and myelinic figures were also present (Fig. 167). This was rather striking in comparison with the fibers that, although small in size, appeared to be normal. It would be interesting to speculate on whether or not the under-development of the muscle fibers in this case has any connection with the progressive loss of satellite cells. But I am unable at this point to go into detail.

STUDY OF SATELLITE CELLS IN MATURE AND FETAL HUMAN MUSCLE AND RHABDOMYOSARCOMA*

PATRICK E. CONEN and C. D. BELL

The general ultrastructural appearances of the satellite cell (Mauro, 1961) and speculations concerning its function have been discussed by previous speakers with whom we are in general agreement. We will describe the morphologic features of satellite cells in mature muscle particularly in neuromuscular disorders, compare those satellite cells with morphologically similar cells in fetal muscle, refer to our attempts to identify satellite cells by light microscopy, and search for satellite cells in rhabdomyosarcoma.

MATERIALS AND METHODS

Materials for this presentation have been selected from muscle samples processed here for electron microscopy over the past 8 years. These include 380 human muscle biopsies (as an aid in histopathologic diagnosis), samples from 19 human fetuses of 8 to 16 weeks' gestation and from rat and mouse fetuses (in study of myogenesis), and muscle samples from experiments on rats, rabbits, normal and dystrophic mice and tissue cultures, and 4 rhabdomyosarcomas in children.

Most muscle samples were maintained at resting length during fixation in 5 per cent glutaraldehyde in cacodylate buffer for 2 hours at room temperature. Tissues were cut into small blocks 2 × 1 × 0.5 millimeters with a long axis parallel to the longitudinal axis of the muscle fibers, washed in cacodylate buffer or stored at 4°C until postfixed in Palade's osmium tetroxide for 1 to $1\frac{1}{2}$ hours at 4°C and embedded in Epon by standard methods. Blocks were embedded in the blunt end of a Beem capsule to prepare longitudinal sections, or in a flat mould (Rockwell *et al.*, 1966) to prepare transverse sections.

Suitable blocks and areas within a block were selected for electron microscopy by oil immersion examination of permanent Epon-mounted thin sections usually stained by toluidine blue (TB) (Bencosme *et al.*, 1959) and sometimes periodic acid-Schiff (PAS) (Cardino and Steiner, 1965), periodic acid-silver methenamine (PASM) (Cardino and Steiner, 1965), or paragon (PG) (Martin *et al.*, 1966) and some unstained slides were examined by phase contrast, interference or polarization microscopy.

During general electron-microscopic examination particular search was made for satellite cells by examining all nuclei in fibers. Satellite cells were beneath the basement membrane of the muscle fiber and separated from the myofibrils and cytoplasm of the fiber by the plasma membranes of the 2 cells. Cells fitting this description rarely contained myofilaments.

* Supported by the Muscular Dystrophy Association of Canada and The Ontario Cancer Treatment and Research Foundation.

RESULTS

Our observations on muscle regeneration will not be described here because they were in general agreement with published reports (see under Discussion) and those given at this meeting.

The satellite cell of mature muscle

Because of a special interest in satellite cells we have always searched for and found them in human biopsies. This cell type was not common in most histologically 'normal' biopsies and was more frequently observed in abnormal biopsies. The relative frequency of satellite cell nuclei and myonuclei was not determined for the human biopsies because the numbers and types of nuclei examined were not recorded for each biopsy. In addition, we have examined very few 'normal' human biopsies and these were biopsies from patients with normal histologic and histochemical findings and minimal clinical symptoms presumed to be not neuromuscular in origin. Our control animal biopsies were truly normal, but we have not systematically examined them for satellite cells.

By correlating 0.5- to 1.0-micron sections examined by light microscopy and adjacent ultrathin sections examined by electron microscopy we made the following observations. In muscle, sectioned longitudinally and fixed at resting length, the satellite cell is usually long and thin with very little cytoplasm, but in contracted muscle it may bulge slightly from the side of the fiber. The electron-microscopic appearance of the satellite cell is distinctive because of its location beneath the basement membrane but outside the sarcolemmal membrane of the muscle fiber (Fig. 168). The plasma membranes between satellite cells and the underlying muscle fibers are usually straight, but when the muscle is partially contracted the plasma membranes, and even the satellite cell nuclear membrane, may have a serrated edge (Fig. 169).

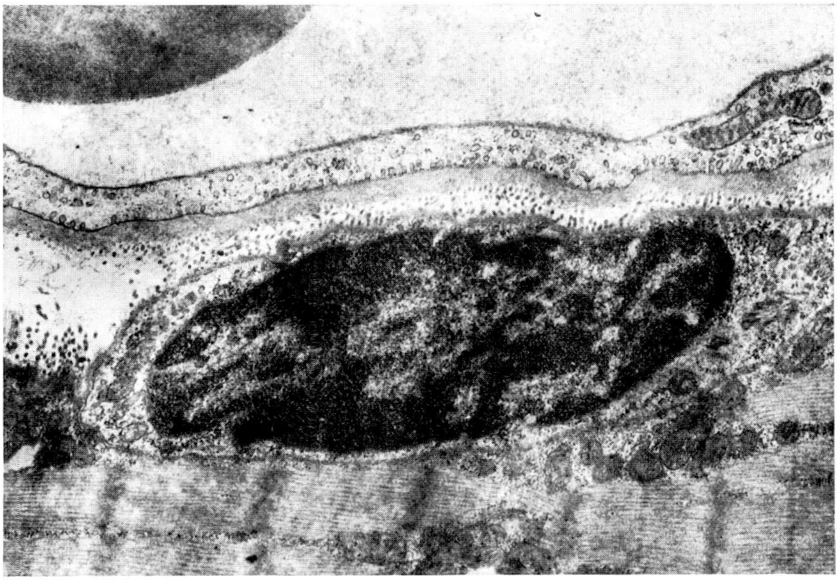

Fig. 168. Satellite cell in the characteristic location enclosed by the basement membrane of the muscle fiber but separated from the fiber. At low magnification this satellite cell might not easily be identified because it is adjacent to a capillary. (Symptomless mother of a boy with nemaline myopathy). × 13,500.

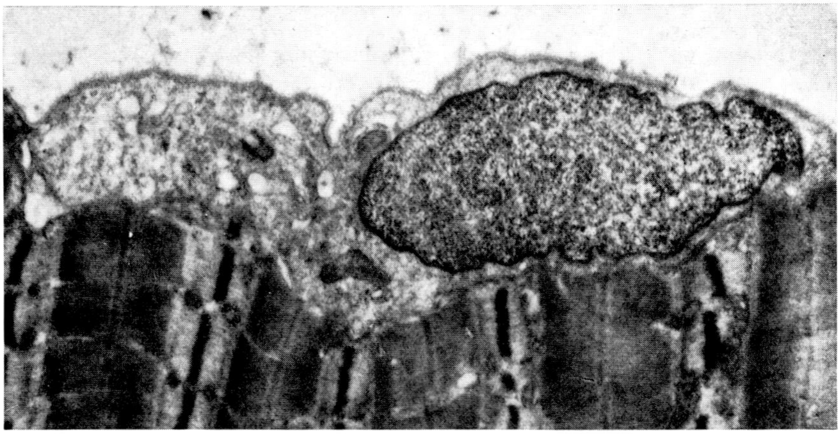

Fig. 169. Satellite cell containing a centriole (left of Figure). The muscle is partly contracted and the plasma membranes at the junction of satellite cell and underlying muscle fiber are serrated as is the nuclear membrane. (Infantile muscular dystrophy). × 7,200.

The nucleus of the satellite cell seldom has an obvious nucleolus and has a peripheral electron-dense arrangement of chromatin compared with the prominent nucleolus and even arrangement of chromatin in the myonucleus (Fig. 170). The nuclear chromatin differences are particularly obvious when glutaraldehyde-osmic fixation is used but not with direct fixation in osmic acid (Fig. 169).

The satellite cell has few organelles and the mitochondria are small and round with a less dense matrix than mitochondria in the underlying fiber (Fig. 171, A and B). These figures also show differences in mitochondria in two adjacent satellite cells. Satellite cells do not contain myofilaments. However, they may appear to contain myofilaments in thicker sections, in cells with scalloped edges and irregular margins, or when there is lack of definition of cell membranes due to the plane of section (Fig. 172). Glycogen is absent or extremely scanty in satellite cells.

Satellite cells may not be recognized on low-power electron-microscopic examination, particularly if their nuclei are round in the plane of the section (Figs. 170 and 172), or are immediately adjacent to a capillary (Fig. 168). Satellite cells in which the section passes through cytoplasm only, without including the nucleus, would have been missed by us since we usually studied nuclei to determine the proportion which were myonuclei or from satellite cells.

Occasionally cells presumed to be satellite cells are in mitosis (Fig. 173) and often there are several adjacent myonuclei. Centrioles are not infrequent in satellite cells (Fig. 169), but we have not seen them adjacent to myonuclei. Some of the structures assumed to be centrioles may be basal bodies of cilia because, although rare, cilia (Fig. 174, A and B) may be found in apparent satellite cells.

Satellite cells in adults may contain electron-dense inclusions (Fig. 175, A and B), interpreted as lipofuchsin and identical to that commonly found as aging pigment in muscle.

Many satellite cells may be found in small areas of a biopsy in a variety of neuromuscular disorders. Elongated cells at the periphery of a necrotic fiber (Fig. 176) pose a particular problem of nomenclature because they may have features of satellite cells, myoblasts, or an intermediate stage (see Discussion). Occasionally most blocks of a biopsy will contain many satellite cells (5 to 10 per cent of nuclei), particularly in young infants. Figure 177 (A and B) shows the light- and electron-microscopic appearance of satellite cells from a young infant with 'muscular hypoplasia'. Satellite cells may also be frequent in older individuals, *e.g.* in a biopsy from an 8-year-old girl with rhabdomyolysis and a symptomless mother of a boy with nemaline myopathy.

The light-microscopic identification of a satellite cell in epoxy-embedded thin sections can only definitely be made by electron microscopy of adjacent ultrathin sections and careful identification of the same cell in both sections. Satellite cells can often be suspected by their nuclear characteristics when studied by light microscopy. Other features are not useful because the membranes separating the cell from adjacent muscle are not visible and the thin rim of cytoplasm in the satellite cell is too small to be recognizable by the absence of glycogen.

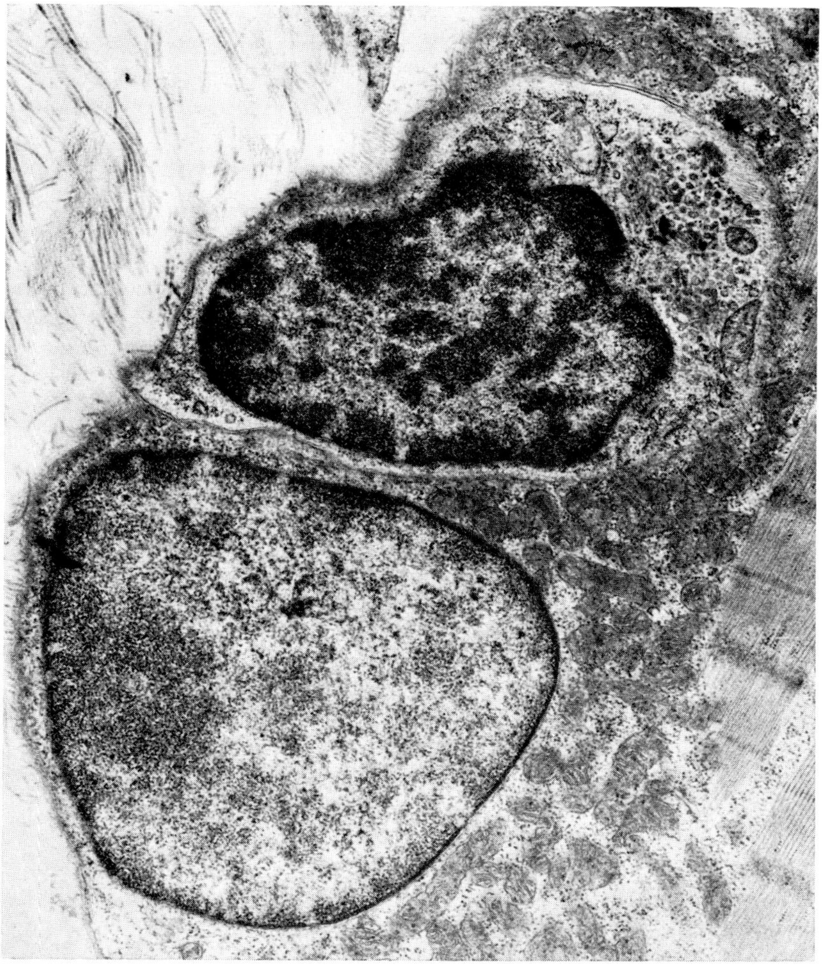

Fig. 170. Satellite cell which might not be easily identified at low magnification because of its rounded shape in this plane of section. The peripheral nuclear chromatin is more dense in the satellite cell than the myonucleus, and the mitochondria are smaller and have less dense matrix. (Duchenne muscular dystrophy). × 16,200.

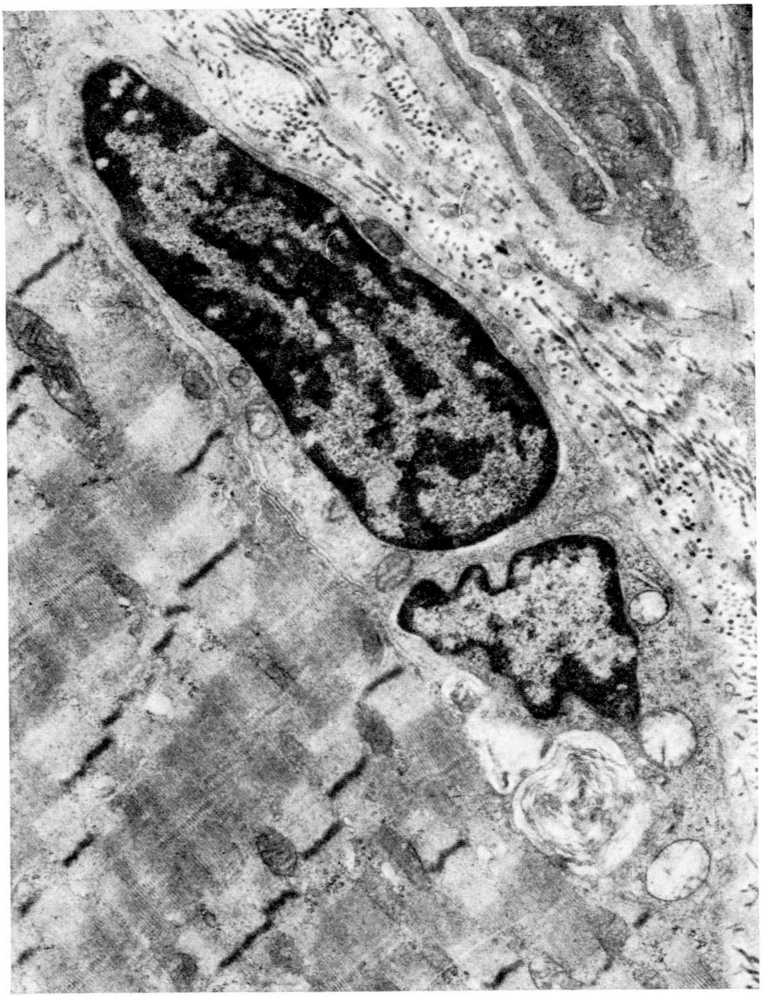

Fig. 171A. 'Twin' satellite cells possibly resulting from recent mitosis. One cell contains large pale mitochondria, and a 'myelin figure' effect possibly due to a glutaraldehyde fixation artefact. (Dermatomyositis). A × 12,000.

Attempts to identify the satellite cell on paraffin sections, 2 to 3 microns thick, using silver impregnation to show a separation between the muscle cell and satellite cell, and PAS to show the basement membrane surrounding both cells, were unsuccessful. Similar studies using PASM technic on epoxy sections have also been unsuccessful.

Satellite cells and myogenesis in man

The increased resolution available with electron microscopy has increased our knowledge

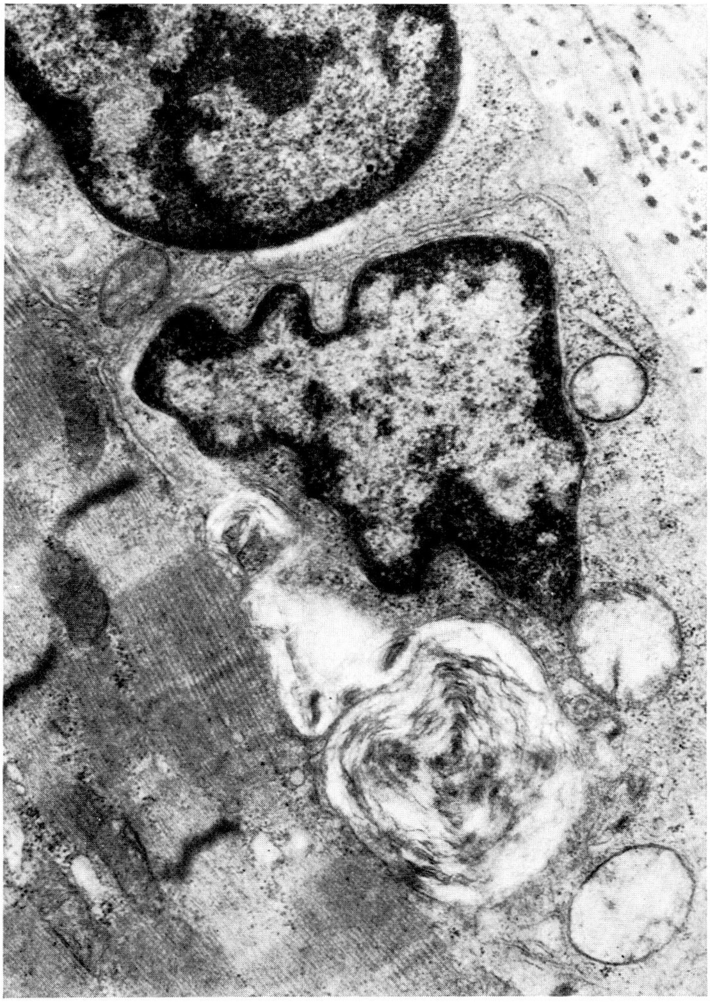

Fig. 171B. 'Twin' satellite cells possibly resulting from recent mitosis. Part of Fig. 171 A enlarged. × 24,000.

of cellular detail in myogenesis, but some problem areas have been made more obvious. Terms used in light-microscopic descriptions of myogenesis have been used differently by various authors (Godlewski, 1902; Tello, 1917, 1922; Boyd, 1960; Sloper and Pegrum, 1967) and the same lack of standard definitions is more noticeable in electron microscopy, particularly if a separate terminology for electron microscopy is to be avoided. A supplemental terminology, using premyoblast, myoblast, premyocyte, myocyte, and satellite cell, was once suggested by us (Conen *et al.*, 1967) but has not proved to be usable.

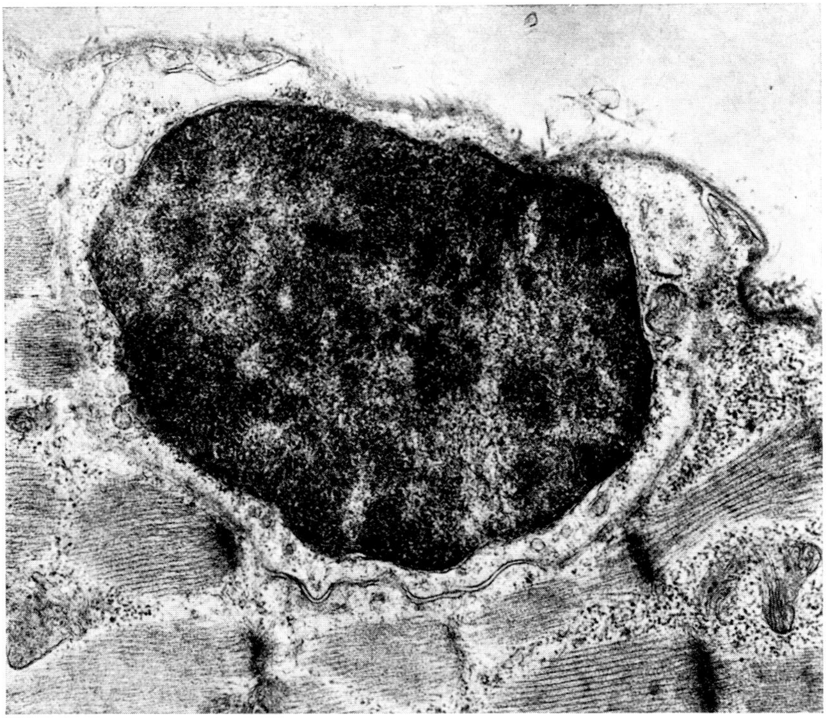

Fig. 172

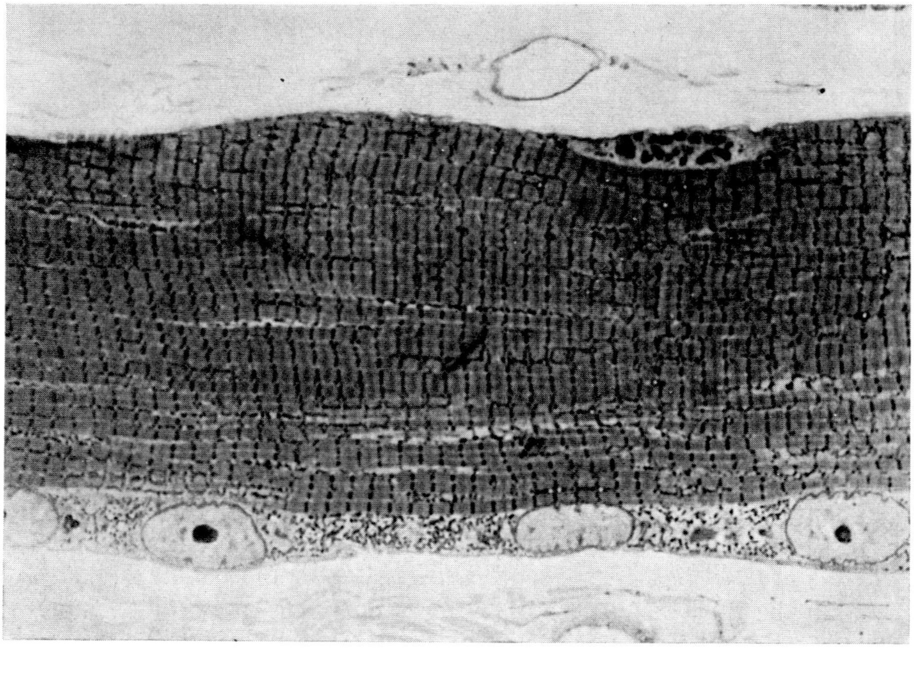

Fig. 1

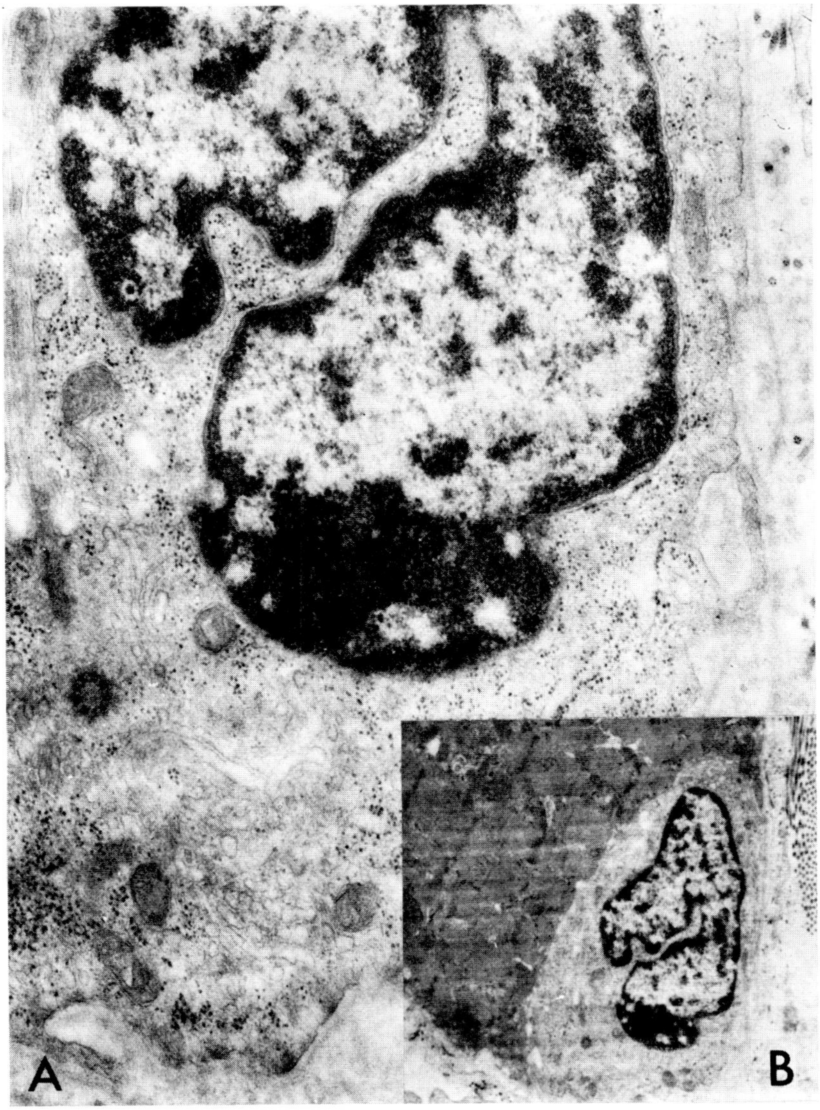

Fig. 174. A 'satellite cell' containing a cilium and basal body (left side of Figure). (Duchenne dystrophy). A, × 28,800; B, × 7,200.

Fig. 172. This satellite cell shows several areas with lack of clarity in the plasma membranes separating it from the underlying muscle. At one area (left side of Figure) myofilaments in the fiber appear to be almost reaching the satellite cell. (Symptomless mother of a second case of nemaline myopathy). × 22,500.

Fig. 173. Light micrograph to show a satellite cell (top of fiber) in mitosis and several sarcolemmal nuclei on the lower border of the fiber. (A second case with infantile dystrophy). × 1,200.

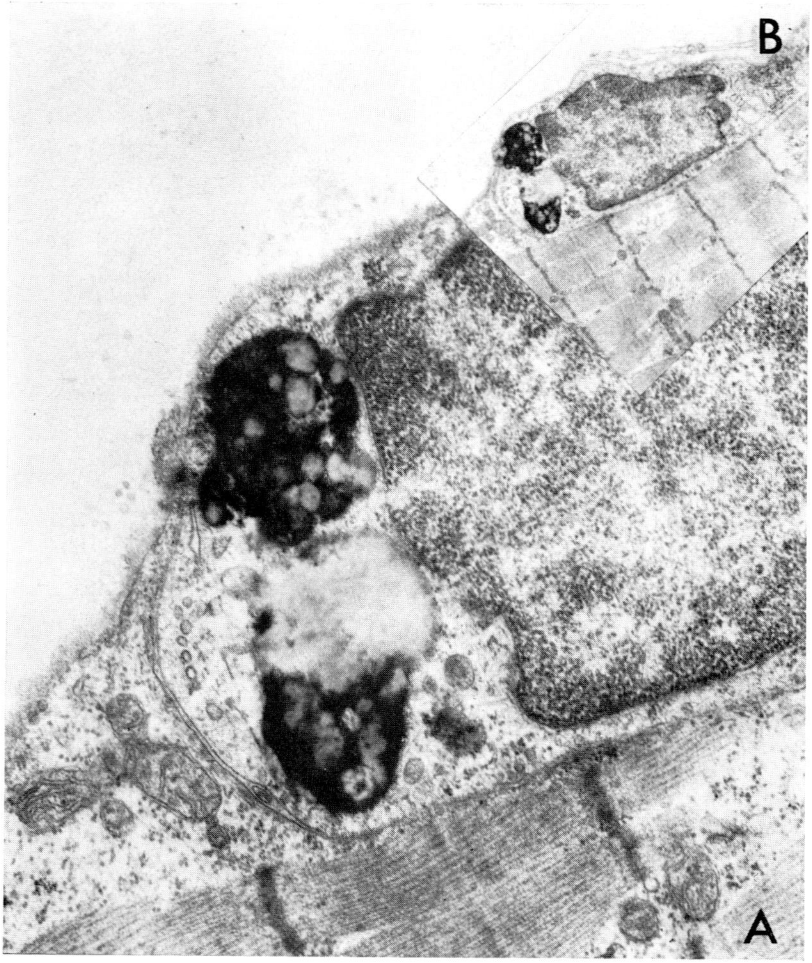

Fig. 175

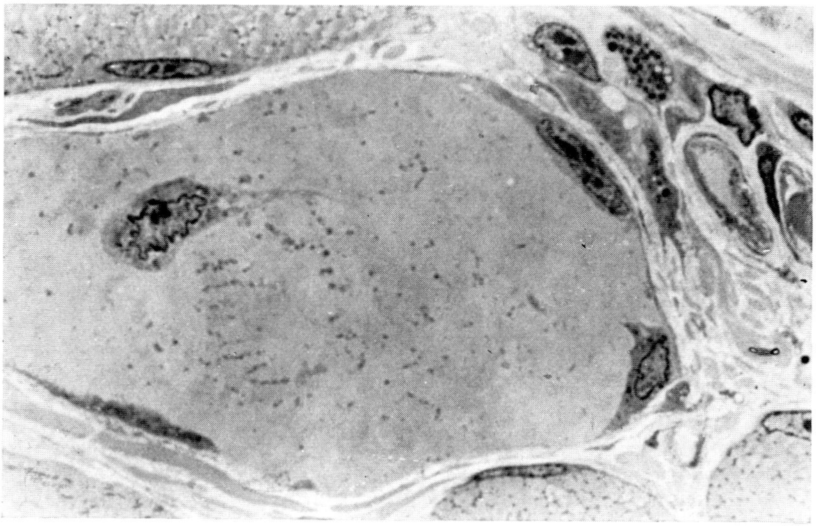

Fig. 176

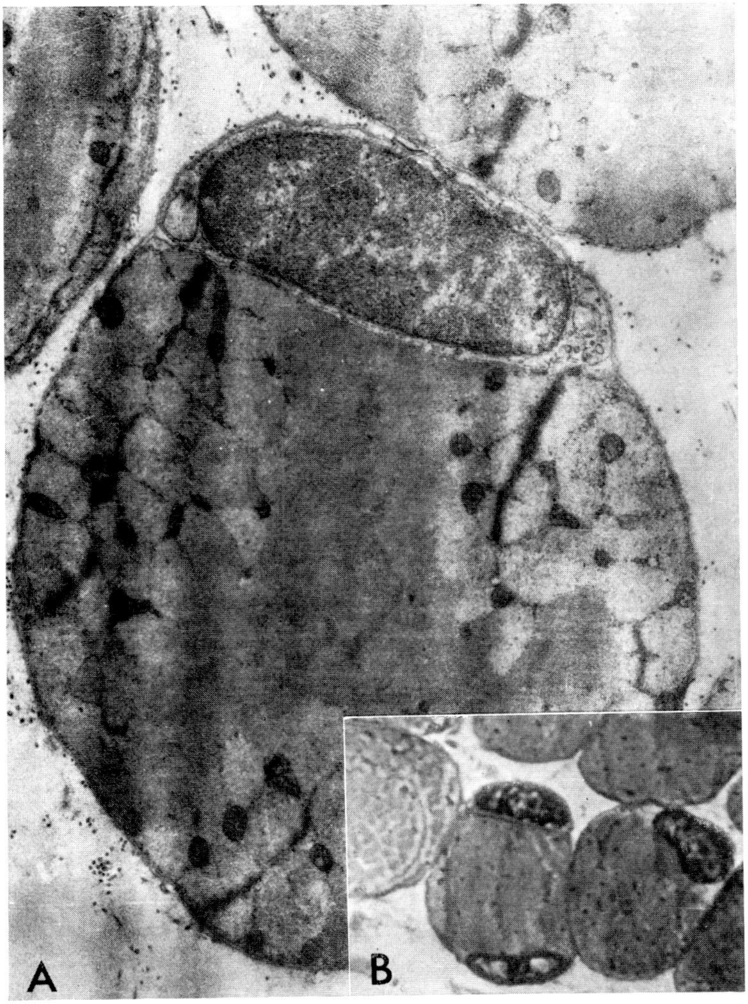

Fig. 177. Satellite cells shown in electron (A) and adjacent light (B) micrographs from a patient with small weak muscles (fiber diameter, 10 to 20 microns). (Muscular hypoplasia).
A, × 5,600; B, × 1,280.

Fig. 175. 'Satellite cell' containing electron-dense inclusions (lipofuchsin-aging pigment). (Same case as Fig. 168). A, × 22,500; B, × 5,400.

Fig. 176. 'Satellite cells' (some myoblast features) limiting the periphery of a necrotic fiber. (Duchenne dystrophy). Toluidine blue; × 11,700.

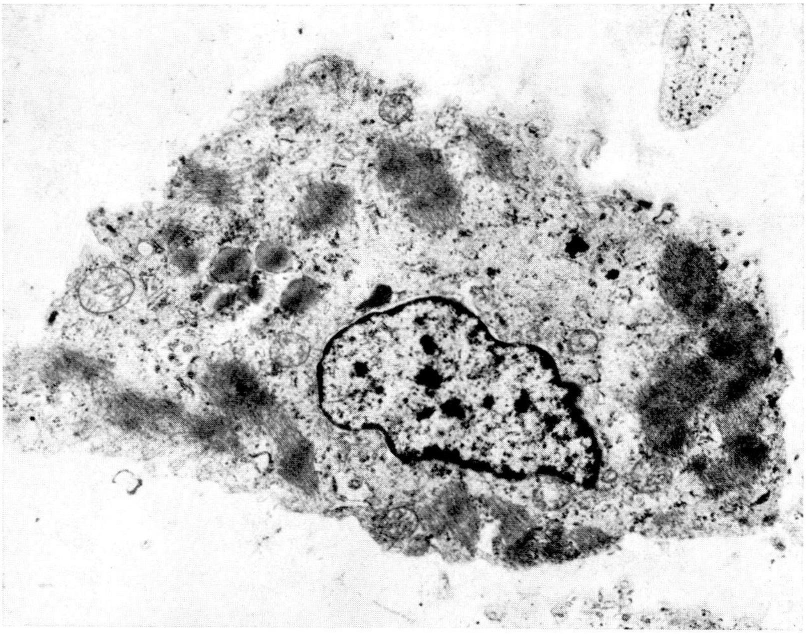

Fig. 178. Apparently single myoblast without surrounding basement membrane. (8-weeks' gestation fetus). × 10,900.

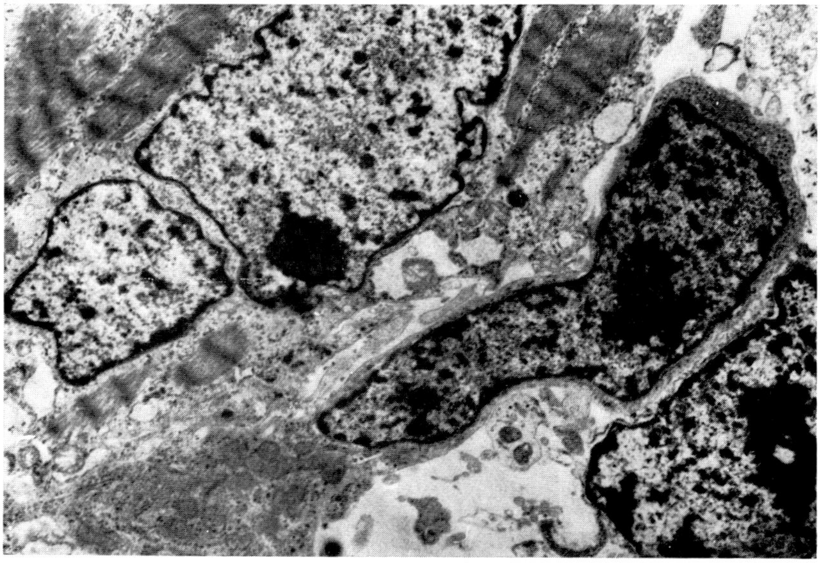

Fig. 179. Myoblasts in early myotubes (upper part of Figure). Below at right of center are 2 cells difficult to identify: probably 'satellite cell' (far right) or fibroblasts (right center). (8-weeks' gestation fetus). × 8,000.

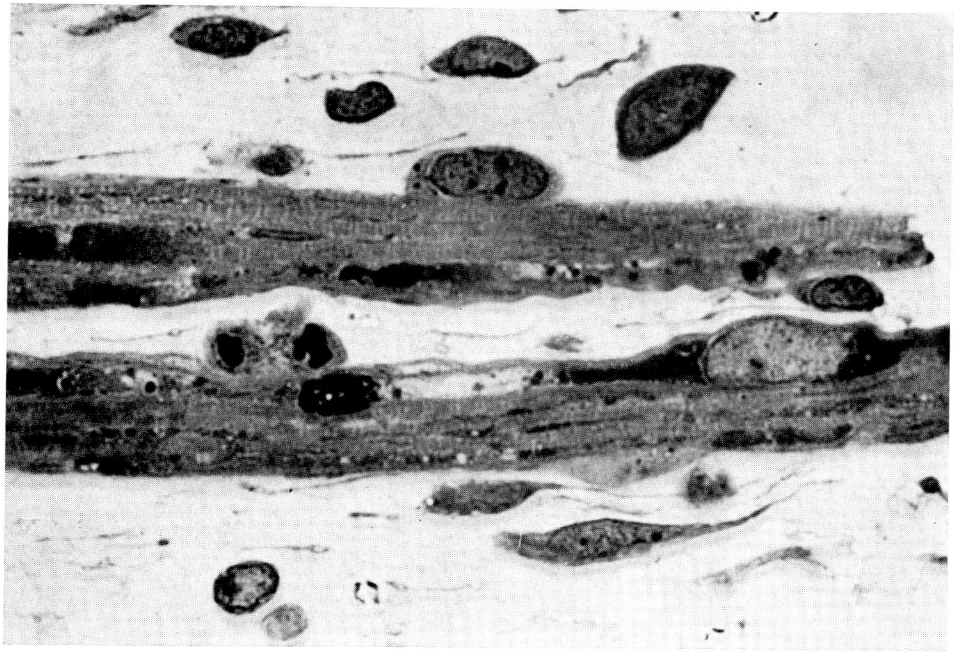

Fig. 180. Light micrograph of 12-week fetus to show probable satellite cells (A: above center of upper fiber, and B: above left of lower fiber). Cell B is in mitosis. Cell A would also appear closely related to the upper fiber if sectioned vertically, but a horizontal section would appear as a single cell. Serial sections through the horizontal section would show the relationship to the fiber if made toward the fiber, but not if made away from the fiber. The large amounts of glycogen in the fiber are densely stained by the PAS-toluidine blue technic. (Male fetus from proven carrier of Duchenne dystrophy). × 1,200.

Myoblast was either an apparently single cell containing myofilaments with or without a surrounding basement membrane (Fig. 178), depending upon the muscle and stage of development, or grouped with others in early myotubes (Fig. 179).

The nucleus was oval, irregular in outline and contained 1 or more small nucleoli. The cytoplasm was abundant and frequently contained closely aggregated bundles of myofilaments which usually had focal, transverse condensations forming rudimentary Z-bands. Mitochondria were spherical and relatively large. Ground cytoplasm contained numerous polyribosomes, particularly in association with myofibrils. The cytoplasm also contained glycogen, dilated cisternae of endoplasmic reticulum and occasionally centrioles.

Nuclei in later myotubes and muscle fibers. These nuclei were vesicular, frequently had large nucleoli and were situated in the center of the muscle fiber with adjacent large mitochondria and aggregates of glycogen and sometimes empty vacuoles (Figs. 180 and 181).

Satellite cells. In most samples of fetal muscle these cells had a similar appearance to those in muscle and were easily distinguished from apparently single myoblasts and nuclei in myotubes (Figs. 182 and 183). This distinction was often accentuated by use of electron-microscopy lead 'stains' which make glycogen electron dense (dark in micrographs) and show a strong contrast between glycogen-rich myotubes and glycogen-poor satellite cells (Fig. 183). However, in early fetal muscle samples or in distal limb-musculature from fetuses of 10 to 14 weeks gestation, many of the muscle fibers and myoblasts did not show a sur-

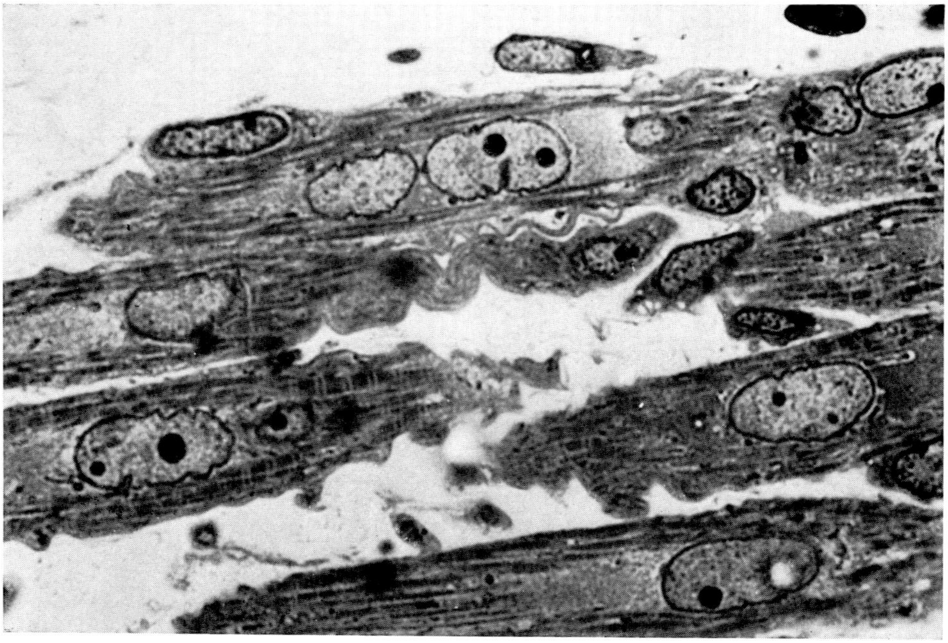

Fig. 181. Light micrograph from the same fetus as Fig. 180 to compare the probable satellite cell (upper left) with nuclei in myotubes. Paragon stain; × 1,200.

rounding basement membrane and consequently one of the hallmarks of the definition of a satellite cell was lacking, *i.e.*, that the cell should be beneath a basement membrane. The term 'satellite cells' has been used here for those cells without surrounding basement membrane which have general characteristics of satellite cells seen in more mature muscle.

Fibroblasts and other mesenchymal cells lacked basement membrane and myofilaments and had extensive, well-developed, rough endoplasmic reticulum and multiple elongated cytoplasmic extensions.

Muscle from human fetuses of 6 to 8 weeks' gestational age showed by light microscopy of the proximal musculature that the majority of muscle fibers were elongated with characteristic cross striations. However, these fibers were intermixed with small stellate or oval-shaped cells which frequently were not striated. The latter cells were characterized by electron microscopy as myoblasts (Fig. 178), or 'satellite cells' and were sometimes difficult to distinguish from fibroblasts (Fig. 179). The lack of, or poorly demonstrable, basement membrane in many of these early fetal samples led to some difficulties in classification of cell type.

Muscle from human fetuses of 10 to 14 weeks' gestational age seldom contained myoblasts without basement membrane. Muscle fibers contained well-formed myofibrils (Figs. 181 and 183) and abundant glycogen (Figs. 180 and 183). These fibers were surrounded by a basement membrane which also enveloped satellite cells (Figs. 182 and 183). These cells resembled those seen in younger fetuses or seen in peripheral limb-musculature. Muscle fibers were frequently aggregated in bundles. In these instances, basement membrane completely surrounded the bundles but was often absent at sites where muscle fibers were closely apposed. Satellite cells were often seen (Figs. 182 and 183), not only beneath basement membranes but also in association with closely arranged muscle fibers. In these latter areas, satellite cells formed rows which longitudinally separated the muscle fibers.

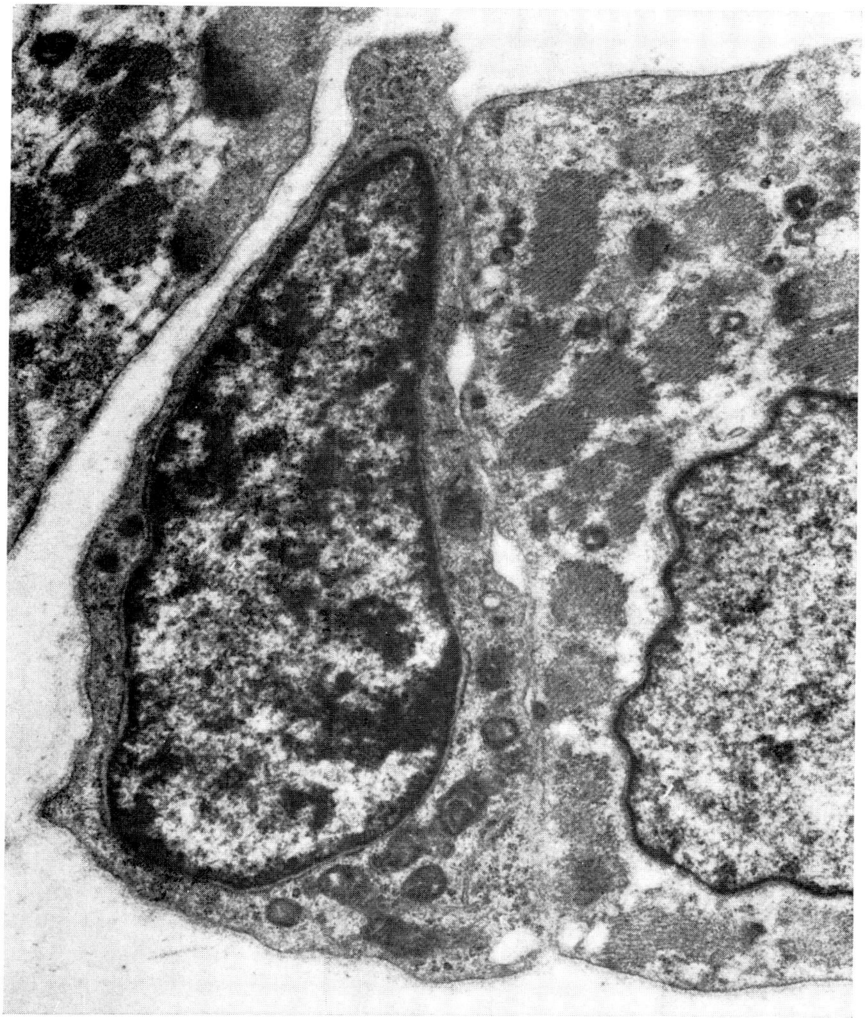

Fig. 182. Satellite cell (left centre of Figure) separated from the underlying myotube but enclosed by the same basement membrane. Compare the peripheral electron density of the satellite cell nucleus with the even staining of the myotube nucleus (far right). (12-weeks' gestation fetus). × 18,900.

Muscle from a human fetus of 28 weeks' gestational age. Satellite cells had the same fine structure but were less numerous than in samples from younger fetuses.

'Fetal'? Duchenne dystrophy. We have recently examined muscle samples from 3 male fetuses, 9, 9, and 13 weeks' gestational age, obtained at therapeutic abortion from proven carriers of Duchenne dystrophy. The serum CPK levels on heart blood samples were elevated in 2 and perhaps all 3 of the fetuses compared with other fetuses. We have not, so far, found any morphologic abnormalities in myogenesis in these fetuses (Figs. 180 and 181), particularly in satellite cells and myotubes.

Rhabdomyosarcomas in childhood. We have examined 4 rhabdomyosarcomas by light and electron microscopy. Three of the samples involved muscle and tumor cells were infiltrating

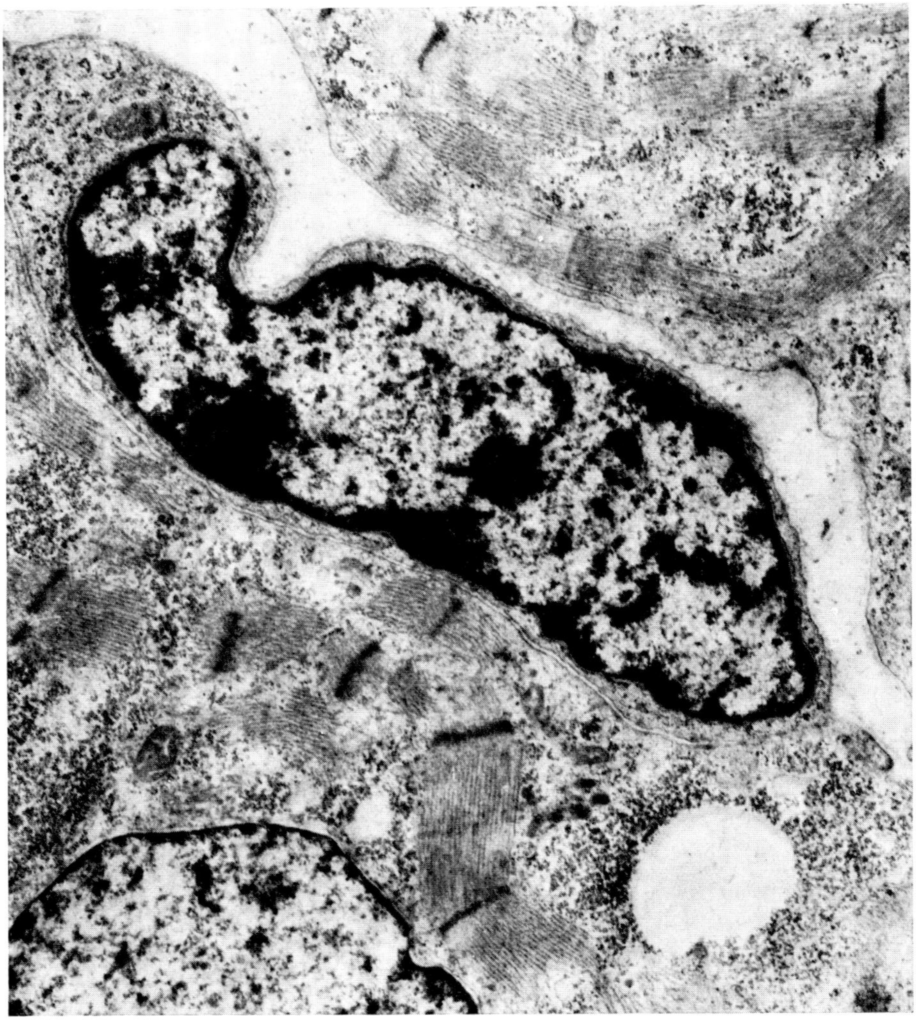

Fig. 183. Longitudinal section to show the well-developed myofibrils and abundant glycogen in the myotube compared with the scant cytoplasm of the satellite cell. Also note the difference in nuclear chromatin in the satellite cell (above) compared to the myotube nucleus below. (12-weeks' gestation fetus). × 21,000.

between normal and altered muscle fibers. The 4th case was a 9-month-old boy who had a Wilms tumor (nephroblastoma) excised which contained many striated muscle fibers in the stroma. He received chemotherapy and radiotherapy for metastatic lesions. During thoracotomy at 2 years of age a single metastatic nodule was removed from the right lung and he is well 5 years later. The nodule (Fig. 184) was composed of groups of well-differentiated myoblasts enclosed within a basement membrane. The cells were not oriented to form myotubes and myofibrils were short and poorly formed. One cell among the myoblasts in the Figure is representative of the few that do not contain myofibrils and may represent

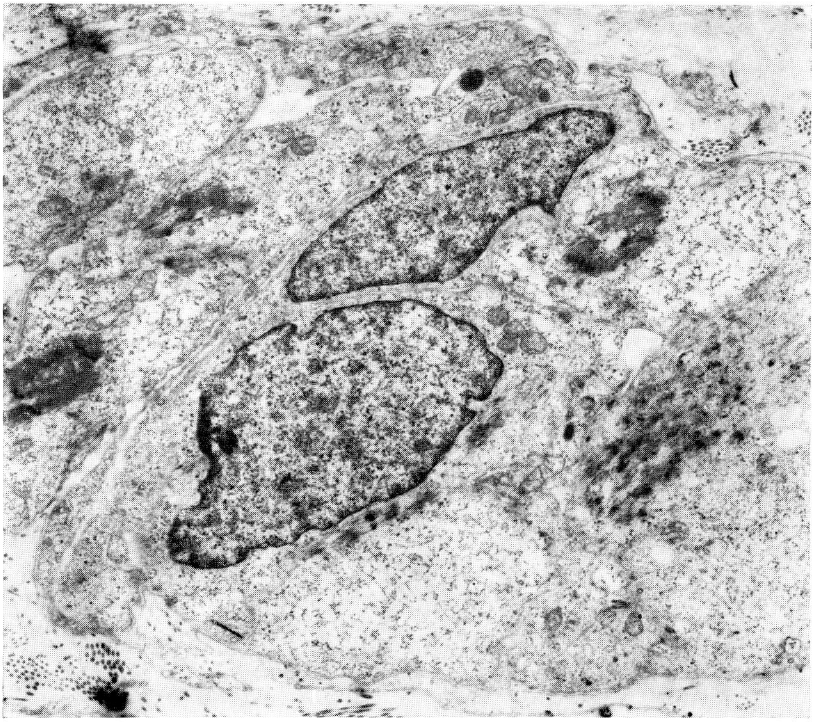

Fig. 184. Electron micrograph of a metastatic rhabdomyosarcoma nodule in lung showing basement membrane surrounding a group of myoblasts which contain poorly oriented myofilaments. One nucleus (upper center) is in a cell with scant cytoplasm which may be a satellite cell (see text). (Rhabdomyosarcoma). × 4,500.

satellite cells, although there may be myofibrils within this cell above or below the plane of this section.

DISCUSSION

During the past century, investigations of the mechanisms of muscle regeneration have been periodically reviewed as new methods of study became available (Godman, 1958; Field, 1960; Betz *et al.*, 1966; Sloper and Pegrum, 1967). Many morphologic studies, using vital stains, histochemistry, tissue culture, spectrophotometry and autoradiography, were limited by the resolution attainable by light microscopy, particularly for identification of cells and spatial relationship of the cells to the sarcolemma. Despite these limitations autoradiography was used by: MacConnachie *et al.* (1964) to show DNA replication and mitosis in nuclei at the edge of fibers; Yaffe and Feldman (1965), and Loeffler (1969) to label one population of cells in hybrid myotube experiments which showed the myotubes were formed by fusion of single cells and not by nuclear replication; Okazaki and Holtzer (1965) with fluorescent antimyosin to show that synthesis of DNA and myosin does not occur simultaneously; Bischoff and Holtzer (1969) to show that fusion occurs in G_1.

The increased resolution obtainable by electron microscopy has provided more information in repeats of classical regeneration experiments, has enabled Mauro (1961) to identify the satellite cell of muscle, and has focused attention on additional problems of interpretation of findings. Sequential studies of regeneration following injury have been repeated with electron microscopy: crushed mouse muscle and ischemic rabbit muscle (Allbrook, 1962); cold-injured rat muscle (Price *et al.*, 1964); heat-injured mouse muscle (Shafiq and Gorycki, 1965); crushed bat-wing muscle (Church *et al.*, 1966) and ischemic rabbit muscle (Reznik, 1969*b*).

These electron-microscopic observations involved considerable study and discussion of the satellite cell of muscle originally described by Mauro and its presence has since been confirmed in several mammals. Mauro also suggested 3 hypotheses for the origin and role of the satellite cell which have been discussed by Lee (1965) and Reznik (1969*b*). These hypotheses are, firstly, that the satellite cells are formed from nuclei and sarcoplasm of muscle fibers; secondly, that they represent dormant myoblasts that were not included in the original myotube; and, thirdly, that they are a wandering (non-muscular) cell that had migrated beneath the basement membrane of the muscle fiber.

In discussion of these hypotheses it is essential, first, to decide whether all cells are of the same type which are in the characteristic location and have many morphologic features of satellite cells. The third hypothesis now seems least likely, particularly since Muir *et al.* (1965), and others, have added additional characteristics of the satellite cell, so that its identification does not depend solely upon its location. Atypical cells, such as those with cilia or lipofuchsin pigment, might be different from other satellite cells, although cilia have been reported in many cell types and aging pigment is common in many cells.

Mauro's second hypothesis concerning the satellite cell has been supported in discussion of several reports of sequential studies of regeneration following the injury. However, Reznik (1969*b*) has critically discussed these reports and used their findings to support his own belief that the satellite cell arises from a previously intact myonucleus by formation of cytoplasmic vesicles, clefts and eventually complete plasma membranes. Reznik and many other workers appear to agree that the satellite cell is a stage in discontinuous regeneration and that this form of regeneration is most frequently observed, although Shafiq and Gorycki (1965) have described discontinuous and continuous regeneration in mouse muscle injured by heat. The weakness of Reznik's argument and that of most of us studying regeneration is that electron micrographs cannot show us the direction of transformation and reliance has to be placed upon the reproducibility of lesions for sequential studies. Unless a time-lapse type of study becomes possible for electron microscopy, we will have to develop other methods of testing the hypotheses.

A basic problem connected with distinguishing between Mauro's first and second hypotheses is concerned with the incidence of the satellite cell in a truly normal population. As described in this communication, human biopsies are almost invariably abnormal because they are taken from patients with neuromuscular disorders; therefore very few 'normal controls' are available to study frequency of satellite cells. In animal controls, the problem is one of adequate sampling because satellite cells are probably so infrequent that the electron-microscopic survey required to determine their incidence would be very laborious except for the bat-wing muscle where 10 per cent of nuclei are from satellite cells (Church *et al.*, 1966). Reznik (1969*b*) has specially searched for them in the mouse gastrocnemius or rabbit tibialis anterior without finding satellite cells in controls. In other cases, the problem could be best investigated if multiple transverse sections of normal muscle mounted on single-hole grids were examined by someone with infinite patience and time so that every nucleus could be counted and scored as in a satellite cell or a myonucleus. Hopefully, an easier way of examining a large number of fibers would use a light-microscopic method for identification of the satellite cell, but such a method has not yet been found. Possibilities for light-microscopic demonstrations of the satellite cell include various pretreatments of the muscle to widen the space between the satellite cell and the underlying muscle fiber and to demonstrate the space by a technic such as Hale's colloidal iron.

If satellite cells are present in muscle at all ages and in all diseases examined, it suggests that skeletal muscle in these conditions retains the ability for regeneration. Some previous theories concerning Duchenne muscular dystrophy have suggested a lack of regenerative ability which may be concerned with a defect in the manufacture of contractile protein or more organized stages of myogenesis if the numbers of satellite cells are normal.

The first two hypotheses of Mauro both imply formation of myoblasts capable of mitosis and production of further muscle cells, and differ only in the origin of these myoblasts. There is now considerable evidence that the satellite cell is capable of mitosis but not the myonucleus. By light-microscopic autoradiography, MacConnachie *et al.* (1964) showed that only cells at the edge of the muscle fiber replicated DNA. Autoradiographic study of DNA synthesis after injury has been performed at the electron-microscope level by Reznik (1969*a*). He showed that satellite cells after cold injury were labeled in 2 to 6 hours and that many nuclei in myotubes were labeled in 12 to 24 hours, which implied origin of the latter from satellite cells. Direct observation of mitoses was made by MacConnachie *et al.* (1964) at the edge of the muscle fiber by light microscopy after colchicine metaphase arrest. Shafiq *et al.* (1968) and we have also directly observed mitoses in thin sections and electron micrographs. Indirect evidence of mitosis in satellite cells is suggested by presence of centrioles and close apposition of 2 satellite cells which might represent daughter cells following a mitosis.

The large number of free myoblasts rapidly produced following injury can result from surviving myonuclei or by division of satellite cells, provided that they do not fuse into myotubes (Sloper and Pegrum, 1967) and do not begin to produce myosin (Okazaki and Holtzer, 1965). Reznik (1969*b*) has suggested that the surviving cells at the periphery of necrotic fibers are myonuclei and not satellite cells as suggested by Church *et al.* (1966), and others, and in human lesions by Shafiq *et al.* (1967). It would seem to us more logical that they are a different cell type *or* a cell at a different stage of the cell cycle and perhaps less vulnerable to various forms of injury than the ordinary myonucleus. From the usual appearance of the 'resting' satellite cell this may be in a G_0 stage by analogy with other cell types.

In myogenesis the sequence myoblast, myotube, muscle fiber, occurs without the 'scaffolding' of an intact basement membrane of the damaged muscle fiber which is usually present in regenerating muscle. Although the 2 processes have been briefly discussed in this communication and in others (Ishikawa, 1966), they may not be directly comparable. In the study of developing muscle it is difficult to prove the existence of single cells with myofibrils because serial sections can demonstrate their connection with other fibers, but their single-cell nature cannot be proven. Other problems already stated concern the necessity of a basement membrane around the muscle fiber because this presently is an important criterion for the definition of a satellite cell, and the difficulty in early stages of myogenesis in distinguishing fibroblast-like cells and satellite cells intimately connected with myoblasts (Ishikawa, 1966). Problems in the clear demonstration of plasma membranes so that myofilaments can be shown to be outside satellite cells can be facilitated if sections are tilted in an electron microscope fitted with a goniometer stage.

Electron-microscopic studies of rhabdomyosarcomas have shown well-formed myofilaments with definite Z-bands in the well-differentiated tumors and poorly formed myofilaments with irregularly oriented primitive Z-bands found in the embryonic type. Immunofluorescence studies for myosin have been positive (Freeman and Johnson, 1968). Cells with little cytoplasm are also present in these rhabdomyosarcomas, as in the case we have described, and may represent satellite cells.

Summary

There have been many exciting investigations in muscle regeneration and these will continue, using increasingly refined technics.

THE USE OF LIGHT MICROSCOPY OF EPOXY SECTIONS IN MUSCLE HISTOPATHOLOGY*

PATRICK E. CONEN and C. D. BELL

We will briefly present our experience with light microscopy of Epon-embedded tissue because of many questions asked during this meeting.

A variety of techniques, such as electron microscopy, histochemistry, enzyme histochemistry, autoradiography, polarization, phase contrast, and fluorescence microscopy, have been helpful in the study of muscle histology, and histopathology. This communication describes the usefulness of light microscopy of 0.5- to 1-micron sections of plastic-embedded material because: (*a*) a considerably larger area can be studied more quickly than by electron microscopy, and in more detail than in sections of paraffin-embedded muscle; (*b*) areas of special interest can be directly compared with electron micrographs of adjacent serial sections; (*c*) several 'special stains' can now be used on plastic-embedded sections; and (*d*) more precise identification and localization of silver grains can be made on autoradiographs of thin sections.

MATERIALS AND METHODS

The materials were the same as those used in the satellite cell study. We will refer to several methods of staining, many of them used by us on various tissues, but on muscle we have only used toluidine blue (Bencosme *et al.*, 1959; Trump *et al.*, 1961; Lynn, 1965), PAS (Cardino and Steiner, 1965), PASM (Cardino and Steiner, 1965), or paragon (Martin *et al.*, 1966), and phase and interference contrast or polarized-light microscopy.

OBSERVATIONS

The thin sections allow study of considerable detail including many organelles and are especially useful for serial sections through an entire cell. The following examples illustrate the detail attainable (Fig. 187) and use of special stains (Figs. 188 and 189) in addition to the examples illustrated in the satellite cell study.

Figure 185 is a longitudinal section of normal muscle stained with toluidine blue (TB) for comparison with Figure 186 which is a low-power electron micrograph of the same block. By oil immersion it is easy to see the Z-lines, A- and I-bands and the mitochondria. Figure 187 shows an artery with surrounding mononuclear cells including a plasma cell and mast cell from a patient with polymyositis. Figure 188 shows a transverse section of muscle from a patient with glycogen storage disease, stained with periodic acid-Schiff-TB, demonstrating the localized areas of glycogen, some of which are metachromatic. Figure 189 shows a paragon stain of longitudinally oriented fiber from a patient with nemaline myopathy

* Supported by the Muscular Dystrophy Association of Canada.

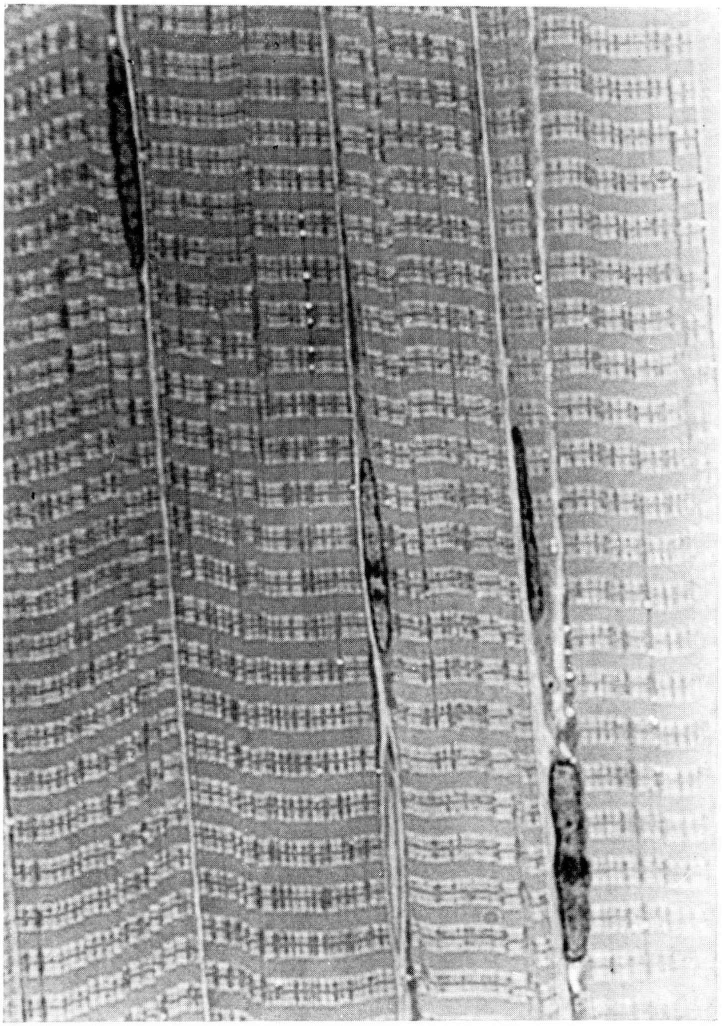

Fig. 185. Longitudinal section of 'normal' human muscle. Toluidine blue; × 1280.

demonstrating the rods which stain red like Z-bands in contrast to the blue colour of the mitochondria and A-bands.

DISCUSSION

Electron microscopy of muscle provides maximal morphologic detail, but the small area that can be studied and the time involved is a considerable disadvantage. Detailed study by oil-immersion microscopy of 1-micron Epon-embedded sections after suitable staining or examination with phase contrast, interference contrast or polarized-light microscopy can help bridge the gap between the detail available in paraffin sections and electron micrographs.

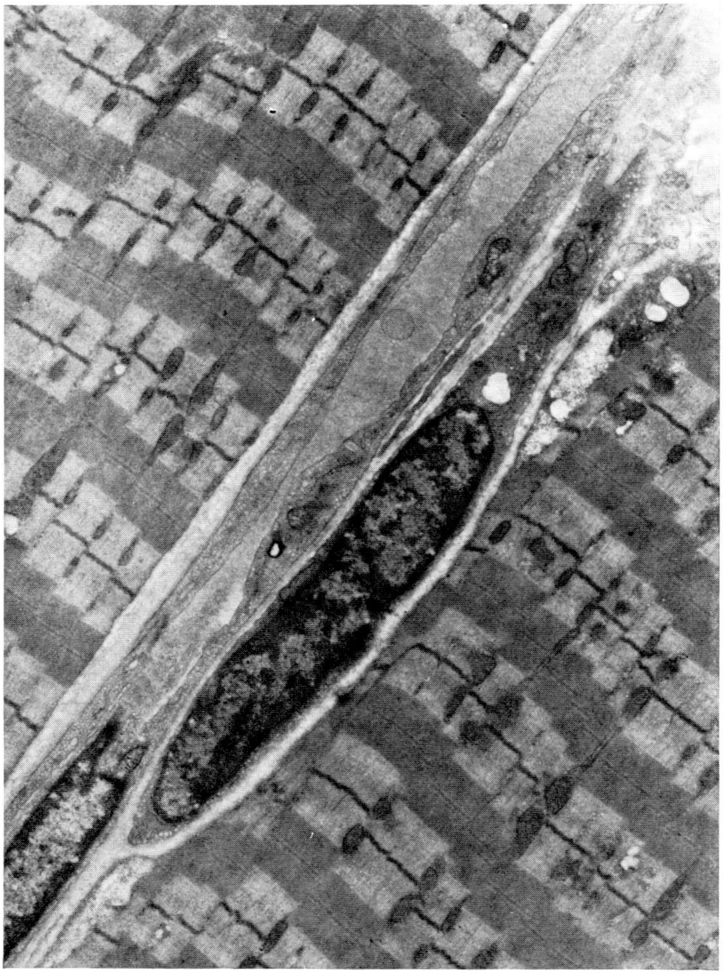

Fig. 186. Electron micrograph of similar area for comparison with Fig. 185. × 4800.

With experience, the number of blocks studied by electron microscopy is reduced and additional time is spent on careful examinations of these thin sections by oil-immersion light microscopy. Examinations of such sections are likely to be even more useful in future because many new staining procedures are being developed and standard histologic stains adapted for Epon-embedded material.

Some of the stains and methods available for light microscopy of epoxy-embedded tissue are: toluidine blue (TB) (Bencosme *et al.*, 1959; Trump *et al.*, 1961; Lynn, 1965), periodic acid-Schiff (PAS) with or without TB counterstain (Cardino and Steiner, 1965), periodic acid-silver methenamine (PASM) (Cardino and Steiner, 1965), paragon (PG) (Martin *et al.*, 1966), Perl's method for iron (Tanaka and Berschauer, 1969), iron hematoxylin and safranin O (Schantz and Schecter, 1965), hematoxylin-eosin (Pool, 1969), Congo-red (Shirahama and Cohen, 1966); after removal of resin from the section, hematoxylin and eosin (H and E) and phosphotungstic acid-hematoxylin (PTAH) (Lane and Europa, 1965),

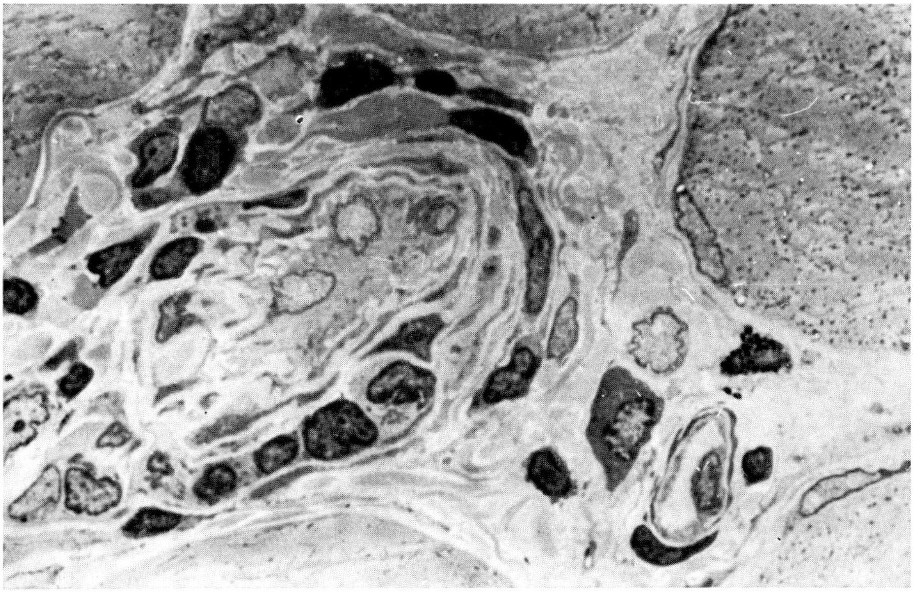

Fig. 187. Section through an artery (left center) to show the endothelium, smooth muscle cells and surrounding mononuclear cells and collagen. A plasma cell and mast cell are on the right of the Figure. Polymyositis. Toluidine blue; × 1300.

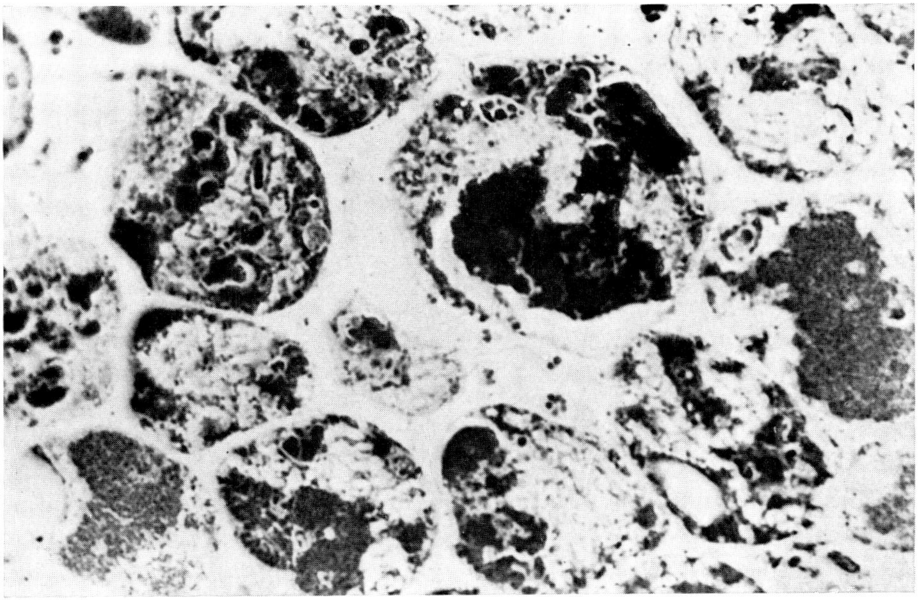

Fig. 188. Transverse section of muscle from glycogen storage disease to show dense PAS-positive areas (some are metachromatic) representing pools of glycogen. PAS; × 1300.

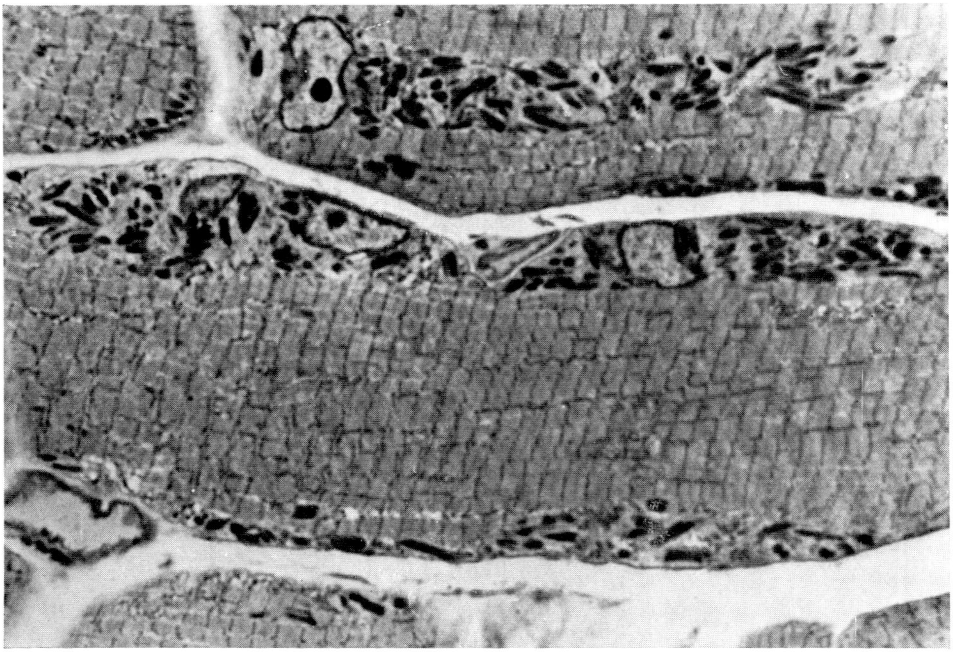

Fig. 189. Semilongitudinal section to show 'rods' in nemaline myopathy. They stain red, the same as Z-bands. Paragon; × 1200.

and selective stains for nervous tissue (Berkowitz *et al.*, 1968) (iron alum hematoxylin, quaternary basic stain, reduced silver stain, modified Giemsa stain). If stained sections are mounted in Epon, resolution is improved and because they fade very little, the sections are permanent.

Thin sections have been prepared from large blocks of epoxy-embedded kidney (Grimley *et al.*, 1965), but this technique has not so far been reported for muscle. Examination of 1-micron sections has an additional advantage by providing considerably greater light-microscopic detail to investigators who do not have ready access to an electron microscope, so that blocks containing areas of greatest interest can be selected and sent to other laboratories for electron microscopy. Probably enzyme histochemical preparations from 50-micron sections of frozen muscle can be embedded in Epon and examined in 1-micron sections comparable with successful methods for acid phosphatase and cholinesterase used on other tissues. Autoradiography using tritiated thymidine has, at this laboratory, successfully been used *in vivo,* and for *in vitro* incubation for several tissues, and would also provide greater resolution to determine the nature of the labeled cells, if applied to muscle.

M. REZNIK: It is very difficult to make a diagnosis of chronic myositis without inflammation in the tissue.

R. K. GILBERT: The study of human muscle is very difficult. I spent a year of my life trying to make diagnoses on the basis of muscle biopsies and I gave up in the end and became interested in the change in the muscle, which became more interesting to me than trying to make the diagnosis.

J. H. VENABLE: Did you do CPK estimations on heart muscle or heart blood?

P. E. CONEN: Heart blood. Although, we had problems getting enough blood, and therefore we also tried placental blood. As controls, blood samples were taken from many other therapeutic abortuses from families without muscle disease. I should point out that the major part of this work was done by the Genetics Department of our Hospital.

A. MAURO: Thank you, Dr. Conen. I think we shall now get on with myogenesis and tissue culture, Dr. Bischoff's presentation.

THE MYOGENIC STEM CELL IN DEVELOPMENT OF SKELETAL MUSCLE*

RICHARD BISCHOFF

INTRODUCTION

The formation of skeletal muscle fibers in development and regeneration is accompanied by proliferation of mononucleated cells. Since myonuclei themselves are postmitotic and do not synthesize DNA nor divide (Lash *et al.*, 1957; Strehler *et al.*, 1963; Shafiq *et al.*, 1968), the rate and quantity of muscle formation depend upon the mitotic activity of the mononucleated myogenic cells that fuse to form muscle fibers. On the basis of labeling experiments, myogenic cells may be divided into 2 classes with respect to their proliferative behavior prior to fusion (Bischoff and Holtzer, 1969): (1) presumptive myoblasts (stem cells) are cells in the division cycle that cannot themselves fuse or make contractile protein filaments but after one or more divisions may give rise to (2) postmitotic myoblasts capable of fusing to form myotubes. The cell cycle separating the presumptive myoblast from the postmitotic myoblast has been termed a 'quantal' division (Holtzer and Bischoff, 1970; Holtzer *et al.*, 1969) in that the behavior of the daughter cells is markedly divergent from that of the parent cell.

Our analysis of this quantal division during myogenesis has been directed towards 4 areas: (1) the temporal relation between the cell cycle phases of presumptive myoblasts and fusion of their progeny; (2) the obligatory requirement for quantal divisions during muscle development; (3) the mode of stem cell propagation during tissue maturation; and (4) the effect of the thymidine analog, 5-bromodeoxyuridine, on the proliferation of stem cells and their transformation into fusible myoblasts.

RESULTS

Muscle differentiation has been studied *in vitro* using aliquots of mononucleated cells obtained by trypsinizing 10-day chick embryo breast muscle. The cells are grown on glass cover slips in Eagle's minimal essential medium with 10 per cent horse serum and 10 per cent embryo extract. In most experiments the cell inoculum was adjusted to give an initial density of 7.5×10^4 cells per square centimeter. Details of the culture technique are given in Bischoff and Holtzer (1968, 1969).

The cells attach to the substrate within 2 hours. Mitotic figures are present throughout the life of the cultures, although the mitotic index is much higher in early cultures (Okazaki and Holtzer, 1966). Multinucleated myotubes first appear at the beginning of the second day, several days before the cells become confluent. The earliest myotubes are highly branched and often have large aggregates of nuclei at the points of bifurcation (Fig. 190). Later, as the myotubes elongate, the nuclei become more evenly distributed.

* This work was supported by grants to Dr. Howard Holtzer from the National Institute of Child Health and Human Development (HD00189, 5T01 HD00030-03) and the National Science Foundation (GB5047X).

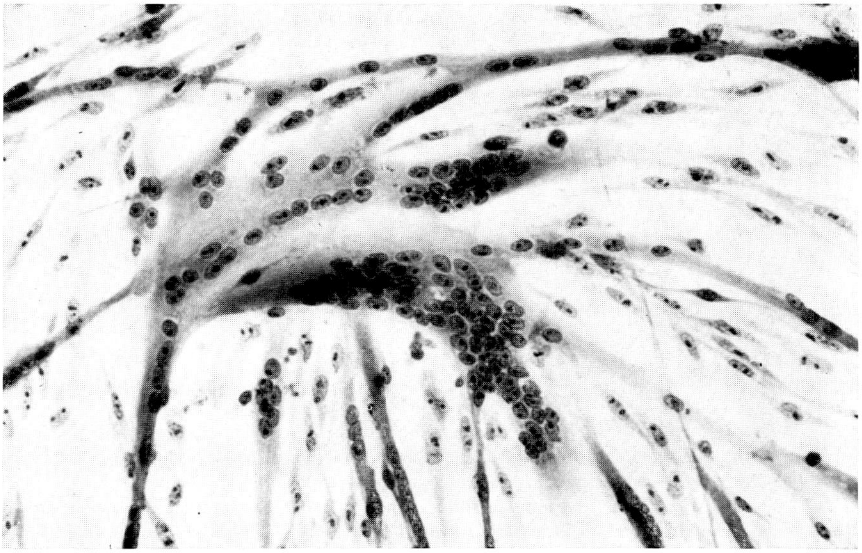

Fig. 190. A 48-hour culture showing accumulation of myonuclei at the point of confluence of several immature myotubes. × 280.

Contractile protein filaments are not present in the mononucleated cells but first appear in the earliest myotubes (Ishikawa *et al.*, 1968). By 3 or 4 days, when spontaneous contraction begins, most of the myotubes are packed with myofilaments. Accumulation of contractile proteins continues as the myotubes enlarge.

Cell cycle and fusion

The cell cycle was analyzed by constructing a percentage labeled mitosis curve (Quastler and Sherman, 1959) following a pulse label of ^{3}H-thymidine (^{3}H-TdR) in 2-day cultures. Twenty-two per cent of the mononucleated cells were labeled after a 30-minute exposure to the label. The duration of the various cell cycle phases, the DNA synthetic period (S), presynthetic period (G_1) and postsynthetic period (G_2) was measured directly from the graph of labeled metaphases. The duration of mitosis (M) was determined by observation of living cells (Fig. 191). The duration of the cell cycle phases *in vitro* is similar to that obtained by Marchok and Herrmann (1967) from 9-day leg muscle *in ovo*. Apparently the timing of the growth-division cycle in these cells is an intrinsic property well buffered against environmental changes.

Since only mononucleated cells incorporate ^{3}H-TdR following a brief exposure, the appearance of labeled nuclei in myotubes during subsequent growth in the presence of cold thymidine is evidence that fusion of the labeled cells has occurred. Accordingly, we examined 2-day-old cultures to determine the *earliest* time after labeling at which labeled nuclei could be detected within myotubes. Since mononucleated cells often lie above or beneath myotubes, careful examination is required to identify labeled myonuclei. This problem is considered in detail elsewhere (Bischoff and Holtzer, 1969). A scattering of labeled nuclei was found in the myotubes at 8 to 10 hours after labeling. The number of labeled myonuclei increased as the labeled stem cells progressed through the division cycle, and, by the end of the experiment (22 hours), labeled nuclei were abundant in the myotubes (Fig. 192).

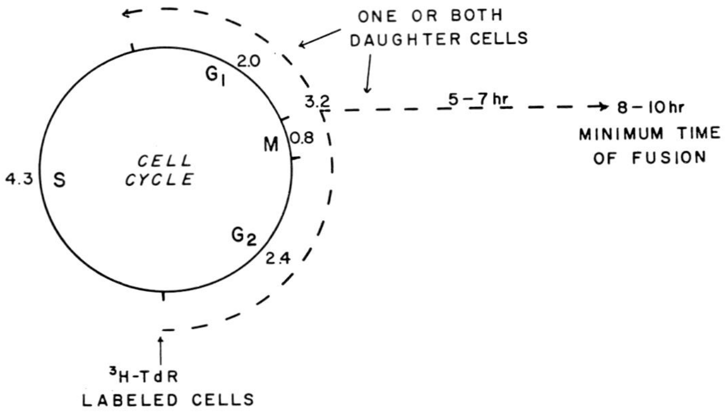

Fig. 191. Diagram showing the relation between the timing of the generation cycle phases (times given in hours) and fusion of myoblasts in 2-day cultures. The duration of mitosis, from nuclear membrane breakdown to reconstruction, was measured by observation of living cells. The duration of S, G_2 and G_1 was obtained by the labeled mitosis method. Cultures were examined for fusion of labeled nuclei at 2-hour intervals following a 30-minute pulse of ^{3}H-TdR. Mononucleated cells that incorporate the label at the end of S (dashed line) will complete mitosis and enter G_1 about 3.2 hours later. Since labeled nuclei are not found in myotubes earlier than 8 to 10 hours after labeling, those cells destined to fuse spend a minimum of 5 to 7 hours in the diploid state (G_1) before fusing. In contrast, cells destined to divide again enter DNA synthesis 2 hours after completing mitosis.

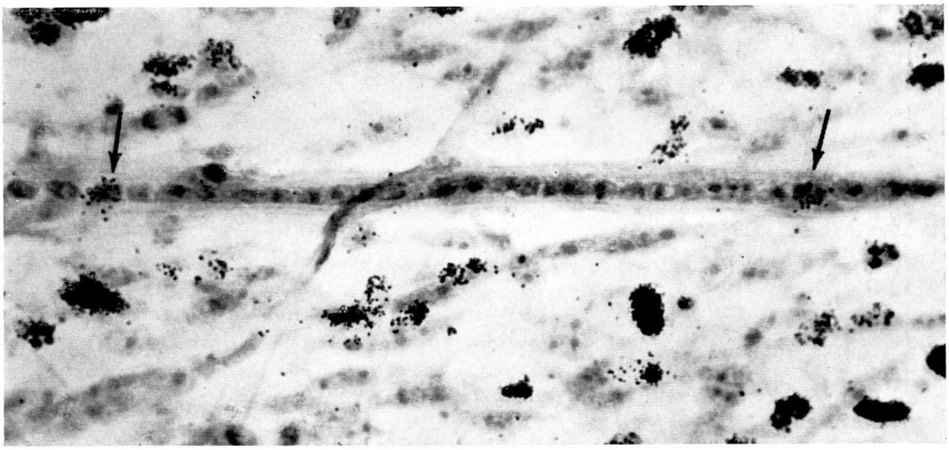

Fig. 192. Radioautograph of 2-day culture sacrificed 16 hours after a 30-minute pulse of ^{3}H-TdR. Most myotubes contain at least several labeled nuclei (arrows) by this time. × 500.

When the minimum interval between DNA synthesis and fusion is compared with the cell cycle timing, the following picture emerges (Fig. 191). Replicating presumptive myoblasts do not fuse during S, G_2 or M. Following mitosis, one or both daughter cells withdraw from the division cycle and acquire the capacity to fuse after a lag period of from 5 to 7 hours in the diploid (G_1) phase. That withdrawal from the cell cycle precedes fusion is shown by the fact that those cells which continue to proliferate re-enter S only 2 hours after emerging from mitosis. Thus, the first recognizable event heralding the transformation of presumptive myoblast to myoblast is the failure to synthesize DNA on schedule.

It should be emphasized that the 5- to 7-hour period between mitosis and fusion is a minimum value and may be longer depending upon contact between competent myoblasts. Mononucleated cells in mitosis adherent to myotubes are common in the cultures. Such cells would have opportunity to fuse immediately upon completing mitosis. That they do not suggests that fusion does not proceed by passive engulfment of mononucleated cells by myotubes, but requires at least several hours after mitosis to develop appropriate surface properties in the mononucleated cell.

One can predict from the 5- to 7-hour delay between mitosis and fusion that the early cultures contain a population of postmitotic mononucleated cells. Indeed, the existence of such cells is reflected in the discrepancy between the expected and observed labeling index. On the basis of calculated cycle times, the labeling index should be 45 per cent if all mononucleated cells are dividing*. The actual labeling index is only 22 per cent, indicating that some mononucleated cells included in the total counts were not in the division cycle. Such non-replicating cells need not all be myoblasts in preparation for fusion. Marchok and Herrmann (1967) found a similar disagreement between the expected (56 per cent) and observed (20 per cent) labeling index in 9-day chick embryo leg muscle *in vivo*.

Fusion of labeled myoblasts has been studied in other systems, but cell cycle times were not measured. Reznik (1968*b*) reported a minimum interval of 6 hours between ^{3}H-TdR injection and fusion of labeled cells in regenerating mouse muscle. In lizard muscle developing *in vitro*, myogenic cells fuse within 12 hours after incorporation of ^{3}H-TdR (Cox, 1968).

Obligatory requirement for quantal divisions

A central question arising from the quantal division hypothesis is whether a cell *must* pass through a division cycle in order to initiate totally new biosynthetic activities. A critical test of this point would be to hold myogenic stem cells in G_1 under environmental conditions favorable for fusion and then challenge them to form myotubes in the absence of further division. Since the majority of myogenic cells in the initial inoculum divide at least once before fusing (Bischoff and Holtzer, 1969), we asked whether these stem cells can fuse directly from G_1 or whether an additional round of the cell cycle is required for their transformation into fusible myoblasts. To hold the stem cells in G_1 we utilized the phenomenon of density-dependent inhibition of division. As cultured cells reach confluence, there is often a decrease in the proportion of cells in the division cycle (Todaro *et al.*, 1965; Stoker and Rubin, 1967). Under these conditions cells withdraw from the cell cycle in the G_1 or diploid phase (Nilausen and Green, 1965).

Accordingly, primary cells were plated at high and low density (45 *vs* 7.5 $\times$ 10^4 cells per square centimeter) and incubated for 20 hours. This should be sufficient time for 1 or 2 divisions (generation time = 9.5 hours) to occur in the sparse cultures. To minimize differences in nutrient depletion and maintain identical conditions of culture, the cells were allowed to settle on coverslips in small dishes for 2 hours, then both groups of cells were transferred to large dishes so that the sparse and dense cultures shared the same medium. At 20 hours, before fusion had begun, both series were dissociated separately with trypsin and replated at high density to test for the proportion of fusible myoblasts. We found a striking difference in the number of myotubes formed by the 2 populations of cells during the subsequent culture period (Fig. 193). Cells from the initially sparse cultures formed a rich network of myotubes, while cells from the dense cultures formed only scanty attenuated myotubes. To rule out the possibility that myogenic stem cells were eliminated in the dense

* Labeling index $= \frac{\text{cells in S}}{\text{total cells}} = \frac{\text{time in S}}{\text{cycle time}} = \frac{4.3}{9.5} = .45$

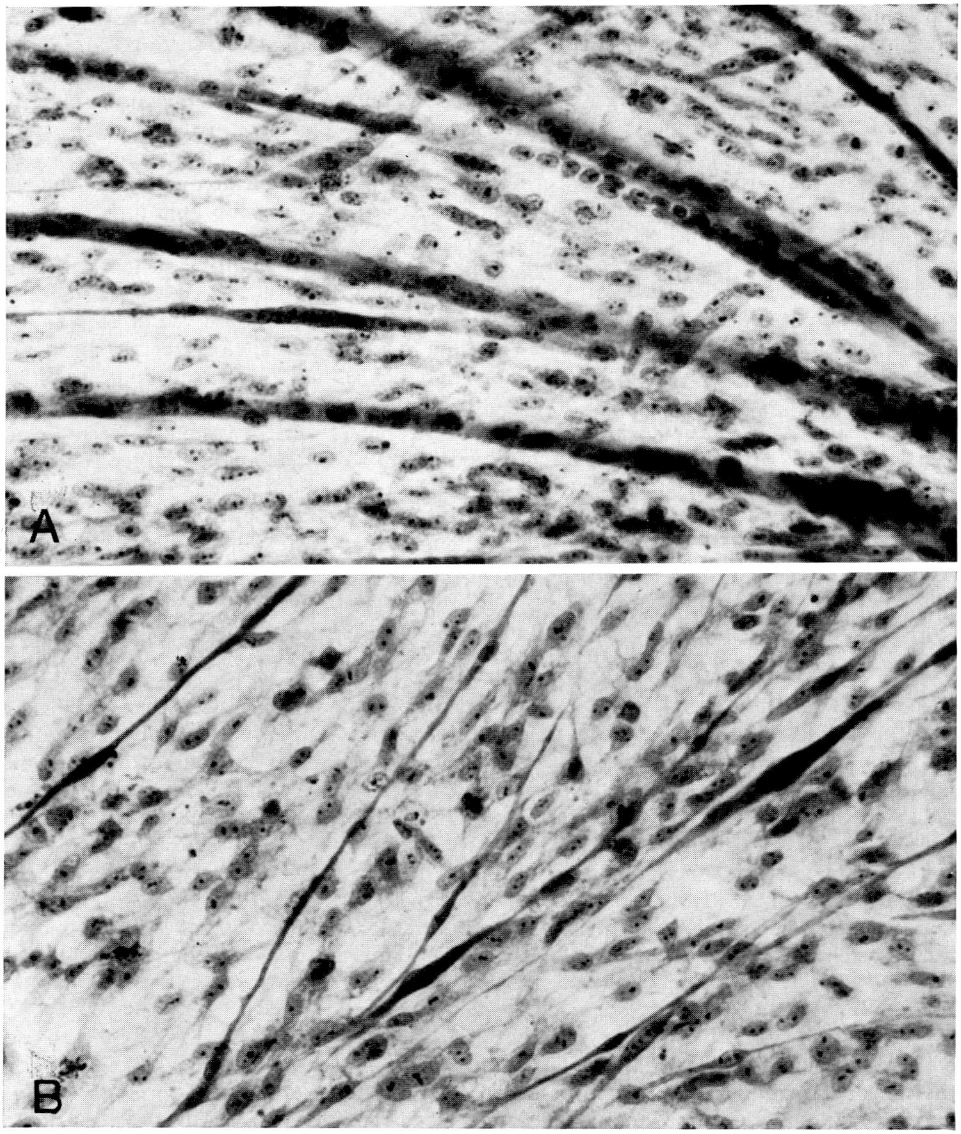

Fig. 193. Effect of cell density of myotube formation. Fresh cells were cultured for 20 hours at high (45×10^4 cells per square centimeter) or low (7.5×10^4 cells per square centimeter) density, then subcultured, and both series were replated at high density. The cells shown here have been in culture for a total of 3 days. $\times$ 500. A: Large myotubes formed by cells cultured at low density for the first 20 hours. B: Attenuated myotubes containing few nuclei formed by cells cultured at high density.

cultures, cells were grown at high density for 20 hours, then replated at low density. Such cultures formed abundant myotubes, indicating that the stem cells persisted in the dense cultures but were prevented from forming myotubes. Since the sparse and dense cells were exposed to the same medium, and crowding, of itself, does not prevent fusion, the only apparent difference between the 2 series is the greater opportunity for division in the sparse cultures.

To determine the difference in replication between the sparse and dense cultures, we measured the collection rate of proliferating cells at 2 points in the cell cycle, DNA synthesis and mitosis. Cultures were set up at high and low density and treated with either ^{3}H-TdR or Colcemid for 4-hour intervals over a 20-hour period. The results (Fig. 194) show a

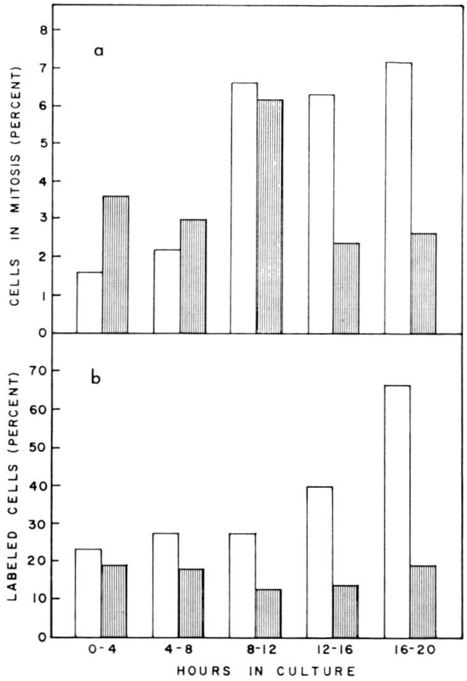

Fig. 194. Effect of cell density on proliferation. Cells were grown at high (filled columns) or low (open columns) density for a 20-hour period. Colcemid (10^{-6} M) or ^{3}H-TdR (0.25 microcurie per milliliter) was added to triplicate cultures for successive 4-hour intervals and the percentage of cells in mitosis and DNA synthesis was determined. 1000 cells were counted in each dish.

substantial reduction in the frequency of labeled cells and cells in mitosis in the confluent cultures compared with the sparse cultures. The initial depression of cell proliferation in both culture situations probably reflects cell damage sustained during tissue disaggregation. Measurement of transit at 2 phases of the cell cycle permits an estimate of the point at which arrest occurs. A significant difference in the proportion of cells in DNA synthesis is evident sooner than is a difference in the proportion of cells in mitosis. Such behavior suggests that cells are able to complete the division cycle in effect upon removal from the embryo, but are subsequently blocked in G_1 in the crowded cultures. This conclusion is also supported by cell cycle analysis of fresh pulse-labeled cells added to pre-existing 4-day cultures (Bischoff and Holtzer, 1969).

In summary, these experiments indicate that myogenic stem cells are incapable of fusing during all stages of their division cycle. Environmental conditions that promote fusion are ineffective when acting upon stem cells held in the diploid state (G_1), the stage from which myoblast fusion normally occurs. One or more rounds of the cell cycle (quantal divisions) appear to be essential for the nuclear-cytoplasmic reprogramming that permits the transformation from stem cell to fusible myoblast.

Propagation of stem cells

Developing muscle *in vivo* and *in vitro* represents a self-limiting system. As the tissue matures, cell proliferation subsides and the number of myofibers becomes constant (Leblond, 1964; Chiakulas and Pauly, 1965; Marchok and Herrmann, 1967; Bischoff and Holtzer, 1969). We have sought to explain this process in terms of the balance between proliferative divisions, which perpetuate stem cells, and quantal divisions, which give rise to postmitotic myoblasts, thereby removing stem cells from the population (Bischoff and Holtzer, 1969; Holtzer and Bischoff, 1970). Regulation of this balance, as development proceeds, may involve attributes of the differentiated myotubes and surrounding extracellular matrix acting during the course of stem cell divisions.

We first attempted to estimate the proportion of myogenic stem cells remaining in the tissue at various stages of maturation. It was found that trypsin subcultures prepared from successively older primary cultures showed a decreasing tendency to form myotubes in the second generation cultures (Bischoff and Holtzer, 1969). Subcultures prepared at 5 or 6 days, for example, formed virtually no myotubes. Since the primary cultures at this stage are rich in myotubes, these observations argue against the likelihood that myonuclei can reconstitute mononucleated myogenic cells, at least after trypsin injury. Whether myonuclei can give rise to myogenic cells following other forms of injury (Hay, 1959; Reznik, 1969*b*) remains an open question in our opinion. Definitive evidence on this point would involve selective marking of myonuclei and following their fate through injury and regeneration.

Although stem cells could not be demonstrated in mature tissue using the subculture technique, the possibility existed that a small number of myogenic cells, if present, might not be expressed because of dilution by the non-myogenic cells remaining. To test this, mature primary cultures (4 to 8 days old) were exposed to ^{3}H-TdR for 12 hours. The cultures were then dissociated and aliquots of the labeled cells were added to 2-day cultures containing young myotubes. Some (less than 1 per cent) of the labeled cells were incorporated into the myotubes from the 2-day cultures. Apparently a small population of myogenic stem cells remains in mature tissue. Whether such cells are also present in normal adult tissue has not been determined.

Four-day cultures contain many broad myotubes packed with myofibrils. Many of the fibers show vigorous spontaneous contractions by this stage. When such cultures were pulse-labeled and analyzed for fusion of labeled cells, little or no fusion was found. The generation cycle of dividing cells in 4-day cultures is essentially the same as that in the 2-day material (Bischoff and Holtzer, 1969). Since some dividing myogenic cells are still present at this period, it seemed possible that something was interfering with the capacity of these cells to fuse *in situ*. To examine this point, we tested the ability of mature myotubes to incorporate fresh mononucleated cells. One-day cultures, prior to fusion, were exposed to ^{3}H-TdR for a time sufficient to label 90 per cent of the cells. These cells were then dissociated and allowed to settle on top of 4- or 5-day cultures. The ratio of fresh cells to pre-existing cells was about 1:10. Radioautographs were prepared at intervals up to 3 days after adding the labeled cells and examined for the presence of labeled nuclei in myotubes. We found

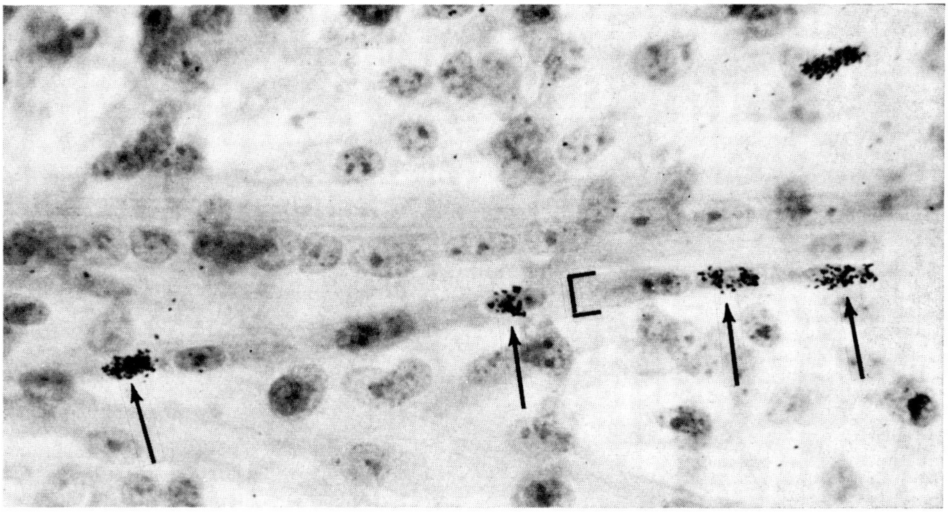

Fig. 195. Radioautograph of 6-day culture which received ^{3}H-TdR-labeled fresh myogenic cells 2 days before sacrifice. Labeled nuclei (arrows) are found within the small myotube (bracket indicates diameter) but are prevented from fusing with the adjacent large myotube. × 800.

many instances of fusion in small attenuated myotubes, but the broad, well-developed myotubes failed to incorporate the labeled cells (Fig. 195). Labeled cells appeared to penetrate the tissue mat completely and occasionally adhered to the large myotubes. Their failure to fuse probably did not result from lack of mobility or failure to make contact. Evidently, a block to fusion develops in myotubes at some point during their maturation. This block may be related to the formation of a basement membrane, which first appears on some of the large myotubes at 4 days (Ishikawa and Bischoff, unpublished). Attempts to induce fusion of mononucleated cells with large myotubes by treating the myotubes with collagenase, however, were unsuccessful. Alternatively, the surface membrane may become refractory to fusion at some stage in development, perhaps when contraction begins.

The block to fusion may operate as a device to limit the growth of existing myofibers and prompt the formation of new ones. Increase in number of muscle fibers seems to proceed by steps or 'generations' (Kelly and Zacks, 1969). If the primary generation of myotubes becomes refractory to fusion at a certain stage of maturation, the remaining myoblasts would be led to seek other contacts and initiate a new generation of myotubes.

Effect of BUdR on myogenesis

The regulatory mechanisms governing the transformation of myogenic stem cells to postmitotic myoblasts can be upset by the thymidine analog, 5-bromodeoxyuridine (BUdR). Myogenic cells grown in BUdR do not fuse and do not synthesize contractile protein filaments (Fig. 196). The cells continue to proliferate, however, and retain the capacity to form muscle when removed from the analog and allowed to divide in normal medium (Stockdale *et al.*, 1964; Okazaki and Holtzer, 1965; Coleman *et al.*, 1969; Bischoff and Holtzer, 1970). Bromouracil-substituted DNA does not impair the mechanism determining the myogenic commitment of these cells. BUdR-suppressed myogenic cells do not revert to an uninstructed state but propagate themselves as presumptive myoblasts capable of expressing their muscle

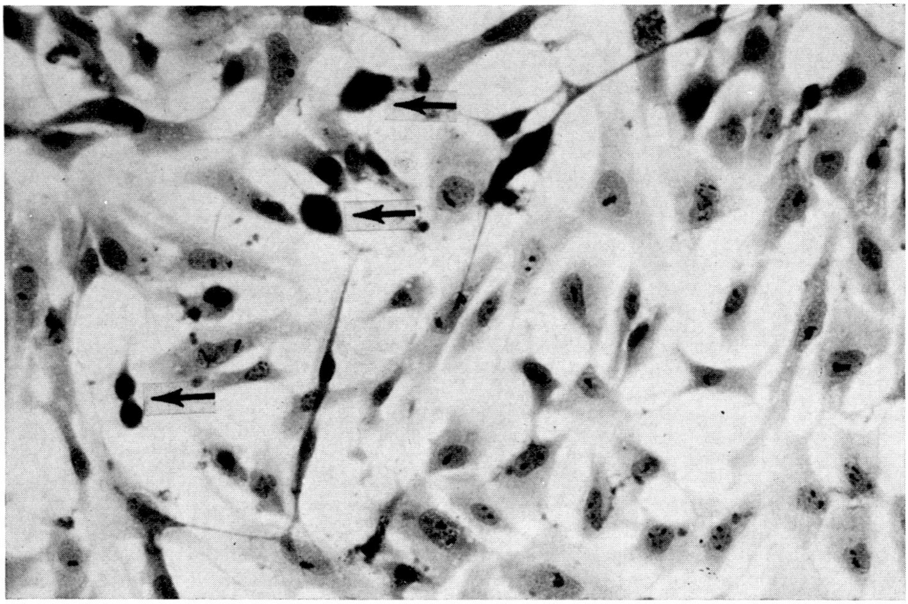

Fig. 196. Fresh myogenic cells grown continuously for 3 days in 50 micrograms per milliliter BUdR. Myotube formation is suppressed and the mononucleated cells become enlarged. The cells continue to proliferate in BUdR and mitotic figures (arrows) are common in the cultures. × 480.

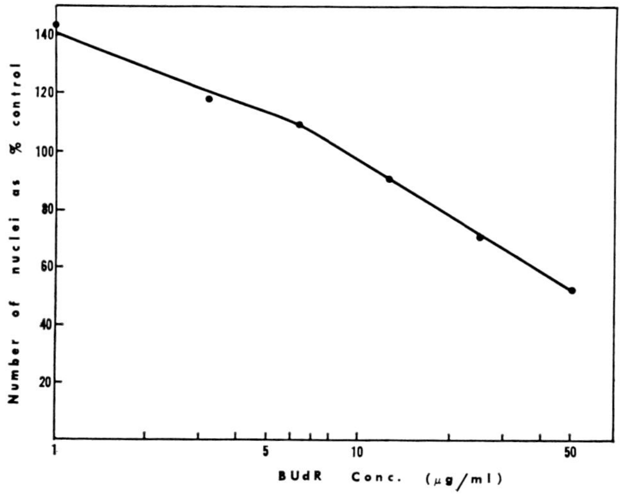

Fig. 197. Effect of BUdR on growth as a function of concentration. Cells were grown for 3 days in various concentrations of BUdR, then the nuclei were isolated with citric acid and counted. The number of nuclei in the BUdR cultures is expressed relative to that found in untreated control cultures. Myotube formation is blocked throughout the range of BUdR concentrations used.

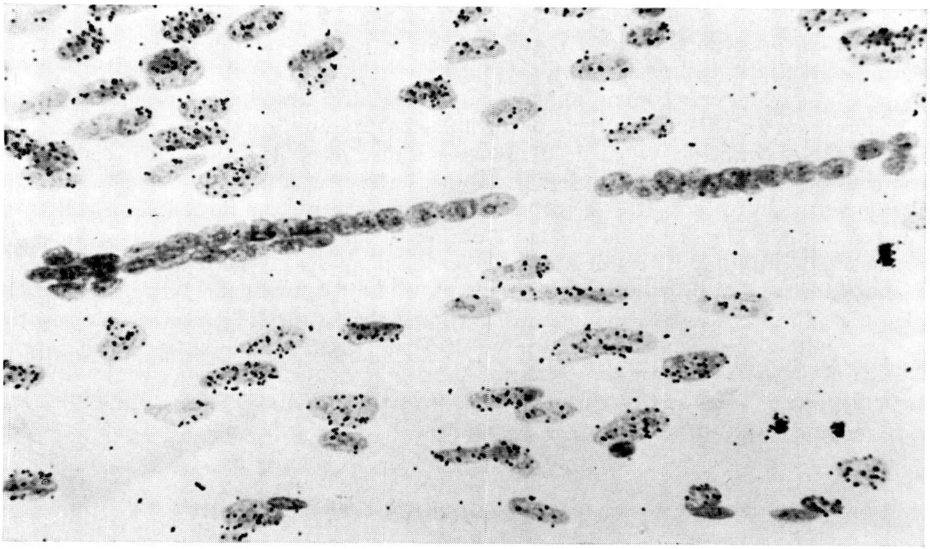

Fig. 198. Inhibition of fusion by BUdR. Two-day cultures were incubated in 50 micrograms per milliliter BUdR, including 0.75 microcurie per milliliter ^{3}H-BUdR, and sacrificed after 48 hours. Cells which have incorporated the ^{3}H-BUdR are prevented from fusing with the myotubes present at the time of labeling. Cytoplasm is not visible in this preparation; however, a myotube is indicated by the aligned nuclei. Feulgen stain; × 500.

phenotype when given the opportunity. In effect, BUdR suppresses quantal divisions but does not interfere with proliferative divisions of the stem cells.

The lag period between removal from BUdR and initiation of fusion in normal medium depends upon the length of time the cells were kept in BUdR. The important element appears to be the number of divisions. The first cells to fuse during recovery, following growth in ^{3}H-BUdR, are those in which the ^{3}H-BUdR has been most highly diluted by division in normal medium (Okazaki and Holtzer, 1965).

Myogenic cells have been kept under BUdR suppression for as long as 25 days and still retain the capacity to recover (Bischoff and Holtzer, 1970). During this time the proportion of myogenic to non-myogenic cells remains essentially constant. There is no overgrowth by fibroblasts. The cumulative production of stem cells in the presence of BUdR is far greater than that found during normal muscle development. These observations suggest that local treatment with BUdR may be of use in stimulating myogenesis in regenerating systems.

Myoblast fusion is blocked throughout a wide range of BUdR concentration. At the lowest effective concentration, 1 microgram per milliliter, the cells are of normal morphology and increase in number more rapidly than do untreated control cells (Fig. 197). With higher concentrations, cells are enlarged and flattened and there is a progressive inhibition of cell production. Preliminary experiments suggest that the growth inhibition by BUdR is not a direct effect of its incorporation into DNA but is probably caused by accumulation of an intracellular pool of the analog which produces a feedback inhibition of nucleotide metabolism.

On the other hand, all available evidence indicates that BUdR acts to block fusion and contractile protein formation via its incorporation into DNA in place of thymidine (Bischoff and Holtzer, 1970; Coleman *et al.*, 1969). We performed 2 types of experiments to determine whether BUdR incorporation for 1 round of DNA synthesis would be effective in blocking fusion (Bischoff and Holtzer, 1970).

Normal 2-day cultures contain immature myotubes and many mononucleated cells that will incorporate ^{3}H-TdR and fuse, following a quantal division, without undergoing another round of DNA synthesis. When ^{3}H-BUdR is used instead of ^{3}H-TdR, the labeled cells are excluded from the myotubes (Fig. 198). In another series of experiments 5-fluorodeoxyuridine (FUdR) was used to block DNA synthesis. Cells were kept continuously in FUdR for 3 days except for a 7-hour period on day 1 during which the block to DNA synthesis was circumvented with either BUdR or BUdR plus excess thymidine. Cells which had synthesized BU-DNA for the 7-hour period showed the typical BUdR suppression, while the BUdR + TdR series formed normal myotubes. Taken together, these experiments indicate that incorporation of bromouracil into the newly synthesized polynucleotide strand of the mother cell prevents transcription of muscle-specific genes in the daughter cells. Although BUdR suppression appears to depend upon incorporation into DNA, the mechanism by which specific genes are inactivated is not known.

DISCUSSION

The generation of functional postmitotic cells from dividing precursor cell populations has been studied extensively in adult renewal systems such as hemopoietic tissue and gut epithelium. The methodology developed for these systems can be profitably applied to the study of growth parameters during embryonic development. The depression of cell production that accompanies functional maturation of most embryonic tissues could result from several conditions acting separately or in concert: (1) lengthening of the cell cycle; (2) decline in the growth fraction, *i.e.* relative number of reproductive cells, as a result of differentiation or death; or (3) conditional withdrawal of some stem cells from the division cycle. Different combinations of these processes probably operate in different tissues.

In skeletal muscle *in vitro* we found no difference in the timing of the generation cycle between newly forming and mature tissue (Bischoff and Holtzer, 1969). Marchok and Herrmann (1967), however, found an increase in the cell cycle duration between 9 and 16 days *in ovo*.

If each stem cell division gives rise to 1 fusible myoblast and 1 stem cell (asymmetric or differential division), then the total number of stem cells would remain constant. The rapid depletion of stem cells and the high rate of fusion between 2 and 4 days in culture, however, suggests that some of the stem cell divisions during this period produce 2 postmitotic myoblasts (quantal division), thereby resulting in the loss of a stem cell. Regulation of the stem cell supply can be explained in terms of a shift in the balance between proliferative and quantal divisions. During the early stages of muscle development most divisions are probably of the proliferative type that propagate myogenic precursor cells and enlarge the stem cell compartment. As the tissue matures, environmental factors act to increase the frequency of quantal divisions, thus reducing the number of stem cells.

The decline in stem cells observed in our cultures is not produced by 'overgrowth' of fibroblasts, since the same effect has been observed in clonal cultures. Trypsin subcultures made from well-developed muscle clones failed to produce significant amounts of muscle in the second-generation cultures (Hauschka, 1968). The nature of the mononucleated cells that do survive in the subcultures is a curious problem since the clone presumably originated from a single *myogenic* cell. Perhaps they represent dedifferentiated or altered stem cells.

Several observations suggest that attributes of the myotubes themselves are instrumental in programming the type of division a stem cell undergoes. The block to fusion that develops in older myotubes may be a signal to dampen further stem cell propagation. A recent report by Yaffe (1969) also underlines the inhibitory effect of myotubes on the maintenance of stem cell supply. Yaffe demonstrated that myogenic cells from newborn rats can be kept in culture for many months without losing the capacity to form muscle, providing the cells are serially subcultured *at the onset of fusion*. Allowing the cells to fuse and form myotubes tended to prevent recovery of stem cells in the subcultures. The behavior of BUdR-suppressed myogenic cells also supports the view that stem cell propagation is not an inherently limited process (Bischoff and Holtzer, 1970). Stem cells continue to produce more stem cells for at least 3 weeks when myotube formation is blocked by BUdR.

Observations on the fate of stem cells in cultured muscle may have some bearing on the source of myogenic cells during regeneration. Although stem cells become greatly diminished in number during myotube formation, they do not disappear entirely. A small number of stem cells persists in the tissue after fusion has ceased. If such reserve stem cells are present in adult muscle, they may be stimulated to undergo repeated proliferative divisions following destruction of muscle fibers by injury. The location in the tissue of these cells is not known; however, satellite cells (Mauro, 1961; Ishikawa, 1966), sometimes in mitosis, have been observed in cultured muscle (Ishikawa and Bischoff, unpublished).

Our major findings on the relation between stem cells and myoblasts are summarized in Figure 199. The progeny of stem cell divisions are subject to 2 fates depending upon en-

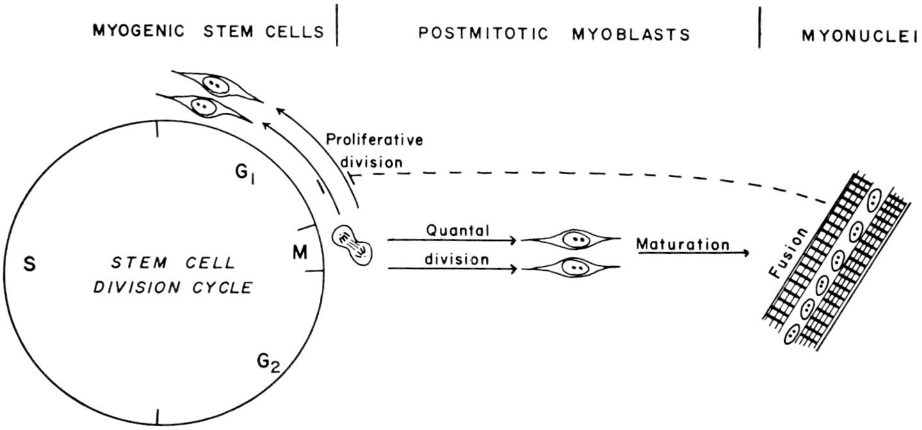

Fig. 199. Proliferative *vs* quantal divisions during myogenesis (see text for explanation).

vironmental conditions. The daughter cells may re-enter the division cycle to constitute another generation of stem cells (proliferative division) or they may withdraw from the mitotic cycle (quantal division) and, for the first time, acquire the capacity to fuse and form myotubes. As the tissue matures, some factor (indicated by the dashed line in Fig. 199), probably arising from the differentiated myotubes or surrounding matrix, tends to suppress proliferative divisions in favor of quantal divisions and results in depletion of the stem cell supply.

J. R. COLEMAN: Can I make a comment here for clarification? You have demonstrated that 8 hours past S or, roughly, 5 hours past the beginning of G_1, the cells become competent to fuse. Have you attempted to estimate at what point they decide to leave the mitotic cycle and become competent to fuse? That transition must be made, I would think, sometime in the 5-hour period between the end of one S-phase and the time at which the next S-phase would ordinarily begin.

R. BISCHOFF: Since the average G_1 is 2 hours, those cells that continue to divide begin DNA synthesis 2 hours after completing mitosis. In contrast, myogenic cells destined to fuse fail to synthesize DNA and remain in a postmitotic-prefusion state for several hours before they become competent to fuse. Apparently, withdrawal from the cell cycle comes before fusion, not as a consequence of fusion.

J. R. COLEMAN: The other interesting question would be what happens to a cell that makes that transition but does not fuse!

R. BISCHOFF: There is some evidence that myogenic cells that withdraw from the cell cycle, but are prevented from fusing, can go ahead and synthesize myofibrils as mononucleated myoblasts. Myotubes become refractory to fusion after 5 or 6 days of culture. A small proportion of highly elongated mononucleated cells capable of contraction appears in these older cultures. These cells may represent postmitotic myoblasts that were unable to fuse. In addition, we find that about 1 per cent of cells grown continuously in BUdR fail to incorporate the analog and such cells can also synthesize contractile proteins in the mononucleated state.

J. R. FLORINI: Do you have any idea of what else the cells might do? So many people here only consider DNA synthesis. Does RNA synthesis not become more active during this period while preparing for fusion?

R. BISCHOFF: We have done radioautography following a 30-minute incorporation of ^{3}H-uridine in 3-day cultures. Large myotubes had fewer grains above them than did the mononucleated cells, but I would interpret this as resulting from absorption of emissions by the thickness of cytoplasm overlying the myotube nuclei. If one compares mononucleated cells with small or flattened myotubes, then there is no apparent difference in grain density. Also, a longer labeling period that permits the RNA to reach the cytoplasm eliminated the difference in grain density between the large myotubes and the mononucleated cells.

J. R. COLEMAN: I would like to disagree with that slightly. Given the difficulties mentioned, we have looked, largely as a consequence of Yaffe's report that multinucleated fibers incorporate less uridine than mononucleated cells, in our cultures with tritiated uridine and we indeed see that label over multinucleated fibers is lower than over the mononucleated cells. But we also, in the early cultures, feel quite secure that the fibroblastic cells are considerably more densely labeled than the bipolar cells.

So really, all the other parameters that complicate this conclusion aside, there are generally fewer grains over the bipolar cells and over multinucleated fibers than there are over the fibroblastic cells.

D. ALLBROOK: Dr. Holtzer, is it also possible to stain myosin through all that thickness and visualize the incorporated isotope?

H. HOLTZER: One can do a double labeling experiment. Cultures are pulsed with ^{3}H-thymidine, fixed at intervals, and then the radioautographs stained with fluorescent anti-myosin. Dr. Okazaki did this in our laboratory and it turns out that we can find a double label in cells 10 hours after they were in the thymidine. They do not show up earlier.

R. BISCHOFF: I would just summarize by saying that the balance between myogenic stem cells and postmitotic myoblast changes as the chicken matures or develops, such as in more mature muscle or older cultures, where stem cells are somehow lost.

Nevertheless, if one examines the older cultures carefully, one can find a few stem cells remaining, which might be analogous to a reserve cell. What these cells are doing, I do not know, but there are some of these.

M. E. KAIGHN: I have done an experiment which bears upon this point, using the clonal technique of Konigsberg. Primary cultures of skeletal muscle cells derived from 12-day chick embryo thigh muscle were subcultured after 1, 2, 3 and 4 days at clonal densities. The fraction of total colonies that formed muscle was scored. I found that, within 2 days, this fraction decreased from about 50 per cent at day 1 to less than 0.5 per cent at days 2 and 3, although the overall cloning efficiency remained constant at about 15 per cent.

THE SOURCE OF MYOGENIC CELLS FOR CULTURE*

JOHN R. COLEMAN

I should like to report on some experiments performed by Dr. Annette Coleman and myself on the source of the cells which undergo myogenesis in culture. The first of these was aimed at the relationship between the cell population in the culture inoculum and that in the limb muscle from which it was derived, and the second asked whether skeletal muscle fibers formed *in vitro* are capable of giving rise to mononucleated myogenic cells in the culture environment.

The culture techniques used in our laboratory have been described in detail previously (Coleman and Coleman, 1968). In general, they are similar to those described by Dr. Bischoff except that the cells are derived from 12-day chicken embryo leg muscle and are plated onto a layer of collagen (Hauschka and Konigsberg, 1966; Coleman *et al.*, 1969). In monolayer culture, as contrasted with clonal culture, the major effect of the collagen seems to be an adhesive one, *i.e.*, the cultures can be maintained for several weeks without the cell layer pulling free from the bottom of the dish. It may also influence the ease with which one can recognize the major cell types present in the culture.

Within several hours after inoculation the cells attach to the substratum and 2 principal types can be identified: highly symmetrical bipolar cells and asymmetrical flat cells. In clonal culture, it is the bipolar cells which initiate myogenic clones, whereas the flat cells generally form more nondescript clones (Konigsberg, 1963; Coleman *et al.*, 1969), are presumed to be mainly fibroblasts, and will be referred to below as 'fibroblastic cells'.

Observation of time-lapse movies in our laboratory has shown that these major cell types are quite stable in morphology, at least during the first several days of culture. Bipolar cells round up during division, but immediately afterward the daughter cells reassume the bipolar shape. The fibroblastic cells behave similarly, *i.e.*, they move as flattened cells, round up for division and flatten again immediately afterward. Furthermore, these cell types can be distinguished in the cell suspension prior to inoculation into culture. Oval, nucleated erythrocytes can also be seen but are ignored in the observations which follow since they do not attach and do not contribute viable cells to the culture. Aside from erythrocytes, all cells are spherical in the inoculum and fall into 2 principal size classes: (1) small cells, about 12 microns in diameter, comprising 60 to 70 per cent of the cell population, and (2) large cells, about 17 microns in diameter, making up 30 to 40 per cent of the total cells present. Time-lapse movies made during the first few hours of culture clearly show that the small cells attach to the substrate and become bipolar, whereas the large cells flatten out and become fibroblastic upon attachment.

Cultures initiated with 6×10^5 cells per 60 millimeter dish form a confluent monolayer within 6 to 7 days of inoculation (Coleman and Coleman, 1968). Massive cell fusion takes place between 50 and 60 hours of inoculation, long before the monolayer is complete. As fusion occurs, a large proportion (60 to 70 per cent) of the nuclei in the culture enter myotubes

* The work reported here was supported by P.H.S. grant number HD-00047 from the National Institute of Child Health and Human Development.

and most of the bipolar cells disappear. Subsequently the myotubes mature into muscle fibers and the cultures are progressively overgrown by fibroblastic cells. By the end of 5 or 6 days *in vitro* the muscle fibers are cross-striated; some have peripheral nuclei, many have centrally located nuclei, and most contract spontaneously. Single cells, mainly fibroblastic, proliferate between the fibers and form a dense meshwork around them.

To what extent does the culture inoculum resemble the cell population of the limb muscle?

The technique of preparation of the cell inoculum is presumed to remove multinucleated muscle fibers by selective filtration through fine mesh nylon, thus enriching the mononucleated cell population in the suspension. Since muscle fiber nuclei are postmitotic, it should be feasible to prelabel the embryo with ^{3}H-thymidine (^{3}H-TdR) and demonstrate enrichment of labeled nuclei in the inoculum as compared with the limb.

Three 12-day chicken embryos were injected intravenously through the chorioallantoic membrane with 10 microcuries per embryo of ^{3}H-TdR (6.7 curies per millimole) and incubated at 37.5°C for 1 hour. One leg from each embryo was then fixed, sectioned and subjected to autoradiographic analysis as described earlier (Coleman *et al.*, 1969). The muscle tissue was dissected from the remaining 3 legs and a cell suspension prepared from it. Part of this suspension was smeared directly onto slides and autoradiography performed. The remainder was used to inoculate cultures. These were fixed at 7 hours, by which time most of the cells have attached, and were analyzed by autoradiography.

Table VIII shows that the 2 recognizable cell types are labeled in similar proportions in the inoculum and after 7 hours of culture, before any cell division has occurred. The small cells accounted for 77 per cent of the cell population in the inoculum, whereas 59 per cent of the cells in the 7-hour cultures were bipolar. This difference may reflect inaccuracies in

TABLE VIII

Nuclear label in different cell types in cell culture inoculum and early cell cultures after a 1-hour labeling period in vivo with ^{3}H-TdR

	Total	Labeled	Labeled (per cent)	Total	Labeled	Labeled (per cent)
	SMALL CELLS			LARGE CELLS		
Culture inoculum	1541	178	11.6 ± 1.2	466	60	12.9 ± 1.2
	BIPOLAR CELLS			FIBROBLASTIC CELLS		
7-Hour culture	964	109	11.3 ± 1.9	837	107	12.8 ± 0.4

cell identification or, more likely, differences in the rates at which the 2 cell types attach firmly to the substrate. In any case, it is clear that attachment is not selective with respect to the stage of the mitotic cycle the cells were in when removed from the limb.

Table IX shows the incidence of labeled nuclei in the leg muscle, the inoculum, and 7-hour cultures irrespective of cell type. The differences are not statistically significant and show only that the cell population used to initiate the cultures was similar to the cell population in the leg muscle. However, if one takes into account the proportion of nuclei in multinucleated fibers in the leg muscle, which is about 13 per cent (Marchok and Herrmann, 1967), the slight enrichment of labeled nuclei in the inoculum and 7-hour cultures is exactly what would be expected (1 to 2 per cent) if the fibers were eliminated.

The impression that multinucleated fibers contribute negligibly to the inoculum is strength-

TABLE IX

Nuclear label in leg muscle, cell culture inoculum and early cell cultures after a 1-hour labeling period in vivo with ^{3}H-TdR

	Nuclei counted (total number)	Labeled nuclei (number)	Labeled (per cent)
Leg muscle	2164	231	10.7 ± 0.6
Culture inoculum	2007	238	11.9 ± 1.1
7-Hour culture	1801	216	12.0 ± 0.7

ened by another observation from our laboratory, namely, that muscle fibers formed *in vitro* do not fragment to any significant extent when trypsinized under conditions similar to those used in processing the leg muscle.

Can multinucleated muscle fibers give rise to cells which are myogenic *in vitro*?

In an earlier publication (Coleman and Coleman, 1968) we reported that the fibroblastic overgrowth of myogenic cultures could be prevented by adding FUdR (10^{-7} M) to the cultures at 2 days. Such cultures consist only of scattered single cells and numerous muscle fibers which mature and remain healthy for as long as we have maintained them – 3 weeks. Will such fibers, if damaged, produce myogenic cells, *i.e.*, will they regenerate? We performed 2 very simple experiments. In the first, we marked 20 areas on the bottom of a 150-millimeter Petri dish containing a 10-day culture which had been exposed to FUdR on days 3 and 4, and maintained in normal medium thereafter. In half of the marked areas, the fibers were cut with a scalpel and the other half were left undamaged as controls. Daily observation for the ensuing 8 days showed occasional fibroblastic cells proliferating both in experimental and control areas, but no myogenic activity was observed and no bipolar cells at all were seen.

In the second experiment, bundles of fibers from similar 10-day cultures were scooped up with forceps, rinsed in saline to remove the few adhering single cells, and placed in several Petri dishes containing complete medium. In half the plates the fibers were minced with scissors, the others serving as controls. Nothing grew in the subsequent 8 days, not even fibroblasts.

The experiments are far from conclusive, but it seems very unlikely that muscle fibers can generate myogenic cells under these culture conditions, even when damaged.

J. H. Venable: In your observations, do you always find that bipolar cells, after mitosis, give rise to bipolar cells?

J. R. Coleman: Well, I would give a qualified 'Yes'. It is sort of a loaded question, I think, but from following numerous cells in time-lapse movies, I would say 'Yes'. But if you ask whether bipolars always divide and give rise to bipolars, I will say 'I do not know'.

I. R. Konigsberg: I think you 'qualify' a little too much! Have you ever seen a bipolar cell divide and not produce bipolar cells?

J. R. Coleman: No, but I am standing up here and you are not! (Laughter). I have never seen a bipolar cell divide and then stretch out into something like a fibroblast.

Unidentified speaker: In these cultures, is collagen used as a substrate?

J. R. Coleman: Yes, yes . . . Everything we do is on collagen.

UNIDENTIFIED SPEAKER: Gentlemen, what prompted you to choose, let us say, 10-day-old chick embryos? . . . Why not use 3-week-old embryos or baby chicks?

J. R. COLEMAN: We have not tried using baby chicks. We have tried a few times with embryos as old as 16 and 18 days and we get a much poorer cell yield from muscle and we get very little myogenesis from the cells that we get.

It looks like it is roughly correlatable with the number of single cells that are left in the myogenic compartment versus the proportion of nuclei that are in multinuclear fibers.

A. MAURO: Dr. Allbrook, are you getting anywhere with the transplant experiments on adult muscles?

D. ALLBROOK: I have used adult muscles in 2 series of experiments:

1. Finely chopped living fibers about 100 to 150 microns long were implanted isotopically. There are about 100 nuclei in each such segment, and according to our calculation some 5 satellite cells per segment. The entire macroscopic muscle site was filled with these fragments. The result after 6 months is: (*a*) mature innervated fibers of normal size; (*b*) fibers of small diameter (Fig. 200); (*c*) collagen and fibroblasts as scar tissue in zones separating groups of muscle fibers; (*d*) fat cells in groups (Fig. 44).

2. Chopped, homogenized and trypsinized material, spun down to contain basement membrane and attached cells (Fig. 43), was grown *in vivo* in diffusion chamber. Fiber formation in the experiments to date has been zero.

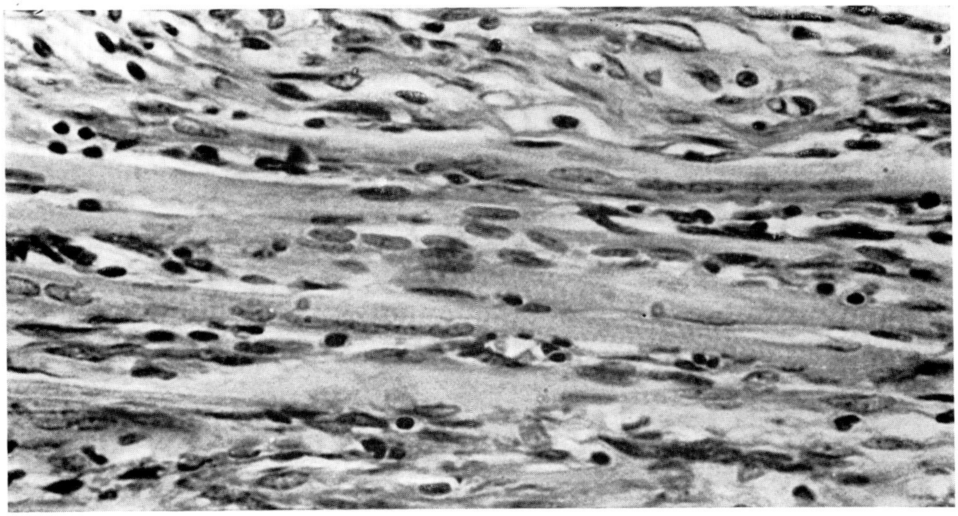

Fig. 200. Six months after implantation of fragments 100 to 150 microns in length, showing many fibers of about 10 microns in diameter, embedded in fibrous tissue.

May I just ask Dr. Coleman, though, something that I do not understand? With the 5-FUdR labeled cells, have I got the significance of this? This stops replication and merely what you are saying is that when you create a situation where there is no mitosis you do not get myogenesis. Is that all?

J. R. COLEMAN: No, essentially what we said is that by giving a 48-hour treatment early in culture after myotubes are formed, we can prevent the mononucleated cell population from growing up and overshadowing what we want to see with the multinucleated fibers.

It does not affect the multinucleated fibers, probably because they are not making DNA and it appears, from looking at older cultures with respect to thymidine kinase, that they cannot phosphorylate 5-FUdR anyway.

H. HOLTZER: I wonder if I could just underline a very important point that was made here.
We tried cutting up myotubes and looking for uptake of ^{3}H-thymidine in the injured cells, the notion being that possibly the nuclei, if they were to come out, could take up thymidine.
After we injured a myotube, added thymidine and sacrificed it within the next 4 to 5 hours, we never found any nucleus in a multinucleated tube taking up thymidine.

I. R. KONIGSBERG: Did you find any cells coming out?

H. HOLTZER: No.

D. ALLBROOK: Would you repeat that, please, Dr. Holtzer? How did you get your myotubes?

H. HOLTZER: We worked on a fibrin clot. We were not plagued by having too many myotubes when we first started this technique. We were very pleased to get a dozen in the whole culture. They stood out by themselves and it was very simple to go in with a scalpel, cut, and then immediately add thymidine, fix after a few hours and look at the injured area.

UNIDENTIFIED SPEAKER: What was the maximum period you waited to get a response?

T. P. STEEN: Maybe the response was slow!

H. HOLTZER: My memory says 6 to 8 hours.

J. R. COLEMAN: Yes, these too, when cut, pull up at their free ends. In this case, fibers that are rounded up and detached float around and generally do not reattach to the substrate.

A. MAURO: Dr. Kaighn, would you like to make any comments on this?

M. E. KAIGHN: We occasionally have seen a few small strips, pieces of fiber with 2 or 3 nuclei. Sometimes they attached and appeared to be viable. However, they usually eventually sloughed off and became necrotic.

I. R. KONIGSBERG: I would just like to make a comment: Five years ago we tried teasing out individual rat fibers, putting them in culture. Assuming that we had done enough damage in teasing, so as to induce degeneration of the fiber . . . we observed the preparation for a week and a half, but the fiber maintained its normal appearance.
Yet, if we took a piece of muscle from the hatched chick, or from a rat or human (I believe this is work that has been done by many people) and put it in a plasma clot, extensive outgrowth was observed. Upon examining the original explant, hollow tubes were seen, indicating that these fibers had degenerated. Thus, it seems that an explant consisting of many fibers behaves very differently from one containing single isolated fibers, so far as regeneration is concerned.

A. MAURO: Now, Dr. Nameroff, would you tell us about your work with myosarcomas?

MYOGENESIS IN A SKELETAL MUSCLE TUMOR*

MARK NAMEROFF

The regulation of cell division is one of the problems which developmental biology shares with pathology. Study of the differentiation of skeletal muscle (Holtzer *et al.*, 1957; Stockdale and Holtzer, 1961; Strehler *et al.*, 1963; Firket, 1958; Okazaki and Holtzer, 1965) has led to the following conclusions regarding cell division in normally developing muscle: (1) muscle precursor cells are mononucleated and may synthesize DNA and divide; (2) nuclei in multinucleated myotubes or muscle fibers do not synthesize DNA and do not divide; (3) muscle-specific macromolecules (*e.g.* myosin) are not found in myogenic cells that are synthesizing DNA or dividing.

One characteristic of tumors is an apparent breakdown or alteration in the regulation of cell division. Are the nuclei in a tumor producing skeletal muscle fibers mitotically suppressed

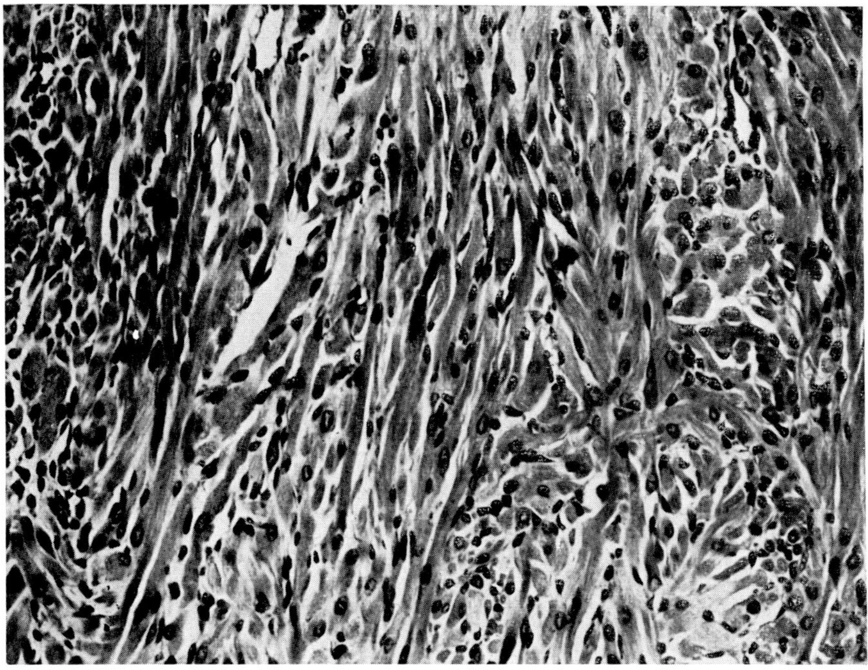
A

Fig. 201, A. Section of 3rd passage tumor. Multinucleated fibers run in many directions. Fibers are approximately 1/3 to 1/2 the diameter of normal muscle fibers. H and E, × 250.

* Supported by a research contract, Project No. 3A-61102B71R-02, from the Medical Research and Development Command, U.S. Army, Washington, D.C., U.S.A.

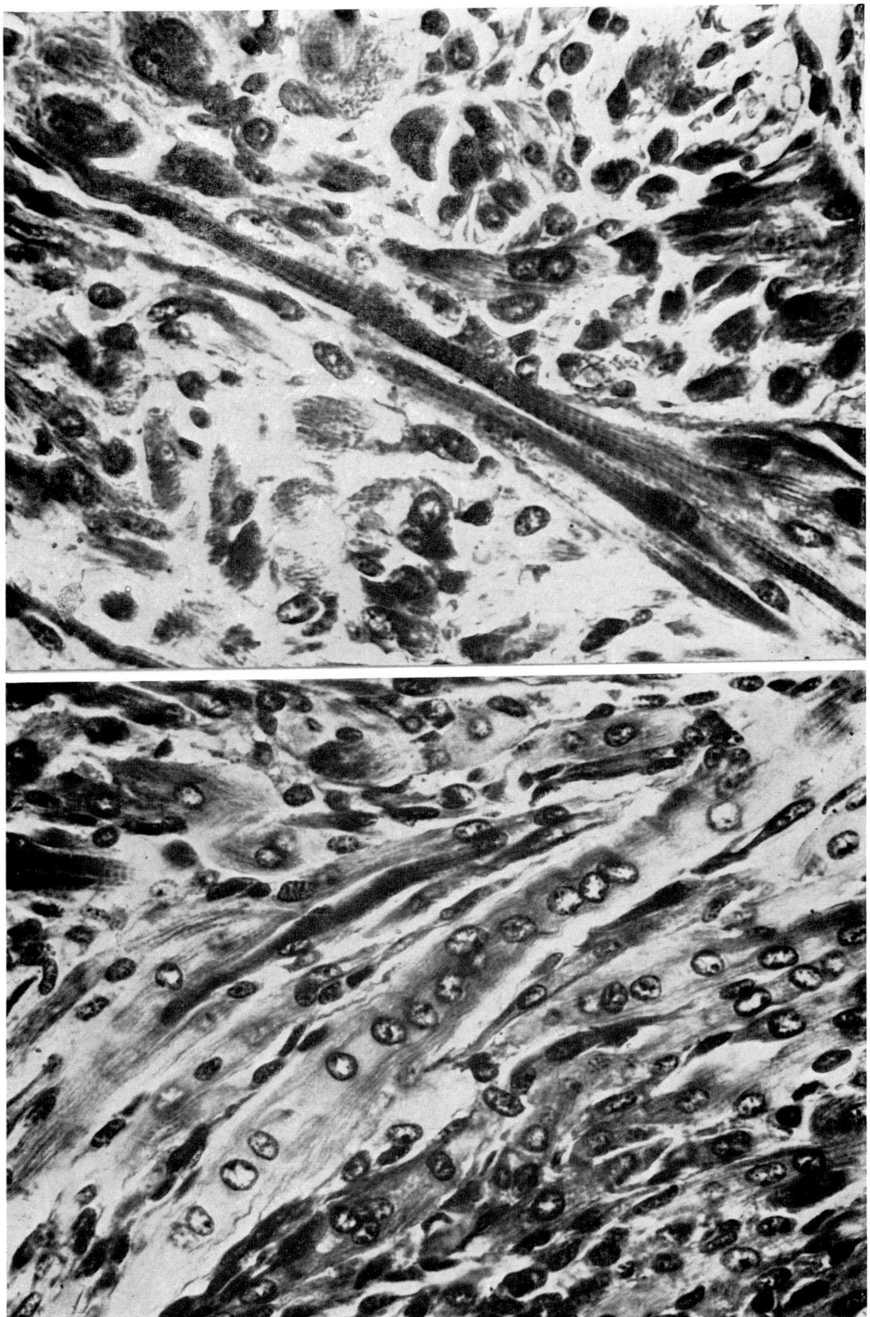

B

C

Fig. 201, B and C. Sections of a 4th passage tumor stained with PTAH. Cross striations are apparent in many cells. × 350.

as are normal muscle nuclei, or is the presumed alteration in mitotic control carried into myotubes so that nuclei therein may continue to synthesize DNA and divide? Studies carried out *in vivo* and *in vitro* on a transplantable skeletal muscle tumor demonstrate that only mononucleated cells divide and that nuclei in multinucleated units are diploid and do not synthesize DNA. Hence, for at least one such muscle tumor, nuclear division in muscle fibers appears to be under the same type of suppressive control that operates in normal muscle fibers (Nameroff *et al.*, 1970).

A tumor consisting of mononucleated cells, multinucleated myotubes, and cross-striated multinucleated muscle fibers occurred spontaneously in the leg of a young female BALB/CAnN mouse. Small pieces of minced tumor tissue were inoculated either subcutaneously or intraperitoneally into female mice of the same strain. In all cases new tumors arose and were histologically indistinguishable from the original (Fig. 201). Transplantation has been carried out through 6 passages with no change in the microscopic appearance of the tumor. No metastases have been observed. In sectioned material, no mitotic figures were observed in multinucleated units. Mitoses were noted only in apparently mononucleated cells. Electron-microscopic observations (Reznik *et al.*, 1970) were in agreement with the light-microscopic findings: no myofibrils were observed in cells containing mitotic figures. In addition, virus-like particles were found in multinucleated cells. Since the muscle fibers in the tumor were randomly oriented, any section of tissue could contain transversely cut fibers which would appear to be mononucleated cells. Hence, it was possible that some mitoses that seemed to be in mononucleated cells were in transversely sectioned myotubes. To investigate this possibility, tissue cultures were initiated from 4th or 5th passage tumors. Mononucleated cells, prepared as described previously (Nameroff and Holtzer, 1969), were plated on glass coverslips or on coverslips coated with collagen (Hauschka and Konigsberg, 1966). On glass coverslips, the cell density did not increase during 2 weeks of cultivation and few myotubes formed. However, autoradiography with ^{3}H-thymidine indicated that many cells were synthesizing DNA (35 to 40 per cent of the cells incorporated the label in a 2.5-hour exposure). Apparently, dividing cells from this tumor adhered weakly to a bare glass substrate, floated off, and were removed when the medium was changed. On collagen-coated coverslips, however, cell density increased and multinucleated myotubes began to form on the 5th to 6th day *in vitro*. Apparently, collagen promoted the attachment of myogenic cells to the substrate. Mitotic figures were found in mononucleated cells but not in any multinucleated unit. Autoradiography with ^{3}H-thymidine demonstrated that nuclei in myotubes did not incorporate the label, whereas many mononucleated cells were heavily labeled (Fig. 202). Feulgen cytophotometry on fixed cultures indicated that all nuclei in myotubes were diploid (Fig. 203).

Myogenesis in this tumor appears to proceed as in normal tissue. Once a nucleus enters a multinucleated unit, it no longer divides. A similar switch from a state in which division is possible to a state in which division is suppressed occurs in normal myogenesis. The difference between normal myogenic cells and tumor myogenic cells is that the tumor cells continue to proliferate *in vivo* under conditions in which normal cells do not divide. Yet tumor cells can give rise to progeny, the nuclei of which are kept from dividing when the cells become committed to forming muscle. Although the nature of the alteration resulting in continued proliferation of tumor cells is not known, it appears that this alteration is lost or 'cancelled' when the cells fuse.

A. Mauro: Thank you very much, Dr. Nameroff.

J. R. Florini: Have you any idea about the formal requirements of the host for this tumor to grow?

M. Nameroff: No. Since this was a spontaneous tumor, and we had only one, we transplanted it into the same strain and the same sex mouse (a female BALB/CAnN mouse).

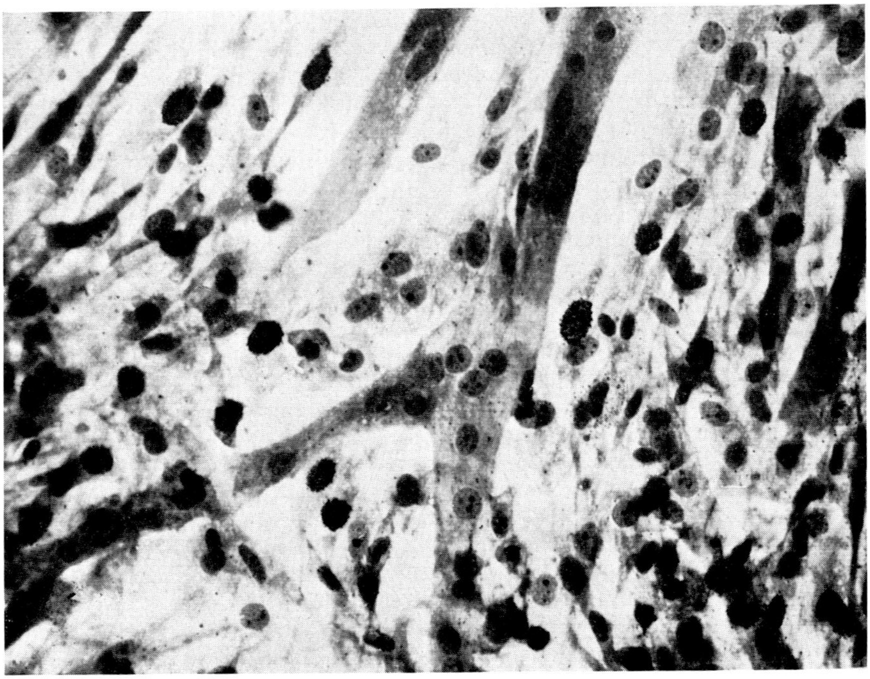

Fig. 202. Autoradiogram of a 5-day culture initiated with mononucleated cells from a 4th passage tumor. ^{3}H-thymidine was given for 2.5 hours before fixation. No nuclei in the myotube are labeled, but many mononucleated cells have incorporated the isotope. × 300.

UNIDENTIFIED SPEAKER: Have you made cultures from this tumor?

M. NAMEROFF: Yes, we have. These cultures are slower than the chick embryo, for example. They begin to form myotubes at about the 5th or 6th day, starting with the same number of cells that we use with the chick culture.

I have subcultured them 2 or 3 times after about 15 days: that is, let the culture go for about 15 days, subculture another 15 days. We have gotten myotubes to form in 2 or 3 subcultures.

I do not want to say anything else. We have not done very much along that line.

M. E. KAIGHN: I wonder if you tried to transmit the culture by means of cell-free extract.

M. NAMEROFF: No, we have not.

M. E. KAIGHN: Because it certainly looks virus-induced.

M. NAMEROFF: Well, I see things there that other people tell me look like viruses. I would doubt that they are causing the tumor.

B. M. CARLSON: Just as a general point of information, is there any good chemical means that you use for myosarcoma?

M. NAMEROFF: There are tumors in the literature which are reported to be rhabdomyosarcoma which have been induced by metal, nickle salts and various other kinds of agents.

You get a tumor in a muscle . . . now I did not want to really get into that. I do not know

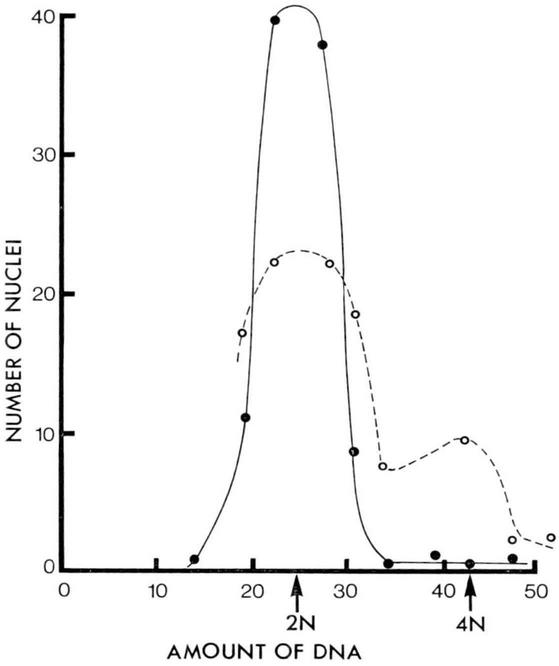

Fig. 203. Histogram of DNA content of nuclei in tumor myotubes (filled circles) and mononucleated cells (open circles) in the same cultures. Cultures were fixed and stained by the Feulgen procedure. The 2N value was determined by examination of nuclei of white blood cells in peripheral blood smears and nuclei in normal muscle fibers *in vitro*. The 4N value was obtained from mitotic figures in normal muscle cultures. Amount of DNA is expressed in the units of absorption displayed on the Barr and Stroud microdensitometer.

what to call this tumor. You say rhabdomyosarcoma! This is a muscle tumor and the things which other people call rhabdomyosarcoma are not this tumor. The pathologists give a name to a particular pattern that they see under a microscope.

UNIDENTIFIED SPEAKER: I think this is a histologically benign tumor. It does not attach to any part of the animal.

M. NAMEROFF: No. It is transplantable but it does not metastasize. It does not kill, at least we did not let it go long enough. I presume it would kill by mechanical effect, compressing a large vessel or loop of bowel.

P. E. CONEN: It is really a rhabdomyoma.

M. NAMEROFF: I do not care what you call it. It is a tumor that is making muscle. I hesitate to say that virus has anything to do with it.

UNIDENTIFIED SPEAKER: Dr. Nameroff made the intriguing suggestion that there may be a number of nuclear switches for turning on myosin/actin production.

M. NAMEROFF: I made the suggestion that there may be a number of switches that turn off DNA synthesis, that is all.

INDUCTION OF DNA SYNTHESIS IN SKELETAL MUSCLE BY ONCOGENIC VIRUSES

M. EDWARD KAIGHN

In the normal course of embryonic development, vertebrate skeletal muscle fibers are formed by fusion of mononucleated myoblasts (Konigsberg, 1965). After fusion, neither DNA synthesis nor nuclear division can be demonstrated in myotube nuclei (Konigsberg, 1965; Marchok, 1966; Okazaki and Holtzer, 1966). Evidence presented in 3 recent reports suggests that, at least in the case of oncogenic viral infection, this repression of DNA synthesis may be reversible (Fogel and Defendi, 1967; Lee *et al.*, 1968; Yaffe and Gershon, 1967). Although different systems were used in the 3 laboratories involved in these studies, strikingly similar conclusions were reached (see Table X). After summarizing the results of these studies, I shall consider the possible nature of the induced DNA.

Fogel and Defendi (1967) found myoblasts from 4 mammalian species to be as sensitive to oncogenic viruses (polyoma and SV-40) as were other mononucleated cells from the same species. This is in agreement with similar findings for the chicken muscle-Rous sarcoma virus (RSV) system (Kaighn *et al.*, 1966). However, multinucleated myotubes were resistant to infection by direct penetration of viral particles. When cultures were exposed to tritiated thymidine (^{3}H-TdR) *before fusion*, autoradiography revealed grains over myotube nuclei beginning about 30 hours after infection. The authors concluded that myoblasts infected prior to fusion, but before morphological transformation, remain capable of forming myotubes (see also Lee *et al.*, 1968) and in so doing induce DNA synthesis in adjacent myotube nuclei.

Yaffe and Gershon (1967) obtained essentially the same results with rat muscle infected by polyoma virus. DNA synthesis, as measured by incorporation of tritiated thymidine, was not observed in myotubes exposed to virus for less than 30 hours. These workers postulated the same mechanism as Fogel and Defendi – fusion of infected myoblasts with existing myotubes. However, they also report that by 50 hours post infection mitotic figures begin to appear in multinucleated muscle fibers. They further state that nuclear division progresses only as far as metaphase, after which the chromosomes tend to form large clumps. This is an extremely interesting observation which has not been made by others working with the same system.

All the work described thus far was done with mammalian cells infected with oncogenic DNA viruses. My colleagues, Drs. Harold H. Lee and James D. Ebert, and I have investigated this problem in the avian-Rous sarcoma virus system (Lee *et al.*, 1968). Certain aspects of the problem are still under investigation in Dr. Ebert's laboratory (Yoshikawa-Fukada and Ebert, 1969*a*) and in Dr. Lee's laboratory (Lee, 1969, personal communication). I shall first outline the results of our studies with the avian system and then attempt to correlate them with the observations reported for the mammalian systems.

The Rous sarcoma virus (RSV) contains RNA rather than DNA as its genetic material. Even so, it has been shown that DNA synthesis is a prerequisite for both productive infection and transformation of chick embryo fibroblasts by RSV (Bader, 1966). Furthermore, various workers had reported that polyoma virus and SV-40 stimulate DNA synthesis in contact-inhibited cell cultures (Dulbecco *et al.*, 1965; Weil *et al.*, 1965; Winocour *et al.*, 1965). With this background, my colleagues and I set out to determine whether RSV infection could induce DNA synthesis in muscle syncytia which in the normal course of their development had ceased to synthesize DNA. The experiments were carried out as follows:

Clones originating from single myoblasts were exposed to RSV. At the time of infection the colonies consisted of both multinucleated myotubes and single myoblasts which continued to divide and contributed to further growth of the colony. After progressive intervals the cultures were exposed to a 1-hour pulse of ^{3}H-thymidine followed by a 'chase' of 1 hour with cold thymidine. They were then fixed and processed for radioautography. We found that grains first appeared over nuclei of myotubes in close proximity to the point of fusion of infected myoblasts with existing myotubes (about 7 hours). At first the label was only seen over newly added myoblasts. Somewhat later (24 hours post infection), groups of myotube nuclei remote from points of fusion were labeled. By 48 hours, large closely packed clusters of myotube nuclei were labeled. No grains were seen over myotubes of control uninfected cultures.

These observations led us to conclude that DNA synthesis is not irreversibly suppressed in myotubes since we had been able to demonstrate clearly incorporation of acid-insoluble ^{3}H-thymidine into myotube nuclei as a result of RSV infection.

Let us now consider the status of our knowledge of the problem of the reversibility of DNA synthesis in skeletal muscle in the light of these recent studies. It seems reasonably clear that, despite the fact that different systems were used, there is agreement among all 3 groups of workers on several counts.

It appears that oncogenic viruses (DNA or RNA) are unable to infect myotubes directly. Rather, they enter the syncytium by the fusion of a previously infected myoblast with an existing myotube. This interpretation is supported by the observation that the time for appearance of grains over myotube nuclei coincides approximately with the time required for an infected cell to initiate substantial virus production (see Table X and Figs. 204 to 207), although the available evidence does not enable us to determine the time interval between fusion with an infected myoblast and the induction of DNA synthesis.

TABLE X

Systems used to study the action of oncogenic viruses on DNA synthesis in muscle fibers

Species	Virus	Viral nucleic acid	Cell culture	Duration[a] of ^{3}H-TdR pulse (hours)	Time[b] for appearance of grains (hours)	Type of cell-virus inter-action	Reference
Mouse	P.V.	DNA	Primary	8	–[e]	Lytic	Fogel and Defendi (1967)
Hamster	P.V.	DNA	Primary	8	32	N.P.[c]	Fogel and Defendi (1967)
Rat	P.V.	DNA	Primary	8	44	N.P.	Fogel and Defendi (1967)
Human	SV-40	DNA	Primary	8	–[f]	N.P.	Fogel and Defendi (1967)
Rat	P.V.	DNA	Primary	6	30	N.P.	Yaffe and Gershon (1967)
Chicken	RSV	RNA	Clonal	1	24	P. but non-lytic[d]	Lee *et al.* (1968)

[a] Labeling period immediately preceding fixation of culture.
[b] Time after infection before fusion for grains to appear over myotube nuclei.
[c] Non-productive infection – no viral progeny.
[d] Viral progeny formed by 'budding' from cell membrane without lysing cell.
[e] Less than 0.5 per cent myotubes took up label probably due to cell lysis.
[f] Less than 1 per cent of cells infected. Label observed after 4 days.

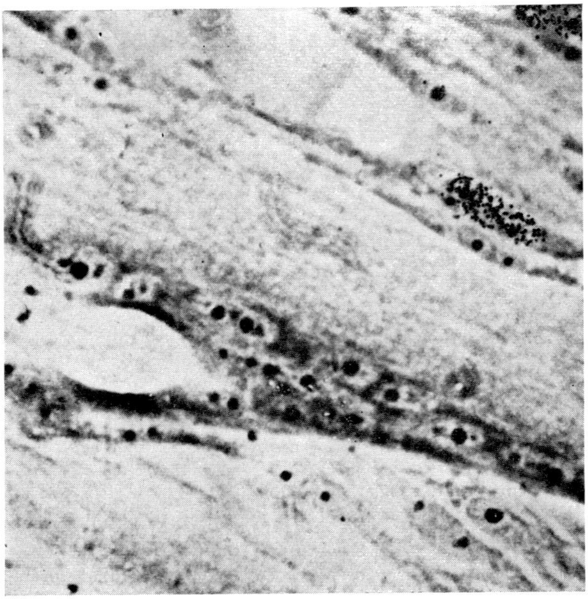

Fig. 204. Seven hours after infection. Myotube is unlabeled, whereas single cells are.

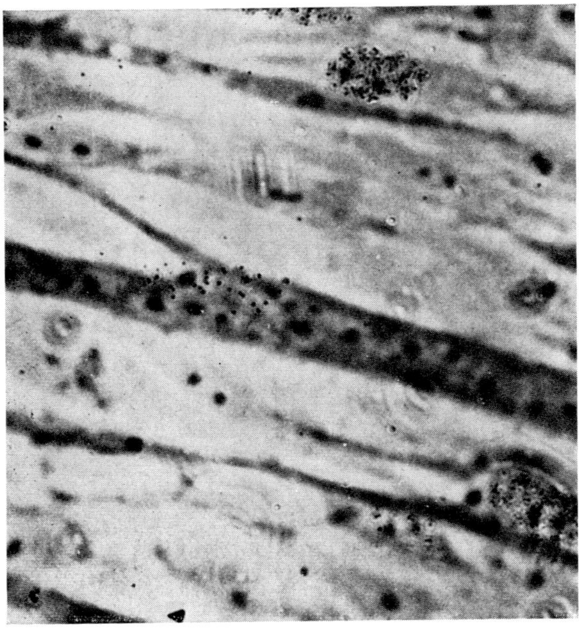

Fig. 205. Light labeling in myotube in region of branching point. Note heavier label over nuclei of single cells; 24 hours after infection.

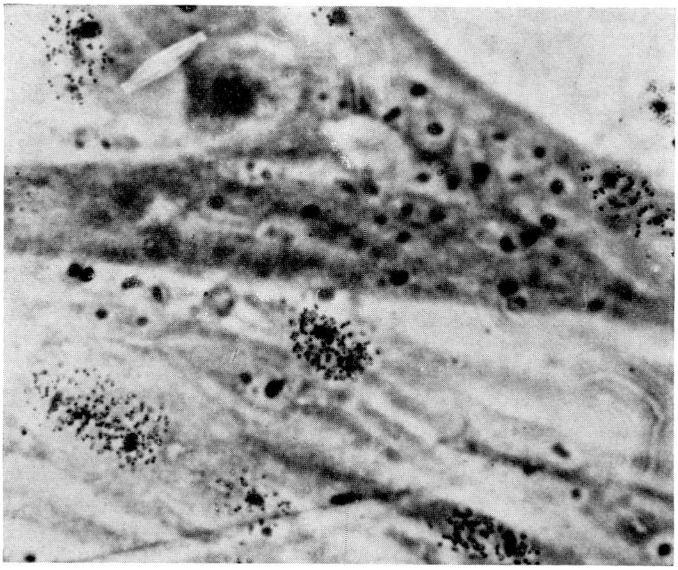

Fig. 206. Similar to Fig. 205. Note label in single myotube nucleus and heavier label in adjacent single cells.

Fig. 207. Part of a myotube in which all nuclei have silver grains. A myoblast with heavy label is out of focus, so that the grains appear light instead of dark.

In the case of polyoma virus (DNA), the available evidence indicates that the ^{3}H-thymidine is incorporated into cellular rather than viral DNA. It is possible to make this distinction in the case of polyoma virus because of the different properties of circular P.V.-DNA. Yaffe and Gershon showed that DNA extracted from infected muscle behaved like cellular rather than viral DNA after heat denaturation and chromatography on a methylated albumin-

Kieselguhr column. The involvement of cellular DNA synthesis in both viral replication and in cell transformation by oncogenic viruses is well established. Inhibitors of DNA synthesis and of DNA-dependent RNA synthesis block both of these processes (Temin, 1963, 1964; Vigier and Goldé, 1964). In addition, manipulation of the culture medium (Bader, 1966) or high population density (Todaro and Green, 1966), both of which decrease the rate of DNA synthesis in a more physiological way, have similar consequences.

It has been proposed by Temin (1964) that the RNA-containing RSV replicates by a sort of reverse transcription; *i.e.*, the RSV-RNA produces a DNA template which in turn is responsible for synthesis of progeny viral RNA. This has been referred to as the provirus hypothesis. Before considering the implications of this hypothesis with respect to the nature of the DNA induced in myotubes, it will be useful to review the more general hybridization studies which have been aimed at elucidating the mechanism of replication of RNA tumor viruses. This method relies on complementary base pairing between radioactively labeled viral RNA and cellular DNA strands. Following renaturation, the complex is treated with RNAase to remove non-specific complexes. The amount of radioactive label complexed to the DNA is then determined. Studies of this type have been carried out in both mammalian and avian systems.

In support of his hypothesis of the formation of a DNA provirus for RSV, Temin (1964) reported finding a novel DNA homologous to viral RNA in infected chick embryo cells which is not present in uninfected cells. However, the difference in the viral radioactivity bound to DNA from infected and uninfected cells was very small because of the low specific activity of the viral RNA used. Furthermore, the total RNA from purified virus rather than the virus-specific 65 S component was used.

In contrast to Temin's initial report, subsequent workers agree that avian tumor virus RNA hybridizes with DNA of infected and uninfected cells to about the same extent. This has been demonstrated for avian myeloblastosis virus (AMV) (Harel *et al.*, 1966*b*; Wilson and Bauer, 1967), for RSV (Bader, 1966; Harel *et al.*, 1966*a*; Yoshikawa-Fukada and Ebert, 1969*b*) and for mouse leukemia virus (Harel *et al.*, 1967). These studies have utilized improved methods to increase the sensitivity as well as the specificity of the hybridization approach. For example, they have examined the relative complementarity of viral RNA fractions of different S values obtained by density gradient centrifugation including the 60-70 S fraction (Bader, 1966; Harel *et al.*, 1966*a*, *b*; Wilson and Bauer, 1967). The specific radioactivity of the virus in these studies was up to 10 times that employed by Temin.

On the other hand, at present it is not possible to rule out completely the provirus hypothesis. Isolation of a rapidly sedimenting RNA from chicken fibroblasts infected with the Harris strain of RSV which is not present in uninfected fibroblasts has been reported (Wilson and Bauer, 1967). This 50-70 S fraction was found to differ from 65 S RSV-RNA isolated from virus particles in that it contained a large fraction of alkali labile (92 per cent) but RNAase-resistant material (56 per cent). This RNAase resistance could be reduced to less than 1 per cent by heating at low ionic strength and DNAase treatment. These data were interpreted as evidence for the possible existence of a DNA-RNA replicative form for the Harris strain of RSV. However, in themselves, they give no information as to the extent of homology between the DNA and RNA of the isolated complex. Clearly more specific information is required. A significant difference between the efficiency of hybridization of RSV-RNA (65 S) with DNA from RSV-induced hamster tumors and DNA from normal hamster embryo cells has been reported (Bader, 1966). This difference held only for the 65 S and 30 S RSV-RNA's but disappeared for the 10 S fraction. It should be emphasized that this comparison was carried out between possibly heteroploid tumor cells and undefined hamster embryo cells from different sources. It would be desirable to carry out this experiment on parallel cultures of diploid cells from an identical source so that differential hybridization efficiency could be more safely ascribed solely to viral infection.

Let us now return to the subject of the nature of the DNA synthesis induced in myotubes by RSV infection. If this DNA were homologous to viral RNA, it would seem easy to isolate and characterize the DNA induced by RSV since uninfected myotubes make none. This approach is complicated because it has not yet been possible to obtain a pure population of myotubes free of myoblasts that actively synthesize cellular DNA. This problem has also been approached by RNA-DNA hybridization techniques. RSV-RNA labeled with ^{3}H-uridine was hybridized with chicken muscle DNA. In a preliminary account of this work it was reported that RSV-RNA hybridized with DNA from uninfected as well as infected chick muscle cultures to the same degree (Yoshikawa-Fukada and Ebert, 1969*a*). If the DNA induced were in fact a DNA-provirus, we would expect that there should be a difference in the pattern between infected and uninfected cells. Furthermore, the hybrid should sediment with an S value at least as high as 35 S. It has been recently reported that RSV (Duesberg, 1968) and AMV (Erikson, 1969) 65 S RNA break down into components sedimenting at 35 S on heating. Temperatures employed in the hybridization process usually exceed those required for complete transition to the 'subunit' form. The hybrid obtained with chicken muscle DNA sedimented as 3 S rather than 35 S RNA (Yoshikawa-Fukada and Ebert, 1969*b*). This could be due to partial degradation by residual RNAase in the incubation mixture. Variable sedimentation values for viral RNA-DNA hybrids have been reported (Harel *et al.*, 1966*a*, *b*, 1967; Wilson and Bauer, 1967; Yoshikawa-Fukada and Ebert, 1969*b*).

In addition to the lack of difference in the hybridization patterns between infected and uninfected cells, DNA extracted from cells of other species yield similar patterns with RSV-RNA; for example, HeLa, calf thymus, L-cells, BHK 21 and hamster liver (Yoshikawa-Fukada and Ebert, 1969*a*). Recently, it has been found that RSV-RNA hybridizes with oncogenic viral DNA adenovirus and SV-40 as well as with a variety of cellular DNA's (Yoshikawa-Fukada and Ebert, 1969*b*). Moreover, the homologous sequences in both cellular and oncogenic viral DNA's appear to be similar on the basis of base composition and melting experiments.

What light do all these observations throw on the nature of the DNA induced in myotubes by RSV? The evidence suggesting a common or, at least, closely similar nucleotide sequence

TABLE XI

Relative uptake of ^{3}H-thymidine into the nuclei of mononucleated and multinucleated components in muscle clones

	Grain counts per nucleus
Infected cultures	
Multinucleated myotubes	21
Mononucleated myoblasts	>60
Transformed cells	17
Control cultures	
Multinucleated myotubes	None
Mononucleated myoblasts	>60

between cellular and oncogenic viral DNA's and RSV-RNA makes it tempting to speculate that synthesis of this segment of DNA is induced by RSV infection in myotubes. Although direct experimental evidence for the provirus hypothesis is minimal, it cannot be excluded as yet. Since RSV infection induces less efficient incorporation of ^{3}H-thymidine into myotube nuclei than into myoblast nuclei (Table XI), it would be interesting to see if a particular pattern of localization could be discerned by the cytological molecular hybridization technique recently published by Gall and Pardue (1969).

Finally, the provocative report of Yaffe and Gershon (1967) raises the question as to whether polyoma virus can stimulate mitosis as well as DNA synthesis in myotubes. This phenomenon has not been reported by others working with the same or any other system and certainly bears confirmation. Unfortunately, the reproduction of Yaffe and Gershon's figures is rather poor. Again, it would be interesting to determine whether DNA synthesis could be localized along the condensed chromosomes within the infected muscle fibers by the Gall and Pardue procedure (1969).

In conclusion, it should be stressed that what I have been discussing is a pathological process brought about by an infectious agent. The response of the muscle fiber to such a pathological stimulus may or may not be related to events which occur during muscle regeneration as considered elsewhere in this monograph.

In summary one can say that:

1. Oncogenic viruses have been shown to induce DNA synthesis in the nuclei of avian and mammalian muscle myotubes by autoradiographic methods.
2. Myotubes are not infected by direct penetration of the virus but rather by fusion with a previously infected myoblast.
3. The nature of the induced DNA is not known at present. Several possibilities are discussed.
4. Because of its far-reaching implications for muscle differentiation, the single report of stimulation of mitosis in rat myotubes should be confirmed.

A. MAURO: Dr. Kaighn, you might start the discussion by reviewing the literature on the Rous sarcoma virus related to tissue culture studies of muscle cells.

M. E. KAIGHN: There are 3 reports in the literature dealing with induced DNA synthesis in multinucleated, syncytial muscle fibers. All 3 groups agree that multinucleated myotubes are not susceptible to direct infection by viral particles but rather the virus gains entrance to the fiber by its fusion with a previously infected myoblast. The only conflict between workers in this area concerns the question of whether stimulation of mitosis in multinucleated myotubes ensues as a consequence of virus infection. Yaffe and Gershon (1967) claim that it does, whereas neither Fogel and Defendi (1967) nor Lee *et al.* (1968) make mention of observing such a phenomenon.

J. R. COLEMAN: Can I clarify one thing? Yaffe has not seen mitosis in multinucleated muscle fibers. He has seen something that looks like chromosome condensation, and that is as much as he has seen.

M. E. KAIGHN: He has never seen, as far as I know, completion of a division. He sees this condensation which he claims are different stages in the mitotic cycle.

UNIDENTIFIED SPEAKER: But he uses the word 'mitotic'!

M. E. KAIGHN: He uses the word 'mitosis' in the title.

A. MAURO: In connection with the theoretical principles which we are trying to establish as a basis for the phenomenon of regeneration, I quote on page 423 from Yaffe's paper: 'These findings indicate that the formation of multinucleated fibers is not necessarily associated with an irreversible change in the capacity of the nucleus to synthesize DNA and undergo mitosis . . .' I emphasize 'undergo mitosis'.

M. E. KAIGHN: O.K. . . . O.K. I know that that is what the paper says, but that is not what the data show!

A. MAURO: I agree with you . . . (Laughter)

M. E. KAIGHN: Besides that, I have not seen the pictures that I understand he circulates around, but the published illustrations are hardly convincing on this point.

I should just like to make one point about stimulation of DNA synthesis in muscle fibers. I think that we have already presented sufficient evidence to claim that RSV stimulates synthesis of DNA if you accept incorporation of tritiated thymidine in an acid-insoluble form as an indication of DNA synthesis.

The design of this experiment is very simple. We infected pure muscle colonies with RSV and at various periods of time after infection, pulse-labeled with tritiated thymidine for 1 hour, followed by a chase for 1 hour in medium containing cold thymidine.

As we already have seen in Bischoff's presentation, within a 2-hour period you would expect to see no label in multinucleated structures after such a pulse and chase.

If you infect cells and then look at this at 7 hours, 24 hours and 48 hours, you begin to see label over the nuclei of mononucleated cells at 7 hours. You see more label over the nuclei of mononucleated cells at 24 hours, and you begin to see a very few cells labeled in muscle fibers.

The pattern of labeling in the muscle fiber and in the single cell is different. There are about 3 times as many grains per nucleus in the mononucleated cells as there are in the intrastrap labeled nucleus.

At 48 hours, there are many more labeled nuclei inside of muscle fibers.

Our experiments rule out the possibility of the cell being labeled with thymidine and then subsequently incorporated into the muscle fiber as the source of the labeled grains.

On the other hand, they do not rule out the possibility that an infected myoblast could fuse with a preformed muscle fiber and stimulate DNA synthesis at sites distant from the point of fusion.

As a matter of fact, we have seen 'puddles' of labeled nuclei with grains, particularly at 48 hours. It is much more prevalent at 48 hours.

M. NAMEROFF: May I see one?

M. E. KAIGHN: Certainly (see Figures 204-207).

M. NAMEROFF: Is there any evidence which rules out that the DNA synthesis that goes on in the nuclei is not viral DNA?

M. E. KAIGHN: I would suspect that it is not viral DNA synthesis because it is an RNA virus. There is ample evidence in the literature that host DNA synthesis is stimulated by avian tumor virus infection and, in fact, the ability of a cell to synthesize host DNA is a prerequisite to its productive infection or to its oncogenic transformation.

UNIDENTIFIED SPEAKER: Does Rous not have an associated DNA virus?

M. E. KAIGHN: No. The associated viruses are RNA viruses, and I did not mention that because it is a very complicated story.

H. HOLTZER: Is there any evidence to indicate that the induced DNA necessary for viral replication is the DNA that would then lead to a normal mitosis of that nucleus? . . . Or is this virally induced DNA always made by the host?

M. E. KAIGHN: We have seen no evidence (and we looked at many many cells) of any kind of replication in infected muscle fibers, although there is ample evidence of induction of DNA synthesis.

I would imagine that the DNA synthesis represents the turning on perhaps of a portion of the host genome, which may be relevant to some viral function, but certainly not enough to lead to cell division.

J. R. COLEMAN: Yaffe, I think, has shown there is no detectable polyoma in DNA made by the cultures. He certainly, in that experiment, has not shown what kind of DNA the multinucleated fibers are making.

L. P. ROWLAND: Could you tell by nucleic acid hybridization?

J. R. COLEMAN: Yes . . . Yes, you could, but if you think that there have been a lot of disputes here in our discussion about autoradiographs, we should not choose to get into this. (Laughter). I assure you, the same kinds of discussion occur with respect to DNA-DNA hybridization and almost as much to DNA-RNA hybridization.

H. HOLTZER: To me this is an interesting question. The Harris kind of hybridization I do not think has been alluded to. This involves fusion of heterologous cell types with the aid of U.V.-inactivated Sendai virus.

There, are you inducing DNA synthesis, again in a meaningful sense, or are you inducing the uptake of some thymidine into some insoluble TCA-precipitable material?

There is no evidence to my knowledge in the Harris kind of induction experiment that in fact meaningful DNA is indeed being synthesized!

J. R. COLEMAN: It may well be that incorporation of infected mononucleated cells into a fiber destroys the controls we have been talking about.

I think one thing that should be emphasized in all 3 of these situations is that they are pathological conditions and that the fibers die!

That raises the question of why they are dying. Is it because the normal control of DNA synthesis has been interfered with, or is it because of some other aspect of infection? One interpretation of these experiments could be that resumption of DNA synthesis by these cells is fatal.

M. E. KAIGHN: Multinucleated fibers are unaffected morphologically for several days after infection even though one can demonstrate incorporation of ^{3}H-thymidine by their nuclei.

On the other hand, if you follow them long enough, they will all slough off the surface of the dish and die and I think that Yaffe mentions the same thing.

A. MAURO: Dr. Nameroff will now present some recent work which is relevant to the previous discussion.

ACKNOWLEDGEMENTS

Dr. Kaighn gratefully acknowledges the helpful suggestions of Drs. H. H. Lee, A. M. Prince, and J. D. Ebert. He would like to thank Dr. Ebert for showing him his manuscript in advance of publication, and the editors of *International Journal of Cancer* for permission to reproduce published material.

INTERFERENCE WITH MYOGENESIS*

MARK NAMEROFF and HOWARD HOLTZER**

Mononucleated myogenic cells liberated by trypsinization from the embryonic breast muscles of the 11-day chick will multiply and fuse on many types of surfaces when cultured at high density *in vitro*. These surfaces or substrates include glass, polyethylene, cellulose acetate, paraffin, collagen, quartz, plasma clots, weakly acidic and strongly acidic cation exchange resins, and weakly basic and strongly basic anion exchange resins (Nameroff, unpublished). Such observations suggest that the surfaces of myogenic cells are rather non-selective in terms of the kinds of substrates that will permit the cells to attach, migrate, divide, and fuse. However, muscle precursor cells are known to have very selective surfaces since they will fuse only with other mononucleated myogenic cells or early myotubes (Okazaki and Holtzer, 1965, 1966), and then only in the G_1 phase of their mitotic cycle (Bischoff and Holtzer, 1969). Therefore we decided to examine the effects of *cellular* substrates on both proliferation and fusion.

Cellular substrates were essentially non-dividing cultures of chondrocytes, liver cells, and muscle that had been initiated with 0.5×10^6 cells on plasma clots in 35-millimeter plastic Petri plates. These cells were obtained from 11-day chick embryos by trypsinization (for details, see Nameroff and Holtzer, 1969) and were grown in a medium consisting of 7.7 parts of Eagle's MEM, 1.0 part of horse serum, 1.0 part of embryo extract, 0.1 part of penicillin-streptomycin mixture, and 0.1 part Fungizone. The initial inocula multiplied to cover the dish and formed sheets of tissue which were several cell layers deep after 5 to 7 days *in vitro*. At this time, 0.5×10^6 freshly liberated myogenic cells were plated onto the cellular substrates. Control cultures were myogenic cells on plasma clots, and 0.5×10^6 myogenic cells together with an equal number of freshly isolated chondrocytes or liver cells on plasma clots. All cultures were inspected daily, under phase-contrast optics, fixed, and stained after 10 days. The medium was changed daily.

Cultures of myogenic cells on plasma clots formed myotubes which exhibited spontaneous contractions by the 3rd day (Fig. 208). In cultures initiated with equal numbers of freshly liberated myogenic cells, together with chondrocytes or liver cells, muscle fibers formed between areas of cartilage or liver (Fig. 209) (Shimada, 1968). But myogenic cells failed to form myotubes on cellular substrates (Fig. 210). To determine whether or not the cells attached to cellular substrates and, if so, whether or not they survived, freshly liberated myogenic cells were labeled by incubating cultures in ^{3}H-thymidine (specific activity 3.0 curies per millimole, Schwarz BioResearch) at a concentration of 0.5 microcurie per millimeter of medium for 48 hours. The labeled cells then were plated on substrates of chondrocytes, liver, or muscle. Each day, for 6 days, several cultures were fixed and at the end of the experiment all were coated with photographic emulsion. The autoradiograms were developed after exposure for 1 to 2 weeks.

* Supported in part by grants from the National Science Foundation and the U.S. Public Health Service, and in part by a research contract, Project No. 3A025601A822, from the Medical Research and Development Command, U.S. Army, Washington, D.C., U.S.A.

** Recipient of Research Career Development Award 5 K3 HD-2970.

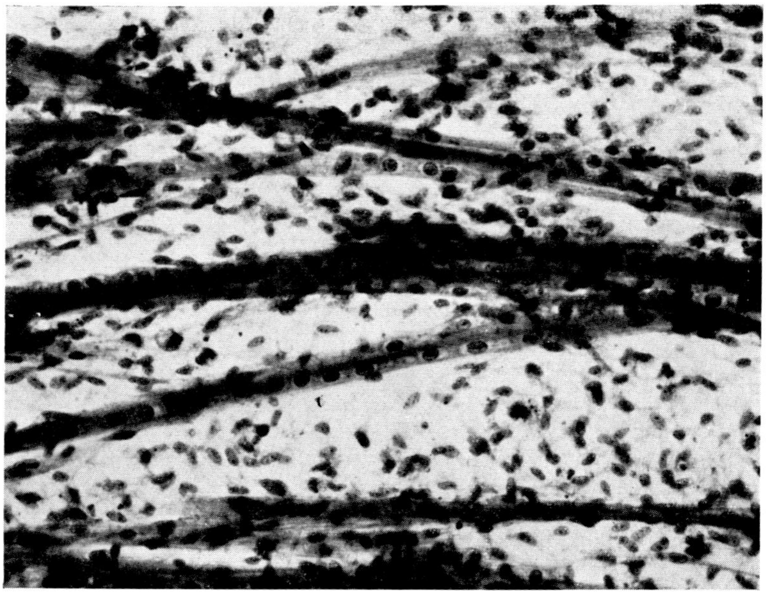

Fig. 208. Muscle fibers formed from mononucleated myogenic cells cultured on a plasma clot; 6 days *in vitro*. × 230.

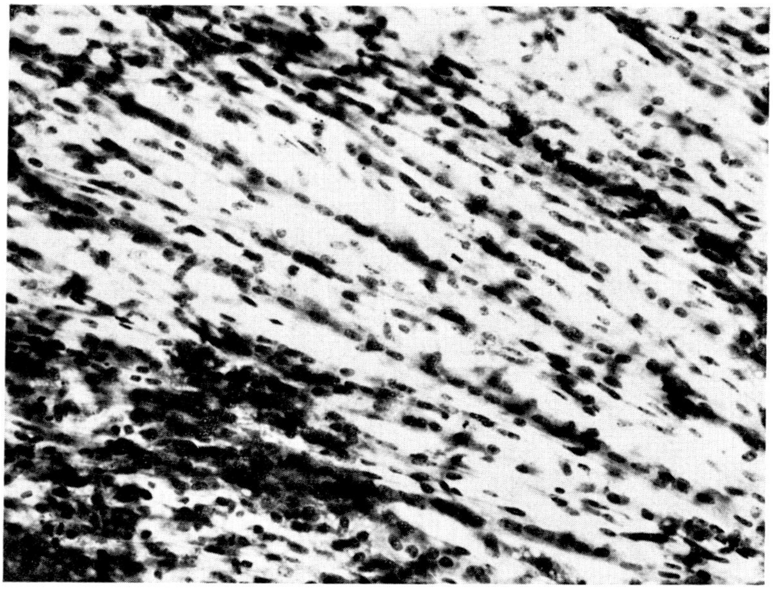

Fig. 209. Myotubes in a culture initiated with equal numbers of myogenic cells and liver cells; 4 days *in vitro*. × 225.

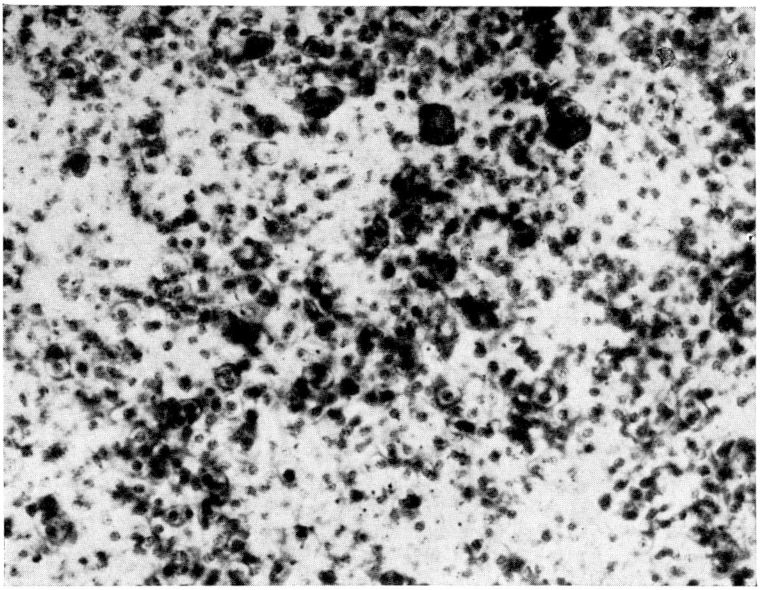

Fig. 210. Absence of muscle differentiation on a chondrocyte substrate. Capsules are apparent around some chondrocytes; 5 days *in vitro*. × 225.

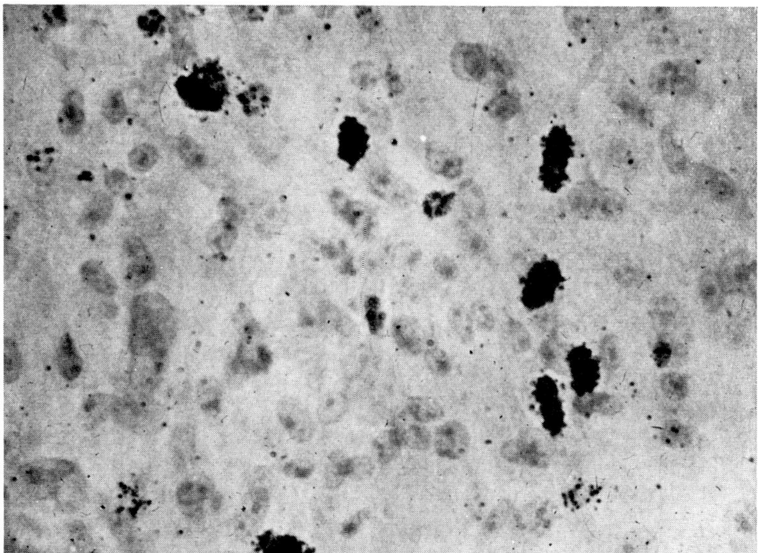

Fig. 211. Autoradiogram showing labeled myogenic cells on a chondrocyte substrate 5 days after plating. × 380.

Approximately the same number of labeled cells attached to the cellular substrate as to the clot substrate in the first 24 hours. The number of cells with grains over their nuclei increased with time in the clot controls. The average grain count per nucleus decreased with time in controls. No increase in number of labeled cells in cultures set up on cellular substrates (Fig. 211), or decrease in average grain count per nucleus (Fig. 213), was observed. However, within 24 hours a few labeled myogenic cells formed short, narrow myotubes on the cellular substrates (Fig. 212). These myotubes did not increase in length or number throughout the experiment. Labeled myogenic cells in controls, subcultured on plasma clots, also formed myotubes within 24 hours. Precultured, labeled cells are different from freshly liberated myogenic cells (Okazaki and Holtzer, 1965); they are larger and the fact that they will fuse

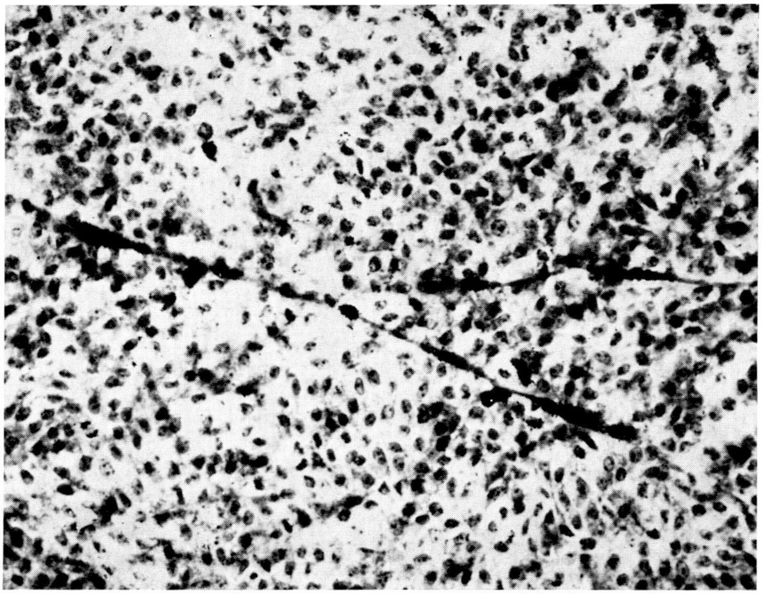

Fig. 212. Short myotube formed by labeled myogenic cells on a liver substrate; 2 labeled mononucleated cells are also visible. × 225.

within 1 day after subculture indicates that growth *in vitro* increases the fraction of the original population capable of rapid fusion. Some of these cells probably fused on the cellular substrates during the 1st day.

The foregoing experiments demonstrated that myogenic cells attached to cellular substrates and remained attached for at least 6 days. The observation that the grain count per nucleus and the number of labeled cells remained constant suggested that mitosis was inhibited on cellular substrates. We confirmed this by exposing cultures of myogenic cells alone, cellular substrates alone, and cellular substrates, with an overlay of myogenic cells to ^{3}H-thymidine, for 4-hour 'pulses' each day. Cultures were sacrificed after each pulse and the thymidine incorporated into trichloroacetic acid-precipitable material was counted in a scintillation counter. The results shown in Figure 214 indicated that DNA synthesis was inhibited on cellular substrates.

The multiplication of myogenic cells may have been suppressed on cellular substrates by

Fig. 213. A: Number of labeled nuclei per 43 × field. Ordinate on left. B: Percentage of labeled cells with more than 50 grains per nucleus. Ordinate on right. C: Percentage of labeled cells with 5 to 50 grains per nucleus. Ordinate on the right.

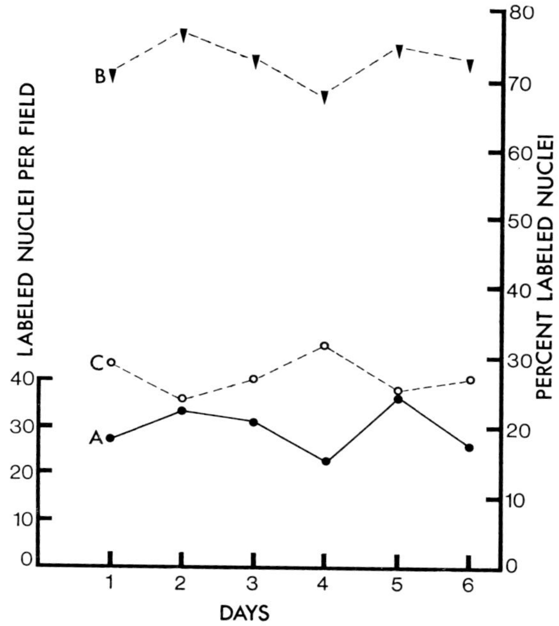

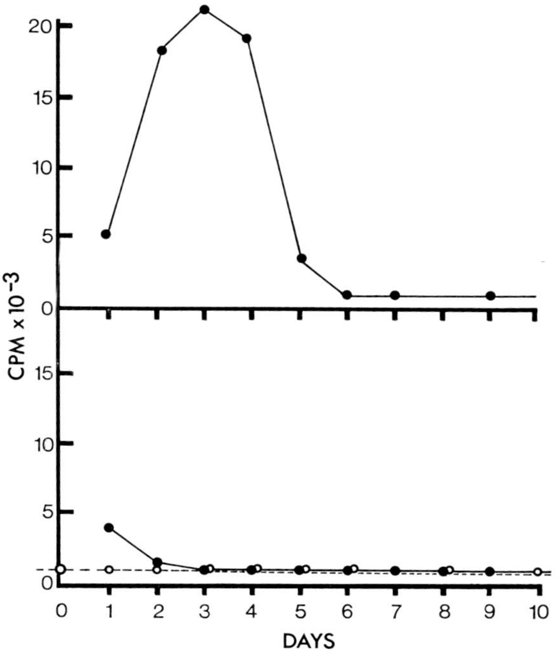

Fig. 214. Top: ^{3}H-thymidine incorporation in a normal muscle culture. Each point represents the amount of isotope found in acid-insoluble material after a 4-hour exposure. *Bottom*: ^{3}H-thymidine incorporation on cellular substrates. Filled circles represent incorporation in cultures of myogenic cells on liver, cartilage, or muscle substrates. Open circles represent incorporation of the isotope by the substrates alone. Ordinate is expressed as counts per minute per culture.

an alteration in the medium caused by the cells in the substrate. Alternatively, direct contact between the myogenic cells and the cell surfaces on interstitial materials in the substrates could have inhibited mitosis. The following experiments demonstrated that direct contact was necessary for mitotic suppression and subsequent inhibition of muscle differentiation. Cellular substrates were prepared in 60-millimeter Petri plates by inoculating the dishes with 1.5×10^6 chondrocytes, liver cells, or myogenic cells. Seven days later, a circular millipore filter (type HA, 47 millimeters in diameter) was placed in each culture over the confluent cell layer. Control cultures were set up with filters placed on plasma clots or directly on the plastic surfaces with no clots. To each dish 1.5×10^6 freshly liberated myogenic cells were added. After 2 to 10 days the filters were removed, fixed, and stained. Mitotic figures were observed in many mononucleated cells on the filters in early cultures. Myotubes formed on the surface of the filters away from the underlying cell layers (Fig. 215).

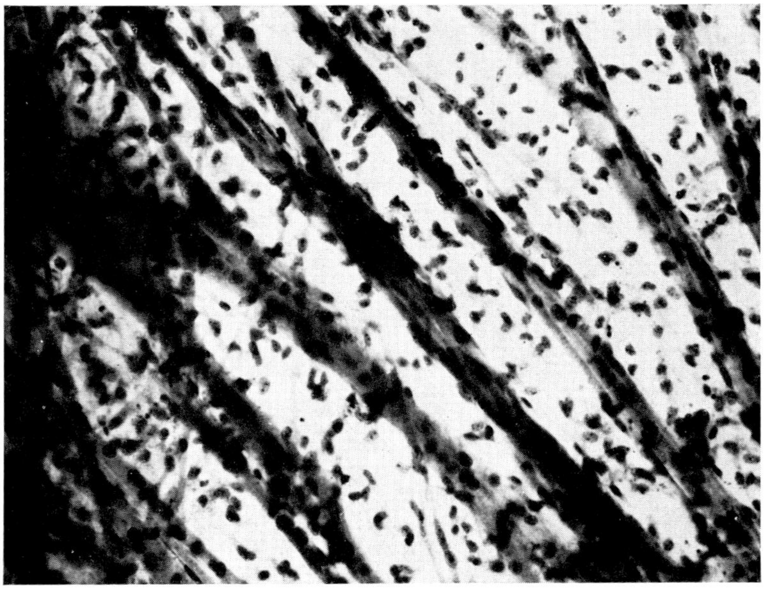

Fig. 215. Muscle fibers on a millipore filter placed over a cellular substrate. × 230.

In this experiment, contact between myogenic cells and cellular substrates was prevented, but the medium was free to exchange between the cells on the filter and the substrates below. An attempt was made to maximize any possible effect on myogenic cells caused by an alteration in the medium. Freshly liberated myogenic cells were cultured on plasma clots in medium in which cellular substrates had been incubated for 24 hours; the medium was changed daily. Myotubes formed in all cultures. In a third experiment, a circular cut approximately 1.5 centimeters in diameter was made in each of a series of chondrocyte and liver substrates. The entire depth of tissue inside the cut was removed. Half the cultures then received 0.5×10^6 fresh myogenic cells; the other half served as controls. Muscle formed in the 'holes' but not on the surrounding tissue (Fig. 216). These 3 experiments suggested that contact was required for mitotic inhibition and subsequent suppression of muscle differentiation, and that an altered medium was most likely not primarily responsible for the observed interference.

Fig. 216. Myotubes in 'hole' cut in a chondrocyte substrate. × 14.

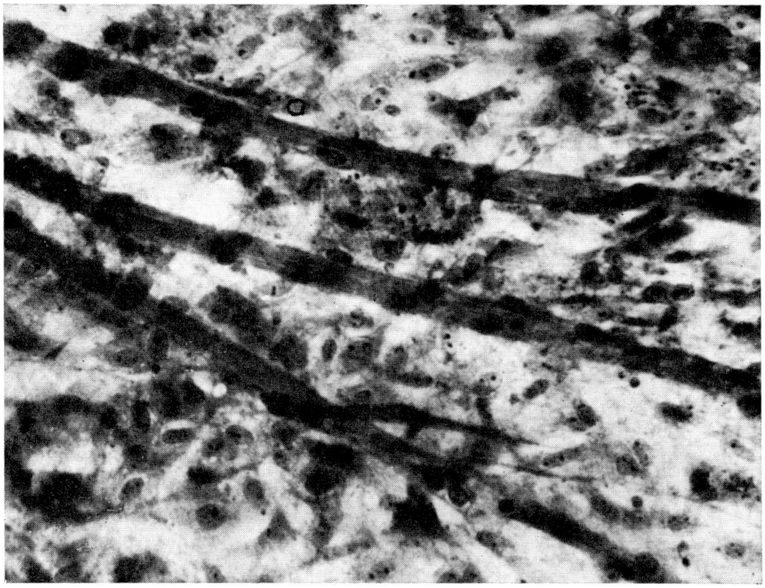

Fig. 217. Demonstration of the reversibility of interference. These myotubes formed from mononucleated cells, which had been on a cellular substrate for 3 days, and then were subcultured onto a plasma clot. × 250.

Interference with myogenesis was shown to be reversible by removing myogenic cells from cellular substrates with trypsin 1 to 5 days after inoculating and subculturing them onto plasma clots. Muscle formed in the subcultures (Fig. 217). This experiment demonstrated that the added myogenic cells were not killed on cellular substrates and that at least some of the mononucleated cells that attached to the substrates were muscle precursors.

Up to this point the experiments described demonstrated that myogenic cells did not divide on cellular substrates and, probably as a consequence of mitotic suppression, did not form muscle. The observation (*cf.* above) that myogenic cells, which were precultured for 2 days and then subcultured onto cellular substrates, formed a few short myotubes within 24 hours suggested that an underlying cell layer did not directly interfere with fusion. To test this hypothesis, a series of cellular substrates was inoculated with increasing numbers

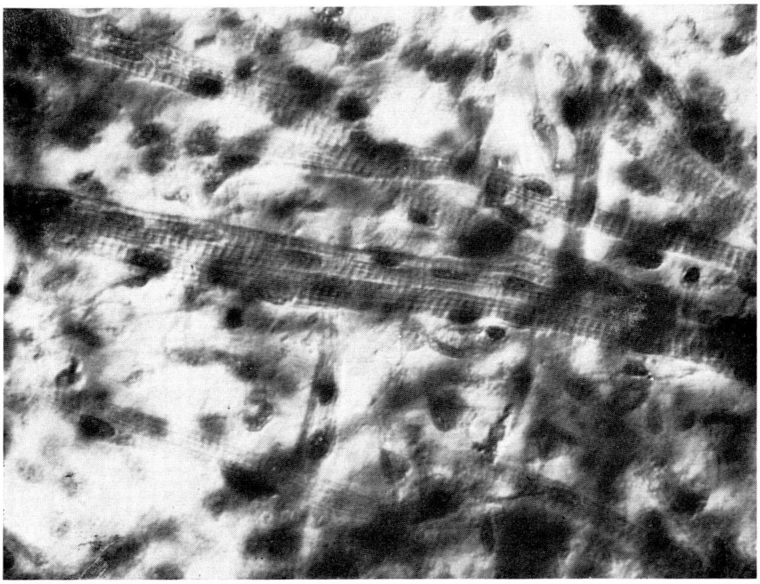

Fig. 218. Muscle fibers in a high density plating (5×10^6 cells) on a chondrocyte substrate. Nomarski optics (courtesy of Dr. J. Griffin, AFIP). $\times$ 315.

of fresh myogenic cells. Since, in a population of such cells, some will have reached the developmental stage at which fusion can occur without an intervening mitosis, it was supposed that in a very dense culture the number of contacts between cells capable of fusion would be increased. According to this hypothesis, if the substrates interfered with fusion, no myotubes should form in spite of the large number of cell-to-cell encounters. When more than 2×10^6 fresh myogenic cells were added to cellular substrates contracting muscle fibers were observed in the cultures (Fig. 218). These fibers were shorter and fewer in number than in controls. Although it is possible that an underlying cell layer could not suppress large numbers of added cells, it would appear that cellular substrates did not interfere directly with the processes of fusion in cells which were mature enough to fuse.

Fusion between competent cells can occur only if the cells establish contact. On cellular substrates contact would occur if the cells moved about on the surface of the underlying layer or if they happened to fall onto the surface very close to one another at the time of

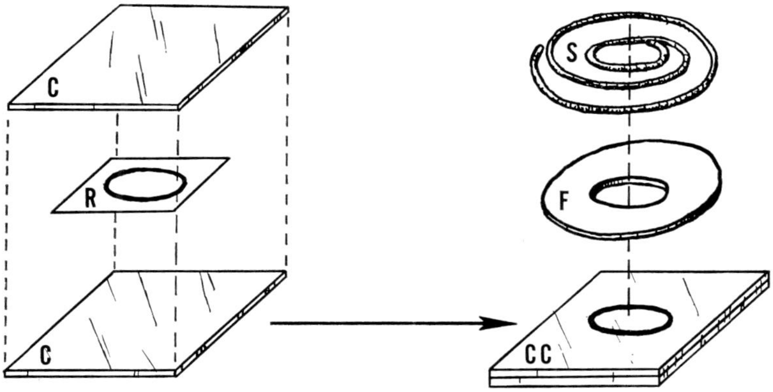

Fig. 219. Method of preparation of ring marker coverslips. Black ring (R) is mounted with Permount between 2 coverslips (C). The resulting sandwich (CC) is coated with a plasma clot and used as a substrate for chondrocytes. After confluence is reached, a millipore filter (F) with a hole in it is placed on the cellular substrate and held in place by a stainless steel wire (S). Labeled cells then are added and the millipore and wire are removed 24 hours later.

inoculation. Myogenesis might therefore be suppressed if myogenic cells did not migrate on cellular substrates. To determine whether migration was inhibited, the following experiment was performed: Chondrocyte substrates were set up in 35-millimeter Petri plates on 22-millimeter square coverslips marked with a circle in the center of 6 millimeters in diameter. After the cells reached confluence, a millipore filter (type GS, 25 millimeters diameter) with a 6-millimeter hole punched in the middle was placed on each of the cultures

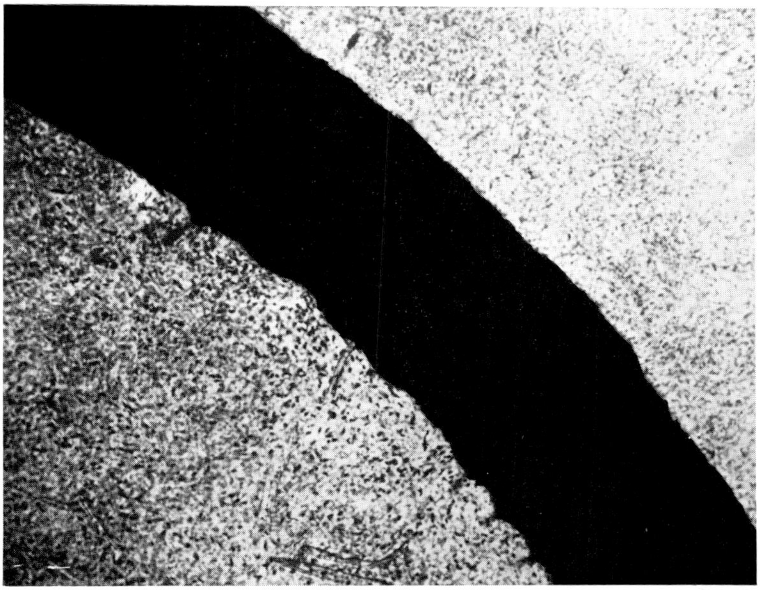

Fig. 220 A. Autoradiogram of movement experiment on a cartilage substrate; 3 days *in vitro*. All labeled myogenic cells have remained inside (to the left) of the ring marker. × 14.

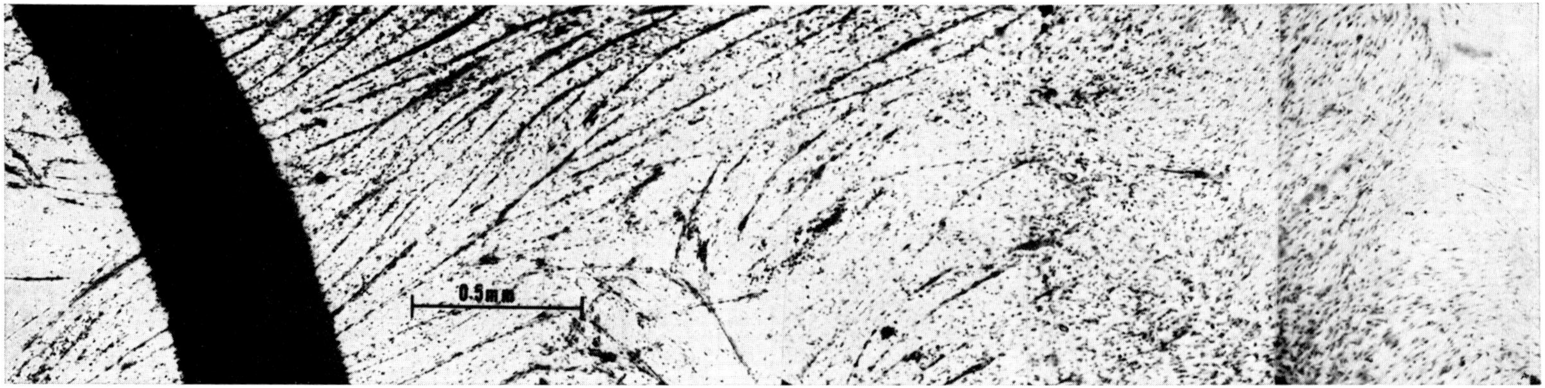

Fig. 220 B. Control on a plasma clot. Mononucleated labeled cells have multiplied, formed myotubes, and migrated far from the area within the ring marker.

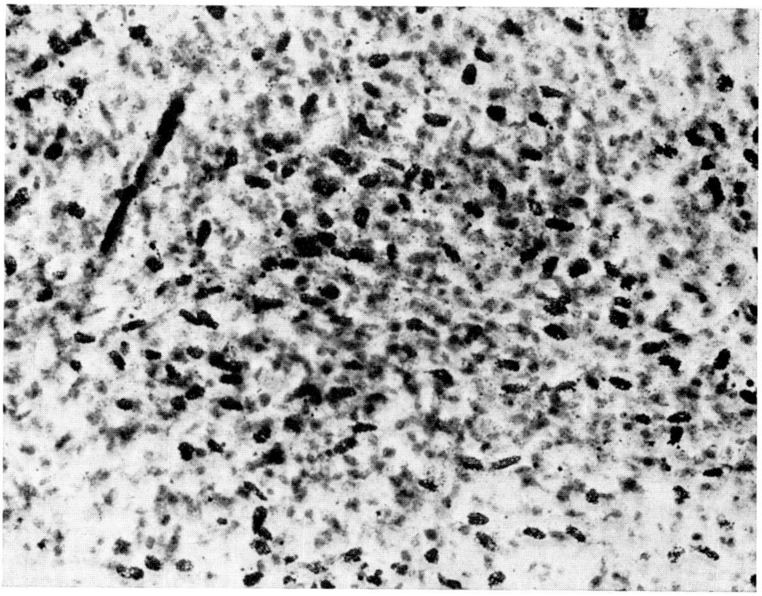

Fig. 220 C. Higher power photo of A. Labeled cells inside ring marker. × 225.

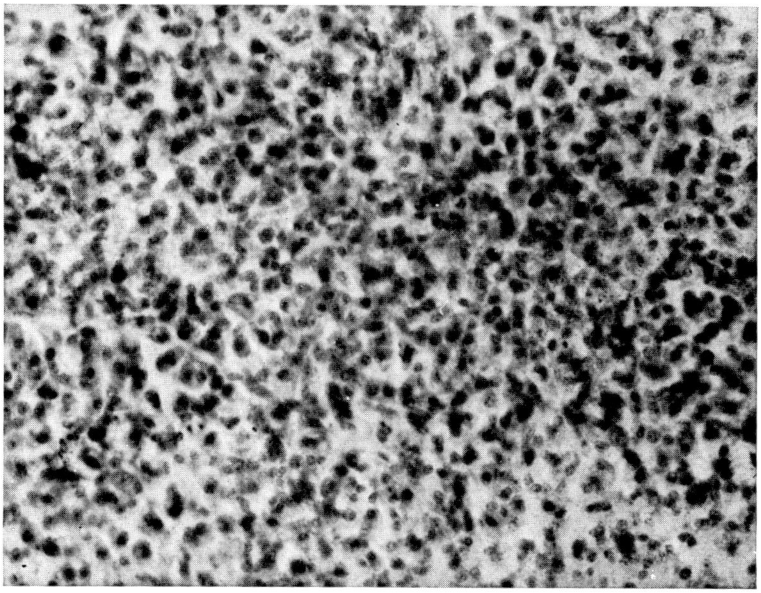

Fig. 220 D. High-power photo of A outside ring marker. No labeled cells are present. × 225.

and was held down by a stainless steel spiral-shaped wire (Fig. 219). The filter was aligned so that the margin of the hole lay over the circle marked on the coverslip. Into each culture, 0.5-1.5 $\times$ 10^6 myogenic cells labeled with ^{3}H-thymidine were plated. The cells attached to the cellular substrate only through the hole in the center of the filter. Outside that area the cells attached to the filter itself. After 24 hours the filters were removed leaving the labeled cells inside the area circumscribed by the circle marked on the coverslip. Several cultures were fixed each day for 7 successive days and autoradiograms were prepared. Two controls were run. The first was identical to the series with chondrocyte substrates, except that the control substrates were plasma clots. In the second control, the filters were left in place for 24 hours on cellular substrates, but no myogenic cells were added. Then the filters were removed and labeled cells were added. The latter control was carried out to determine if the presence of a filter altered the substrates so that labeled myogenic cells either could not attach to them or would divide and form muscle on them. Examination of the autoradiograms revealed that the labeled myogenic cells did not migrate outside of the circle marker when plated on cellular substrates (Fig. 220). In the controls on plasma clots, mitotic figures were observed, myotubes formed, and the cells migrated out of the circle for distances up to 10 millimeters from the edge of the marker. Labeled myogenic cells attached to but did not divide on cellular substrates which had been incubated with millipore filters for 24 hours before addition of the cells. This experiment suggested that in addition to suppressing mitosis, underlying cellular substrates also inhibited the movement of myogenic cells. These 2 effects would complement each other in suppressing muscle differentiation.

It is not known whether living cells are required in cellular substrates to effect interference with myogenesis. Myogenic cells divided and formed muscle on cell layers which had been fixed with absolute alcohol or irradiated with ultraviolet light (Fig. 221). No muscle formed on substrates treated with 2.5 per cent glutaraldehyde, but washing may not have removed all of the aldehyde. These experiments were aimed at destroying the cells in the substrate without severely altering the extracellular matrix. Although the results do not lend themselves

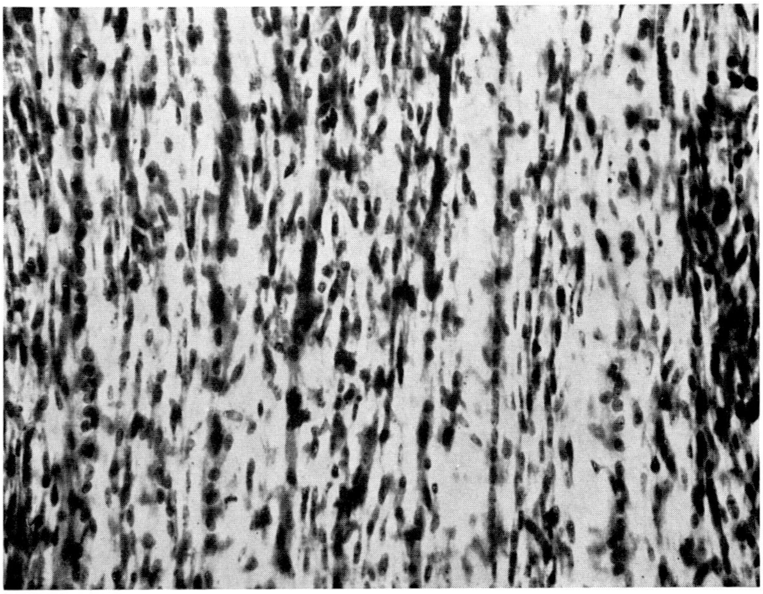

Fig. 221 A. Myotubes on an alcohol-fixed chondrocyte substrate. $\times$ 225.

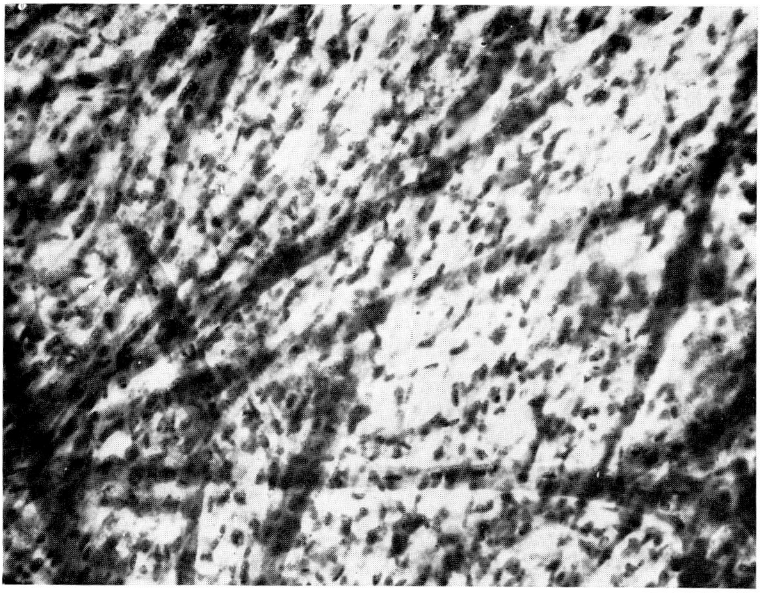

Fig. 221 B. Myotubes on an ultraviolet-irradiated chondrocyte substrate. × 180.

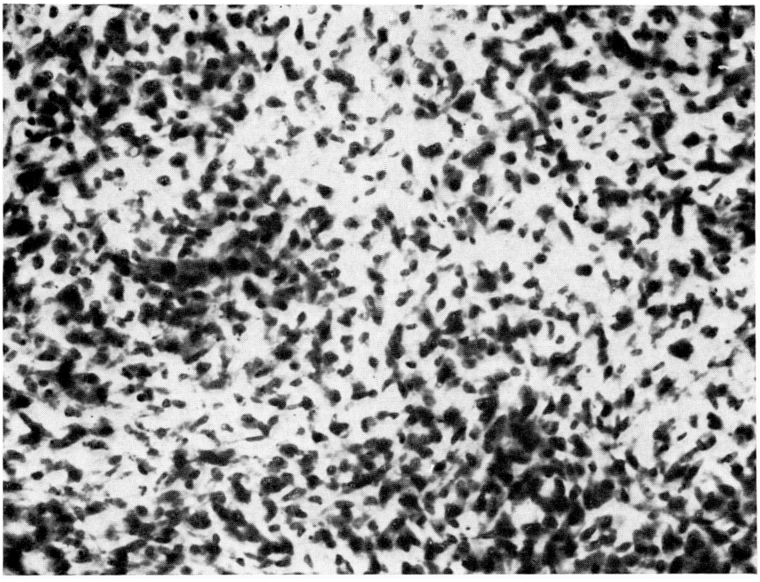

Fig. 221 C. Glutaraldehyde-fixed chondrocyte substrate on which myogenic cells were plated. No muscle appeared. × 225.

readily to interpretation, the following observation may yield a clue concerning the absence of suppression in these cultures. Cellular substrates with ^{3}H-thymidine-labeled myogenic cells added to them were sectioned perpendicularly to the surfaces of the cultures. Autoradiograms were prepared and the position of the labeled cells with reference to the extracellular matrix and cells in the substrate was noted (Fig. 222). Some of the labeled cells were on the surface of the cultures. Many, however, were located below the surface at depths ranging up to several cell diameters. The cells below the surface usually were entirely surrounded by extracellular materials and were not in contact with the surfaces of cells in the substrate. We do not know whether the 'buried' cells actively burrowed down into the underlying layer or were enveloped by material secreted around them by the original substrate cells. On substrates fixed in alcohol or irradiated by ultraviolet rays the labeled cells were not found below the surface of the killed cultures. This observation may indicate that the extracellular materials must surround myogenic cells to suppress their movement and mitotic activity. Hence, although the matrix may be the active material, suppression may require living cells to secrete that material around the myogenic cells.

In view of the proposal that collagen may be specifically required for myogenesis (Hauschka and Konigsberg, 1966), it is worth noting that the extracellular matrices of the tissues, which suppress muscle differentiation, contain collagen. If the contacts established between myogenic cells and cellular substrates are cell-matrix contacts, then either some other material present in the matrix abolishes the postulated collagen effect or collagen is not a specific promoter of myogenesis. Polysaccharides present in the extracellular materials (Nameroff and Holtzer, 1967) may be likely candidates for the role of mitosis and movement inhibitors.

In summary, these experiments demonstrate a reversible interference with muscle differentiation by heterotypic and homotypic cells, their associated interstitial matrices, or

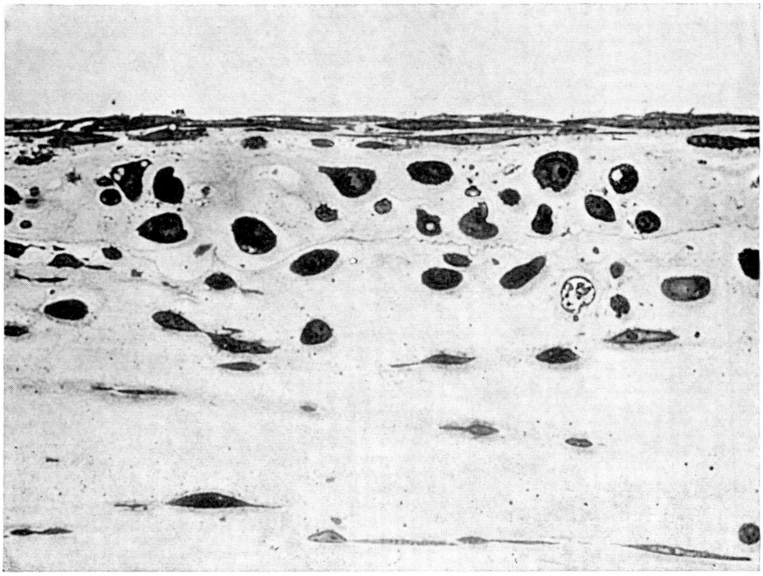

Fig. 222 A. Cross section of a normal chondrocyte culture. Fixed with glutaraldehyde on day 8 *in vitro*. The entire depth of the culture is visible. Note the perichondrium-like layer of flattened cells on the top surface. There has probably been some swelling during processing. × 360.

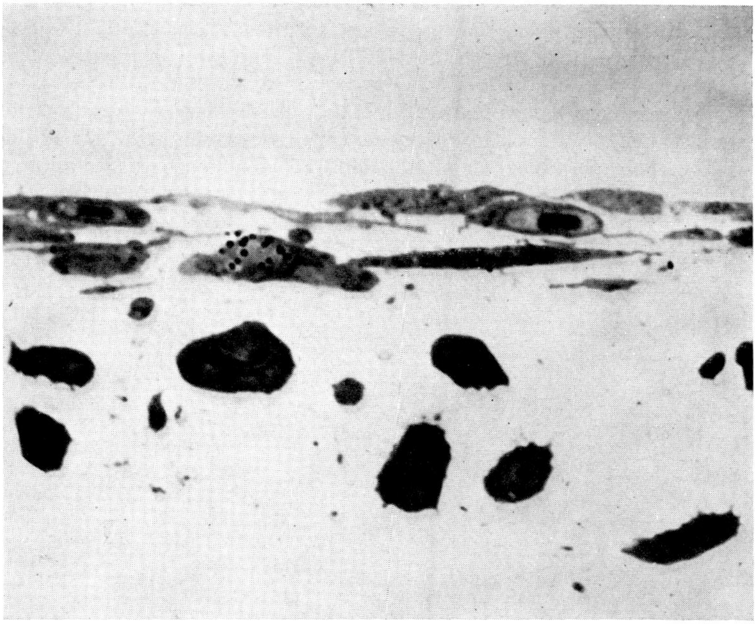

Fig. 222 B. Autoradiogram of labeled myogenic cells on a chondrocyte substrate. Cross section of the culture. The labeled cells are below the surface and are surrounded by extracellular matrix material. They are not in direct contact with the cells in the substrate. × 720.

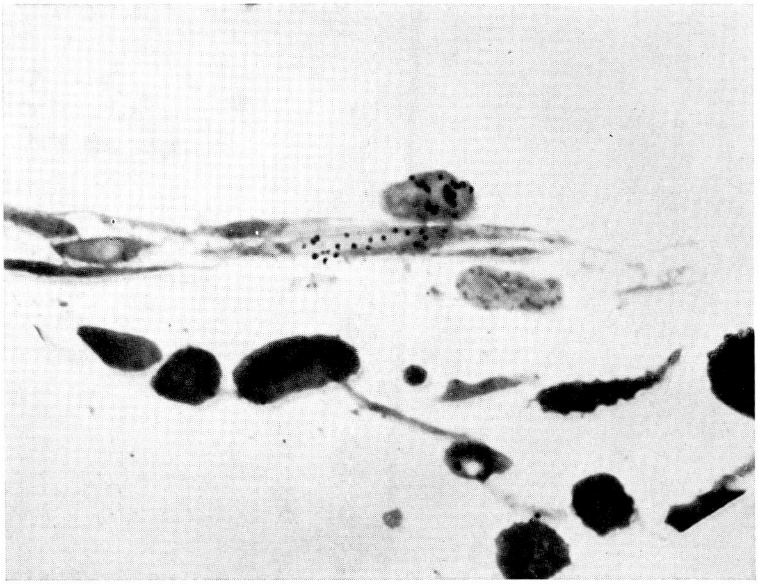

Fig. 222 C. Cross section as in B. One labeled cell is on the surface of the culture and one is in the perichondrial layer. This is an unusual field since few cells were found on the surface after 24 hours *in vitro*. × 720.

both. They are consistent with the hypothesis that such interference acts by inhibiting mitosis and movement of muscle precursor cells rather than by preventing fusion or synthesis of muscle-specific macromolecules (Nameroff, 1968). Inhibition of movement reduces the possibility of contacts between cells that are able to fuse. Inhibition of mitosis also acts to reduce cell-to-cell contacts by keeping the population density low. Mitotic suppression also may keep developmentally immature myogenic cells from giving rise, through division, to progeny capable of fusion. Whether or not interference acts *in vivo* during normal histogenesis is as yet unknown.

THE FORMATION OF PRESUMPTIVE MYOBLASTS IN INJURED AMPHIBIAN SKELETAL MUSCLE*

MICHAEL E. BARBERIE

The search for the origin of new muscle cells in the regeneration of skeletal muscle has centered around 2 possibilities: reserve cells or muscle itself. The latter source has been proposed for the amphibia where the formation of presumptive myoblasts has been described as a result of fragmentation of injured muscle fibers into nucleated and non-nucleated parts (Thornton, 1938; Hay, 1959). The nucleated fragments were reported to be the source of myoblasts for muscle regeneration. To study this mode of possible myoblast formation further, adult salamanders (*Triturus viridescens*) were injured by amputation through the forelimb and allowed to regenerate for 24 hours to several weeks before pieces of injured muscle were removed and prepared for examination with electron microscopy and radioautography. One of the early changes observed in injured muscle fibers was the dilatation of sarcoplasmic reticulum and Golgi cisternae. Swollen sarcoplasmic reticulum tubules of relatively constant diameter were evident throughout the deeper parts of the injured fiber. The diameter of the tubules of one fiber, however, often varied from that of the neighboring fiber, apparently reflecting stages in the progressive swelling of sarcoplasmic reticulum. The sarcoplasmic reticulum was especially swollen near the muscle nuclei where it is possible that the relative sparsity of myofibrils allows more extensive swelling than can be achieved deeper in the fiber where surrounding myofibrils restrict swelling. Serial sections of the area around muscle nuclei revealed vesicles that were thought to result from the fragmentation of swollen sarcoplasmic reticulum. The vesicles were arranged in rows delineating the peripherally placed muscle nuclei from the rest of the fiber. Individual vesicles were often in close proximity to one another and many times appeared to have incomplete membranous walls between themselves, suggesting fusion of neighboring vesicles. Other muscle nuclei were surrounded by large membrane-lined clefts that almost completely separated the nucleus and its narrow surrounding rim of cytoplasm from the fiber; only a few strands of membrane or cytoplasm connected the nucleated fragment to the fiber (Fig. 223). These static images suggested that there was a continuous progression of events near the nucleus starting with the swelling and fragmentation of sarcoplasmic reticulum and Golgi cisternae, and progressing through stages of vesicle fusion until, finally, extensive fusion of vesicles detaches cytoplasm-enclosed muscle nuclei from injured fibers. The detaching nuclei remained large and euchromatic and the narrow rim of surrounding cytoplasm contained numerous free ribosomes, some polysomes, a few small mitochondria and Golgi cisternae (Fig. 223). The nucleated fragments thus resembled cells described by other investigators as presumptive myoblasts (Bergman, 1962; Hay, 1963; Price *et al.*, 1964). In addition to these morphological similarities further suggestion of the possible myoblastic capacity of detached muscle nuclei is offered by radioautography. In the present study correlating thick radioautographic sections with the same cells observed in thin sections showed that when tritiated thymidine was made available to injured muscle fibers for 1 hour, the muscle nuclei that picked up the

* Aided by funds from NIH grant FR 05431-05.

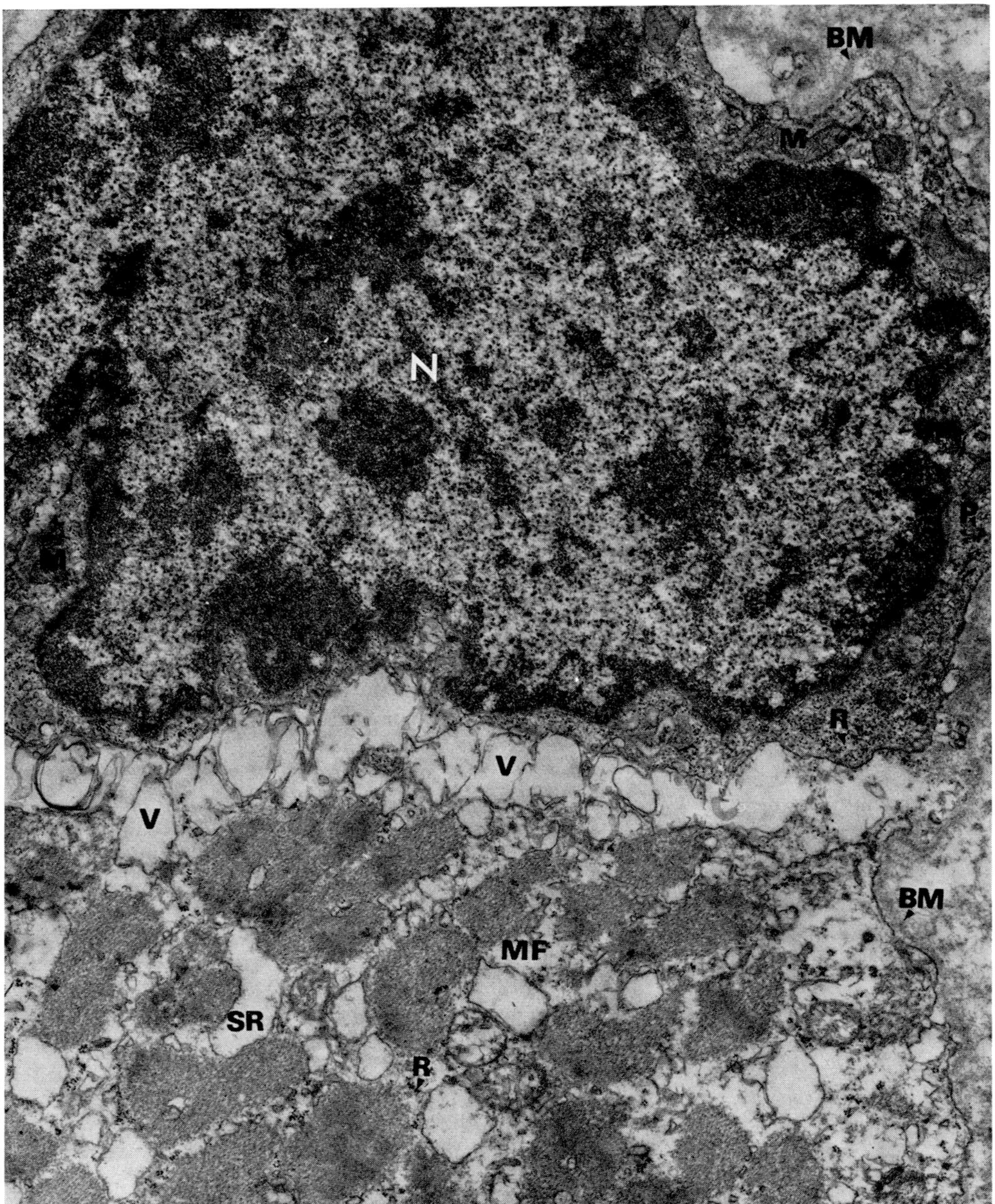

Fig. 223. Electron micrograph of a portion of injured muscle fiber ($\times$ 14,000). A peripherally placed muscle nucleus (N) is almost completely separated from the injured muscle fiber (MF) by a row of fused vesicles (V). The narrow rim of cytoplasm surrounding the nucleus contains mitochondria (M) and ribosomal accumulations (R), including polysomes (P). The basement membrane (BM) of the fiber is evident in places. Swollen sarcoplasmic reticulum (SR) and clusters of ribosomes (R) occupy the interfibrillar cytoplasm.

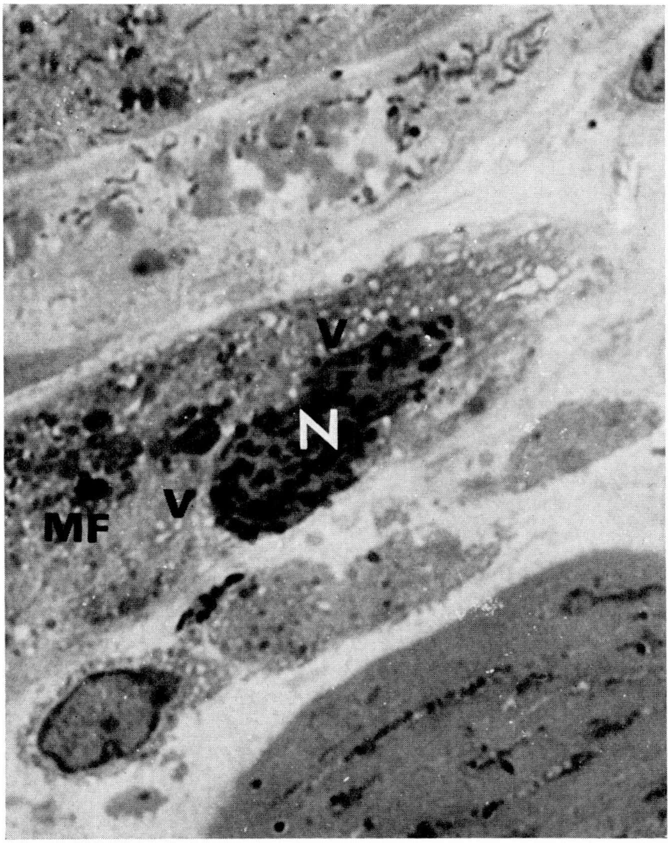

Fig. 224. Radioautograph of injured muscle fiber (× 5,000). A detaching muscle nucleus (N) is heavily labeled with tritiated thymidine 1 hour after exposure. Vesicles (V) surround and delineate the nucleus from the rest of the injured muscle fiber (MF). Surrounding muscle fibers exhibit a variety of post-injury changes.

label (unlike nuclei of normal muscle which never takes up the label) were also surrounded by the type of vesicles which have been described here as leading to detachment. The vesicles surrounding labeled muscle nuclei were often visible in the light microscope (Fig. 224). This uptake of tritiated thymidine indicates that at least some detaching nuclei are actively synthesizing DNA (and therefore are not dying) and gives additional support to the proposal that nucleated fragments are presumptive myoblasts since labeled nuclei appear in myotubes a few days or more after the isotope is made available to injured muscle.

It is interesting that the satellite cell of muscle, which is gaining considerable recognition as a possible source of myoblasts (Shafiq and Gorycki, 1965; Church *et al.*, 1966; and others), was seen only very rarely in normal salamander muscle. Several weeks after muscle injury the number of cells meeting the description of satellite cells was moderately increased, but it was not clear whether these cells were progeny of proliferating satellite cells or whether they represented nucleated fragments of injured muscle that were retained within the basement membrane of the fiber.

Conclusion

In the regeneration of amphibian skeletal muscle, presumptive myoblasts are formed when cytoplasm-enclosed muscle nuclei are detached from the fiber. Detachment occurs when vesicles, derived from swollen membranous systems of the injured fibers, gradually coalesce to form larger and larger spaces that separate muscle nuclei from injured fibers. The appearance of detaching nucleated fragments and their uptake of tritiated thymidine suggest that they are presumptive myoblasts.

A. MAURO: Thank you, Dr. Barberie, for your clear presentation.

E. D. HAY: Well, that puts 3 of us in that school. (Hay, 1959; Reznik, this volume, pp. 133-156).

A. MAURO: Yes . . . I think so. The hypothesis of the fragmenting muscle cell has been very well represented in the last 2 days.

Dr. Venable, would you like to give us your presentation?

J. H. VENABLE: My work has really not been in regeneration, nor in myogenesis, but rather in muscle growth.

I want to show you some of the problems that you run into when you try to do a total quantitative study in a whole muscle.

The basic design of my experiment is to find a muscle which has the normal morphology that you see in the adult but which is still showing myogenesis in terms of mitotic activity of specific cell types.

ACKNOWLEDGEMENT

Dr. Barberie gratefully acknowledges the aid and stimulation generated by Dr. J. David Deck during the course of his study.

TRIAL ANALYSIS OF THE CYTOKINETICS OF A RAPIDLY GROWING SKELETAL MUSCLE

JOHN H. VENABLE and MICHAEL D. LORENZ

It is well documented for several muscles of laboratory rodents that during the early phases of postnatal muscle growth the numbers of cell nuclei are increasing, even though the numbers of muscle fibers are stabilizing (Goldspink, 1962; Enesco and Puddy, 1964; Kochakian *et al.*, 1964). To what extent each of the cell types multiply is not wholly understood, but their total numbers are all enhanced. From what forebearers do each of these cells arise?

The purpose of our continuing study into the cytology of muscle tissue is to develop a statistical representation of the proliferation and fate of 5 types of cell nuclei in a single growing muscle. The nuclei of interest belong to muscle fibers, endothelium, fibroblasts, pericytes (undifferentiated perivascular cells), and satellite cells (Mauro, 1961). The muscle studied is the *m. gracilis* in Charles River, CD-1, mice. The period of growth in question falls between the 9th and 10th postnatal day, when the mice weigh 7 and 8 grams, respectively. Three parameters are measured on each of the test days. The first is the percentage representation of each of the 5 nuclear types, these 5 types accounting for all but 1 per cent of the cells in the muscle. The second is the per cent labeling accomplished in each of these nuclear types by a single 12.5-microcurie dose of tritiated thymidine administered 1 hour before the 9-day reference point. And the third is the total number of nuclei in the muscle as estimated from its DNA content. By multiplying the percentages of each of the nuclear types by the total number of nuclei, all data can be converted to absolute nuclear numbers.

The growing *m. gracilis* is the same muscle studied by Kityiakara and Angevine in 1963. From a thymidine labeling experiment, they came to a conclusion that myoblastic activity is confined primarily to the ends of this muscle. This interpretation is made on the basis of a higher percentage of thymidine labeled nuclei at the muscle's insertions in contrast to its middle. Although their data are accurate, their interpretation is open to question. The fact is that labeled nuclei can be found anywhere within the muscle during the period of postnatal growth. The labeling percentage is higher at the ends only because fewer unlabeled nuclei have had time to accumulate there. The per cent labeling in our study is calculated from randomly selected blocks from all parts of the muscle and is more representative of the total muscle than it is of any particular region.

The technical basis for the data collection comes from 2 similar studies: (1) an analysis of hypertrophy in the *m. levator ani* (Venable, 1966*a*) of mice, and (2) a study of nuclear proliferation in growing rats (Enesco and Puddy, 1964).

The nuclear identifications and enumerations made in the present experiment depend upon light-microscopic examination of sections 0.75 micron thick made from epoxy-embedded muscle (Figs. 226-229). The criteria used for nuclear identifications have been previously published (Venable, 1966*b*). The critical points, however, are illustrated in Figure 225.

A major problem in designing these experiments has been the choice of an optimum age period where the cells can be identified by the means established for adult mice, yet where cellular proliferation remains rapid enough to produce a significant nuclear increase during a 24-hour period. The 24-hour period is chosen because we suspect that within that time the

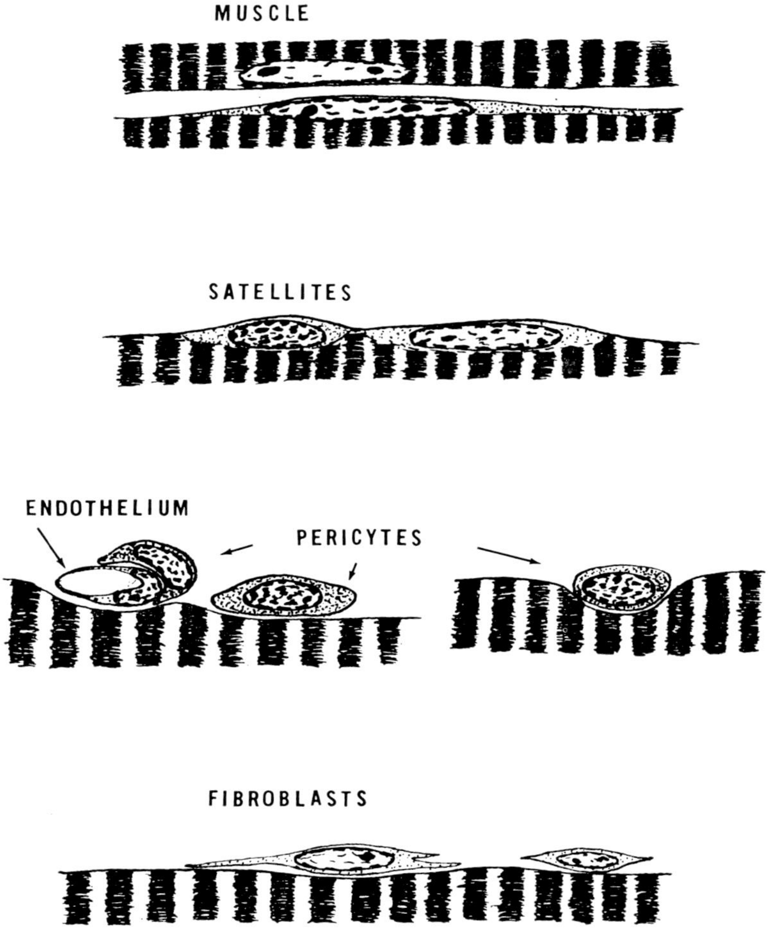

Fig. 225. Types of cells in primary bundles of skeletal muscle.

Muscle nuclei have nuclear envelopes abutting the dark A-bands without intervening space. Chromatin is minimally clumped, primarily at the nuclear envelope. Nucleoli are common and large.

Satellite cell nuclei have envelopes separated from the nearest dark A-band by a clear space approximately 0.4 micron wide (no more than 0.5 and no less than 0.2). This correlates with intervening double plasmalemmae and thin shell of non-fibrillar cytoplasm seen by electron microscopy. Chromatin clumping is abundant and coarse. Nucleoli are small if seen. Silhouettes of satellites blend with contours of the muscle fibers.

Pericytes are morphologically similar to satellite cells but generally show more chromatin clumping, are smaller, are separated from the nearest dark A-band by at least 0.5 micron, and do not accommodate the contours of the muscle fiber as smoothly as the satellite cells.

Fibroblasts show light cytoplasm streaming from their nuclear poles. Their nuclear envelopes are separated from the nearest dark A-band by a space at least 0.5 micron wide. The chromatin clumping is lessened in comparison with satellites and pericytes and a nucleolus is commonly present.

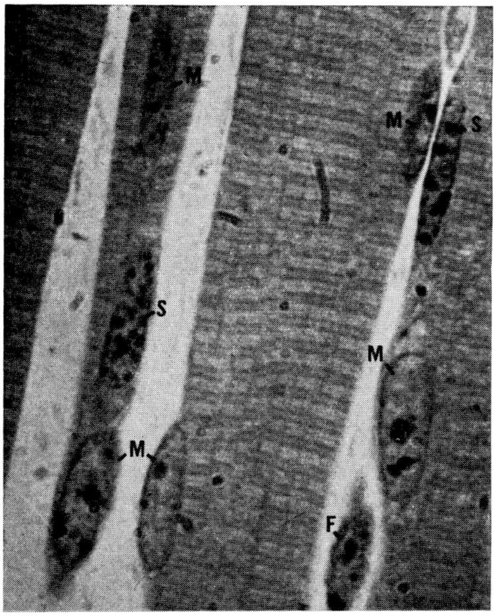

'ig. 226

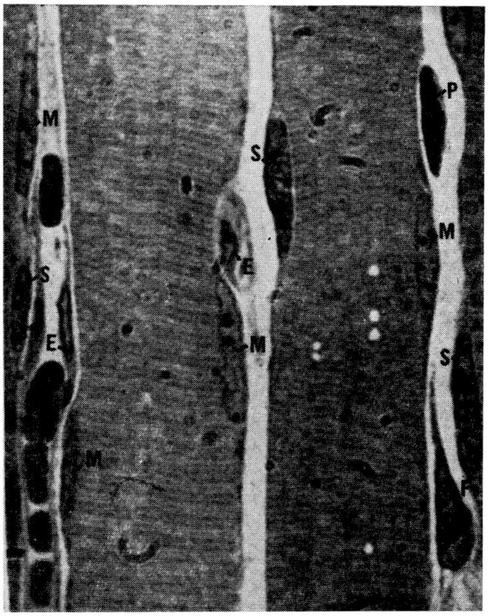

Fig. 227

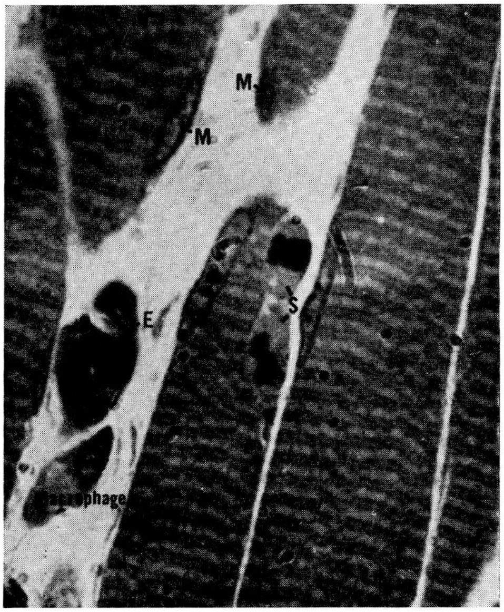

'ig. 228

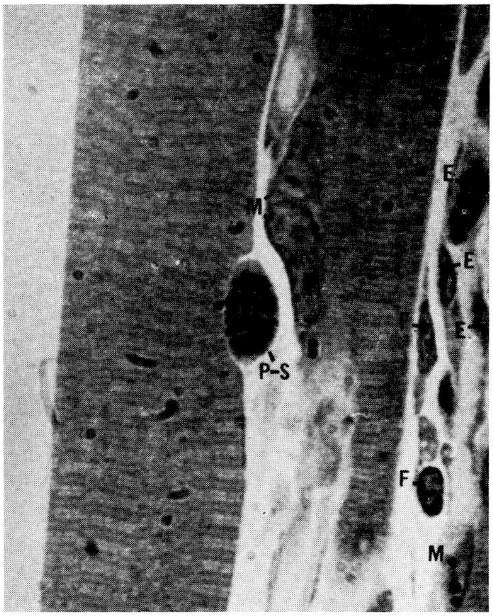

Fig. 229

Figs. 226-229. Nuclear identifications in 0.75-micron thick epoxide sections stained with 1 per cent toluidine blue in 1 per cent sodium borate.

Nuclei labeled M are true nuclei of the muscle fibers. Likewise, nuclei of endothelial cells are labeled E, those of pericytes P, those of satellite cells S, and those of fibroblasts F. The P-S in Figure 229 represents a pericyte-satellite 'transitional', a cell which has the intimate association with the muscle fiber typical of a satellite cell, but bulges free of the muscle fiber's contours as if still free in the endomysial space. This same cell is in prophase and it is possible that the mitotic process causes the unusual bulging of what otherwise would be a satellite cell.

Satellite cell nuclei in these Figures (S) all show a white line between the nuclear envelope and the nearest A-band. Their chromatin pattern is coarser than that of muscle nuclei. Figure 228 shows two daughter cells of a satellite cell just completing mitosis. × 700.

TABLE XII

Per cent encounters of nuclear types in m. gracilis of 7 mice, 9 and 10 days of age

	9 days old					10 days old					
Animal number and body weight	517 (6.8 g.)	518 (6.9 g.)	519 (7.1 g.)	$\bar{x} = 6.95$ g.		520 (8.4 g.)	521 (8.2 g.)	522 (7.8 g.)	523 (7.3 g.)	$\bar{x} = 7.92$ g.	
Nuclear type				Pooled data	Per cent					Pooled data	Per cent
Muscle	75	111	112	298	46.9	83	101	111	75	370	47.2
Endothelium	28	27	48	103	16.2	31	21	40	29	121	15.5
Fibroblast	39	36	39	114	18.0	21	35	23	35	114	14.5*
Pericyte	9	7	16	32	5.0	8	10	18	18	54	6.9
Satellite	13	21	34	68	10.7	26	22	35	23	106	13.5
Satellite-pericyte transitionals	7	6	2	15	2.4	3	0	2	7	12	1.5
Total satellite-pericyte forms	29	34	52	115	18.1	37	32	55	48	172	21.9*
Other nuclei	4 (Schwann)	1 (mast)	0	5	0.8	1	4	2	0	7	0.9
	Total cells 635					Total cells 784					

Pooled data on mitotic figures

Pericytes	2/86	MI = 0.023
Satellite cells	4/174	MI = 0.023
Endothelium	2/224	MI = 0.009
Fibroblast	1/228	MI = 0.004
Muscle	0/668	MI = 0.000

* Significant chi squares

$\chi^2 = 0.011 + 0.298 + 6.394^* + 5.232^* + 0.142$

mitotically active cells experience no more than 2 generation cycles. The greater the number of generation cycles, the more difficult it is to postulate a model to fit the data.

Newborn mice are unsuitable subjects because their muscles contain many variably sized myoblastic cells, both mononucleated and polynucleated, a condition which presently defies interpretation by light microscopy on the basis of the 5 nuclear types found in mature muscles.

By the 4th postnatal day, we begin to find typical satellite cells, fibroblasts, and pericytes. By the 9th day, 99 per cent of the cells can be classified into one of the 5 major types. The satellite cells can be identified at this age as undifferentiated cells nested into the silhouette of normal appearing muscle fibers. These nuclei are separated from the closest myofibrils by a 'white line' no more than 0.5 and no less than 0.2 micron thick. Pericytes and satellite cells have a noticeably dark, condensed chromatin pattern distributed throughout the karyoplasm. If they contain a nucleolus, it is small. In clear contrast, muscle nuclei and fibroblasts have noticeably larger nucleoli and possess chromatin clumped primarily at the nuclear membrane.

J. C. T. CHURCH: Dr. Venable, would you show us the vessel alongside that (undifferentiated) cell you call a perivascular cell?

J. H. VENABLE: The vessel is not there. I do not think undifferentiated cells with the morphology of a pericyte are always located in a perivascular position. They may move. The section can also be cut in such a way that the juxtapositioned vessel lies either above or below the pericyte, out of the plane of section. I differentiate 'pericytes' on the basis of an undifferentiated cellular morphology and a position in the endomysial space, whether or not I can see an adjacent blood vessel. However, most of these cells will be found alongside capillaries.

The morphological criteria applied in the differentiation of the 5 nuclear types at the level of light microscopy correlate specifically with the characteristics seen by electron microscopy, but as yet we have not done comparative differential counts both by light and electron microscopy. Significant mean counts require over 600 individual nuclear identifications on ½ square millimeter sections, a number practically achievable by light microscopy, but a stupendous task when attempted by electron microscopy. Neither have we corrected for differing nuclear dimensions in relation to the thickness of the sections (Abercrombie, 1946) a factor which may change our relative percentage figures somewhere between 1 and 5 per cent (Venable, 1966*b*), raising the figures for the smaller pericyte and satellite nuclei, lowering the figures for the larger muscle nuclei. It is assumed here that these errors are consistent for both the 9th and 10th days and thus cancel out to some extent for purposes of comparison.

Recognizing these errors, we have proceeded, nevertheless, with the light microscopic enumerations and have estimated the percentages of each nuclear type in 3 9-day and 4 10-day old mice and have made preliminary comparisons. Table XII shows the results. There are only 2 significant differences in the percentages of the different nuclei on the 10th day as compared to the 9th day. At 10 days the percentage of fibroblasts has decreased and the percentage of undifferentiated cells, pooled from the classifications of pericytes, satellites, and pericyte-satellite transitions, has increased. The reality of this difference is open to question on the basis that the differentiation of fibroblasts, pericytes, and satellite cells represents the more difficult decisions and one simply could be measuring a change in bias of the observer. But we remain unconvinced that our error in cellular identification explains this difference entirely.

As to mitotic activity, we have been somewhat disappointed at the level of mitosis found in 9-day-old mice (7.0 grams body weight). In this preliminary analysis we find the mitotic index for muscle nuclei to be 0.000, the mitotic index for pericytes and satellites to be 0.023, that for endothelial cells 0.009, and that for fibroblasts 0.004. This is not the amount of

TABLE XIII

Distribution of ^{3}H-thymidine in m. gracilis of 9- and 10-day-old mice

Nuclear type	1 hour after injection				24 hours after injection					
	Per cent	Number of cells (× 10⁶)	Number labeled / Number counted	Labeled cells (× 10⁶)	Per cent	Number of cells (× 10⁶)	Increase in number of cells (× 10⁶)	Number labeled / Number counted	Number of labeled cells (× 10⁶)	Increase in labeled cells (× 10⁶)
Muscle	46.9	9.536	0/298 = 0.000	---	47.2	10.463	0.927	9/370 = 0.024	0.251	0.251
Endothelium	16.2	3.294	7/103 = 0.070	0.230	15.5	3.436	0.142	10/121 = 0.083	0.285	0.055
Fibroblast	18.0	3.660	1/114 = 0.010	0.033	14.5	3.214	−0.446	5/114 = 0.044	0.141	0.108
Pericytes	5.0	1.017	21/32 = 0.030	0.031	6.9	1.530	0.513	5/54 = 0.093	0.142	0.111
Satellites	10.7	2.175	5/68 = 0.074	0.161	13.5	2.993	0.818	18/106 = 0.170	0.509	0.348
Satellite-pericyte transitions	2.4	0.488	2/15 = 0.133	0.065	1.5	0.332	−0.156	2/12 = 0.167	0.055	−0.010
Satellite-pericyte forms	18.1	3.680	8/115 = 0.070	0.257	21.9	4.855	1.175	25/172 = 0.145	0.706	0.469
Others	0.8	0.163	0/5 = 0.000	---	0.9	0.199	0.036	0/7 = 0.000	---	---
Total		20.333*	16/635 = 2.52	0.520		22.167*	1.834	49/784 = 6.25	1.383	0.883

* Number of nuclei on day 9 and 10 are calculated from a DNA growth curve plotted from analysis of muscles from mice of the same stock, body weight, and muscle size. The mean and standard deviation for DNA content on day 9 is 12.2 ± 0.3 micrograms; on day 10, 13.3 ± 0.9 micrograms. These figures are divided by 0.6 picograms DNA per nucleus (a figure for mouse hepatocytes) to estimate total numbers of nuclei. The mean DNA content contains greater error than the standard deviation of figures read off the growth curve indicates since the curve itself is constructed from mean DNA determinations each of which has a standard deviation between 10 and 20 per cent of the value of the mean.

activity for which we hoped, yet it proves to be enough to make a first order approximation of the percentage of each cell type capable of incorporating tritiated thymidine.

In Table XIII we have taken our percentages for each cell type and multiplied them by our estimates of the total number of cells in the muscle (20.3 million cells on the 9th day and 22.2 million cells on the 10th day) giving us estimates of the actual numbers of each nuclear type in the whole muscles. Similarly, the estimated number of labeled cells of each type has been calculated.

The estimates of total nuclei in the muscle are based respectively on a mean of 12.2 micrograms of DNA in the gracilis of 9-day-old, 7-gram mice, and 13.3 micrograms of DNA in the gracilis of 10-day-old, 8-gram mice. The mice providing the DNA figures are of the same age, mean body weight, and breeding as the mice providing the cellular percentages, but are not the same individual animals.

By noting the indicated change in the various populations of nuclei over the 24-hour period some interesting points develop.

One unmistakable result is that no muscle nuclei are labeled initially, but by 24 hours 250 thousand labeled muscle nuclei appear among a total population of 10.5 million, a total population expanded by 927 thousand in the same time period. The labeled muscle nuclei likely come from some cell labeled on the 9th day. The satellite cell would seem to be the most likely candidate (Moss, 1969).

One notes also the change in the total number of labeled nuclei of all types. Approximately 520 thousand nuclei are labeled on the 9th day, half of them belonging to the pericyte-satellite populations (257 thousand), the other half belonging mainly to endothelial cells (230 thousand). A relatively few fibroblasts are labeled (33 thousand) but this is based on the identification of a single labeled fibroblast. One day later, the 10th day, the 520 thousand labeled nuclei have multiplied to 1.3 million nuclei, somewhat more than a doubling. Half of the labeled nuclei are in the pericyte-satellite classifications (706 thousand) while only 25 per cent are endothelial (285 thousand). For these categories this represents a 275 per cent change and relatively no change, respectively. The remaining 25 per cent are, primarily, labeled muscle nuclei (250 thousand) and, secondarily, labeled fibroblast nuclei (141 thousand).

Considering the higher mitotic indices of the pericytes and satellites (0.023) as opposed to the endothelial cells (0.009), one can rationalize the labeling figures by postulating that the labeled pericytes and satellites have completed 2 mitotic cycles, the labeled endothelium less than 1, and that the satellite population has provided the new, labeled muscle nuclei. But, while this provides an answer for the increase in labeled cells, it creates other discrepancies.

If one assumes that labeled satellites can quadruple in number in 24 hours, one is forced to assume at the same time that all satellite cells do not participate in this mitosis. Look at the population totals. There are 2 million satellites on the 9th day. If all of them completed 2 mitotic cycles we would have to account for 8 million of these cells on the 10th day. One cannot account for such an increase since no more than 2 million new nuclei have accumulated from all sources. It seems more rational to assume that only some of the cells identified as satellite cells are mitotically active. The same considerations must be true for pericytes if our population statistics are anywhere within 20 per cent of the true figures. All of the pericytes cannot be mitotically active at any one time. In other words by functional criteria there are more classes of cells than we have accounted for on purely morphological grounds.

The data on fibroblasts gives a disconcerting result, one which throws doubt on the ability to identify fibroblasts. According to our calculation the population of fibroblasts decreases by some 12 per cent, not an easy figure to believe . We suspect then that the changes in percentages for fibroblasts and pericyte-satellite classifications over the 24 hours are not as great as these data indicate.

In summary, we say only that these cytological enumerations represent a first attempt at quantitating the total population kinetics of a single muscle. They are as yet crude and doubtless neither complete nor accurate enough to warrant serious acceptance as a cytological model for muscle growth. Yet we believe quantitative analysis of the cytological populations of muscle either by these or other means holds a key to understanding the cytological events we call myogenesis.

A. T. MILHORAT: How accurate would you require a DNA determination in order to make any sense from your numbers, because DNA determinations can be off by 10 to 20 per cent? (What will such errors do to your figures?)

J. H. VENABLE: In my experience they have not been off that much. On second thought, yes they have. The usual error as measured by the standard deviation of DNA analyses of muscles from 4 mice identical by the usual measures (body weight, age, breeding, etc.) fall between 15 and 20 per cent of the mean. The 12.2 micrograms and 13.3 micrograms DNA used in Table XIII contain that degree of error individually, but these means ($\bar{x}$) fall on a smooth curve covering body weights 1.0 through 13.0 grams. These means ($\bar{x}$) predict an 11 per cent increase in DNA between the 9th and 10th day. The 11 per cent increase in DNA seems an appropriate and most important figure in Table XIII. You can plug in any number of nuclei at 9 days, 1 thousand or 1 billion, and use the 11 per cent increase for the 10th day and arrive at the same conclusions. The true value for the per cent increase in DNA may range from 8 to 14 per cent, I do not know how to determine its error, but even these extremes provide similar conclusions about the shifts in population.

M. E. BARBERIE: I would like to know why you limit your transitional cells, as you call them, to transitionals between pericytes and satellites. Do you not have other differentiating cells which look for all the world like satellite cells?

J. H. VENABLE: Yes, there are other differentiating cell types, such as Schwann cells, but they represent less than 1 per cent of the total population and have been lumped together. (The transitional classification appearing in Tables XII and XIII simply represents undifferentiated cells positioned in regard to the silhouette of the muscle fiber in such a way that we cannot decide whether to call them pericytes or satellites. You see these by electron microscopy also, where a single fused glycolemma (basement membrane) appears between the muscle fiber and what otherwise would be called an indenting pericyte. I do not take this as evidence that pericytes become satellites or vice versa, and for this reason I would not use the term ‘transitionals’. One should keep an open mind as to what cell types might transform into others. We can list the 5 major cell types and postulate transformations interlinking them all. But the only evidence that any of these transitions do exist is the evidence that label moves from the mononuclear populations into the multinucleated skeletal muscle fibers. It would seem logical that the satellite cell is the most direct contributor.

A. MAURO: Dr. Venable, would you like to make some general comments, or even give us a summary of the proceedings of the past two days?

IN SUMMARY

JOHN H. VENABLE

Doubtless it is impossible to draw all the observations and interpretations presented at this conference into a unified whole. Within its theme – Regeneration of Striated Muscle, and Myogenesis – much receives common agreement, yet much sounds strange to one or another of us. We each look through tiny windows at one or more examples of the myogenetic phenomenon but never see the whole. Likely we each make valid observations and rational interpretations of what we see. But most of us fall woefully short of experience with all the nuances of the fundamental process. Conflicts arise because as yet we have not found the definitive experiment or group of experiments that form the basis of a unified theory.

As to the cytology of the regenerative process, we have disagreed primarily over the question of the origin of mononuclear, relatively undifferentiated cells wedged between the glycolemmae (basement membranes) and plasmalemmae of the muscle fibers. The question "From where do these cells arise?" repeatedly draws argument which oscillates between 2 schools of thought. One school would have these cells originate by dedifferentiation of muscle nuclei; another would have them arise from a pre-existing mononuclear population. The former concept stems from the sum of 3 circumstances: the similarity of position of these cells with that of normal muscle nuclei; the sightings of membranous tubes and vesicles around muscle nuclei, making the latter appear to be 'budding off'; and the fact that in the species most often studied by this school, urodeles and adult mice, the number of such cells found in unaffected muscles approaches the limit zero. The primary evidence that specific skeletal muscle nuclei have made the transition into mononuclear cells has been supplied at this conference by Dr. Steen. He has transplanted triploid and thymidine-labeled homografts of muscle tissue into regenerating limbs of axolotl and recovered labeled nuclei in the ultimately regenerated muscle. He even finds some of these labeled nuclei in the new cartilage. This evidence does not seem to clinch the issue, however, on 2 grounds. Firstly, his is a light-microscopic study of relatively thick sections. Thick sections are necessary for the detection of the triploid nuclei, but thick sections prevent optical resolution sufficient to differentiate muscle nuclei from mononucleated cells alongside muscle fibers. And secondly, the muscle homografts include many cell types other than muscle cells. Thus, we are stalemated by technical inadequacies – the specter that haunts us all.

A second school would have these subglycolemmal mononucleated cells arise solely by mitosis from a pre-existing mononucleate population, on the presumption that such cells are always present in normal skeletal muscle. This concept stems from the experience of researchers who have found these mononucleated cells without fail in all the normal muscles they have examined thus far. They admit that the relative numbers of such cells may approach a point where casual identification is unlikely. But they hold that they are always there. Of course, the rebuttal comes quickly that there are microscopists who have never seen these cells in their material, or who have seen them so rarely that it seems facetious to consider them the sole progenitor of the large numbers of mononuclear cells which appear at sites of muscle damage and repair. Again technical inadequacy precludes conclusion. No one can know if such cells are always present. And if they are present, there is no direct measurement of their mitotic rates to prove or disprove their ability to account for the numerous mononucleates seen in muscle regeneration.

There is also the middle ground. Here one accepts the possibility of 2 origins: one by vegetative proliferation from a morphologically similar precursor, and one by dedifferentiation from its closest differentiated neighbor, the skeletal muscle cell. Perhaps both can occur simultaneously. Perhaps the first can occur under one set of circumstances, the second under another.

Leaving this major contention for the moment, let us look at the remarkable number of points of fundamental agreement.

As our chairman, Dr. Mauro, exclaimed during one discussion, it seems that "no one knows of amitotic activity". We may all be relieved that the thought of muscle nuclei dividing by a budding process is absent from our discussions and is hopefully dead. May it rest in peace. No longer do we succumb to the interpretations of pinched profiles of nuclei as indications of division. No longer do we interpret clusters of nuclei in growing myotubes of reorganizing muscle fibers as products of amitosis. The tissue culturists among us have impressed us with the mobility of cells in general and of muscle nuclei in particular, a mobility capable of producing non-random distributions of the nuclear population. The sight of muscle nuclei in aggregation means simply that the nuclei are situated close together. No other interpretation is valid unless there is other quantifiable evidence that the nuclei in the area have indeed increased their number. We apparently agree that classical mitosis is the sole process by which pre-existing nuclei within a muscle system can multiply and that this mitosis is found in mononuclear forms rather than multinuclear forms.

The question of utilization of precursor cells from outside the muscle system during regeneration, being transported there via the blood, is as yet unanswered. Dr. Reznik's use of localized radiation on muscle prior to injury provides the typical evidence that muscle regeneration requires mainly cells which are situated locally rather than those which are imported with the wave of leukocytes indicative of the inflammatory response. But, Dr. Sloper cautions us from the experience of his work that, although imported cells may not be a major source of myogenetic elements, such contributions have not been excluded and we must be cognizant of their influence on statistics of mononucleated populations within any regenerating system.

A point mentioned earlier needs repeating in the context of common consent. In all regenerating systems, considerable numbers of mononucleated cells are present within the confines of the glycolemmae of skeletal muscle fibers. We still lack accurate information on their frequency of appearance in both normal and abnormal circumstances in all but a limited number of muscles in a limited number of species. Despite our minuscule knowledge of their true population size, their presence or absence in the normal muscle system before regeneration seems crucial to any discussion of the myogenetic process. Their significance is signalled by the 2 ubiquitous findings: (1) these are the only cells within the glycolemmae of muscle fibers that show mitotic activity; and (2) thymidine labeling apparent initially in these mononucleated cells subsequently appears in muscle nuclei. There is the inescapable conclusion that a mononucleated cell either contributes some of its progeny or its tritiated thymidine to a population later recognized as muscle nuclei. Students of the myogenetic process cannot avoid focusing their attention on these cells. Just what are they? From where do they come? What can they become? Do they represent a single population?

Before this conference, many of us thought we had unassailable ideas on these points; but a substantial number of us left unsure, less willing to press our views as the truest ones. You find much argument over these cells in the discussions. Besides their origins, many arguments arose over names: "Do you call that mononucleated cell a satellite cell?" – "Do you call that mononucleated cell a myoblast?" – "Do you call that mononucleated cell a potential macrophage?" This problem is a common one in descriptive histology and cytology. It is caused partially by the change in meaning which occurs when a modifying adjective or an adverb is advanced to the status of a noun. For example, when mononucleated cells lying satellite to muscle cells become known as *satellite cells*, our mental concepts narrow even though there has been no change in these cells' morphologic or genetic potentials. We are still talking about relatively undifferentiated mononuclear cells of unknown history and equally unknown future lying in that peculiar position satellite to muscle fibers.

Nomenclature, the plague of our communications! Mononucleated cells positioned beneath the glycolemmae of apparently normal muscle fibers and having relatively sparse undifferentiated cytoplasm are usually called 'satellite cells'. If morphologically similar cells are seen in obviously regenerating skeletal muscle systems, they might be referred to as 'presumptive myoblasts'. If myofilaments are detected in their cytoplasm, the straightforward term 'myoblast' might be applied. If the mononucleated cells are seen where the neighboring sarcoplasmic mass is fragmented and nearby cells contain lysosome-like granules, the term 'presumptive macrophage' might be used. Whether or not all these cells have a single precursor remains unknown.

Dr. Muir provides us with a fundamental discussion of the so-called satellite cells: that is to say, mononucleated cells that fit the description first given by Dr. Mauro in 1961. Since this description comes originally from cells seen in presumably normal skeletal muscles, the term satellite cell, used almost as a hyphenated noun, to some has come to mean 'a normal mononuclear cellular component of muscle tissue'. Dr. Muir, who himself once contributed, perhaps unintentionally, to this concept, wisely cautions us now on this point. We must keep the facts before us that these cells have not been described in investigations of normal skeletal muscle in invertebrates; it is seen rarely in urodeles and axolotl; and, as has been indicated previously, has been missed by competent electron microscopists studying mammalian skeletal muscles. Dr. Muir asks "Can the presence of these cells reflect localized myopathic processes rather than normal states?" He points out also that all so-called satellite cells are not morphologically identical, although they show pertinent similarities. He establishes clear doubt that we can find suitable morphologic criterion resolvable by light microscopy by which we can identify these cells positively and in every instance. The pleomorphism of satellite cells upsets simple thoughts about a single origin or a single cell set.

But, perhaps we dwelt too long on the significance of mononuclear cells in the regeneration of muscle fibers. Other parameters equally important to the initiation, course, and completion of the myogenetic process in the total muscle come into view, but remained poorly defined.

The special ability of urodeles to regenerate whole limbs remains a puzzle. Dr. Carlson shows us that regeneration within minced autografts in frogs and rats differs from that seen in salamanders. We wonder how the basophilic cuffs he sees around peripherally positioned muscle fragments in minced grafts placed in the former 2 species relate to the lack of such cuffs in the case of the salamander, where randomly oriented myoblastic activity is detected throughout the grafted mince. Dr. Aloisi, likewise, notes that muscle tissues of different species show different myogenetic capacities and characteristics. Yet, Drs. Allbrook, Holtzer, and others doubt that the phenomenon of myogenesis at the cellular level is fundamentally any different in any of these species. Between the levels of cellular and histological development, there are many controlling factors of myogenetic expression about which we know little.

It is apparent that further insights into the mode and control of the myogenetic process depends on experiments of more detail and sophistication. The usefulness of yet more morphological observation without additional measurement of correlated dynamic parameters approaches the null point. Cells must be identified functionally as well as morphologically. Dynamic processes within muscle tissues must be traced by appropriate labels, not by arrangements of groups of micrographs, the individual pictures of which cannot be accepted as belonging to a congruous set, and cannot be arranged quantitatively along any positive or negative time scale. Most studies presented at this conference show recognition of this principle.

For example, Dr. Bischoff uses thymidine labeling in culture systems to measure the time myoblasts spend in the various phases of their mitotic cycle. His quantified data coupled with the characterization of the morphological events of regeneration show that the 'fusion' of mononuclear myoblasts into the multinucleated fiber system cannot be triggered by the same 'switch' that halts DNA synthesis – a conclusion which could not be deduced by morphological observation alone. Use of controlled *in vitro* systems continues to provide valuable understanding of the myogenetic process. Dr. Nameroff's study of cultured myosarcoma cells and Dr. Kaighn's studies of the effects of Rous sarcoma virus on the uptake of thymidine in muscle nuclei of chick embryonic muscle are immediate examples. In a similar way, Dr. Church's work with bat wing muscles demonstrates the value of choosing model systems where experimental parameters are easily controlled.

Many of the labeling and *in vitro* studies presented at this meeting draw attention to a shift in interest toward control mechanisms at the cellular level. Dr. Holtzer, Bischoff, Nameroff, and others wonder at the nature of 'switches' directing cells of myoblastic potential to stop their mitosis and begin differentiation. Dr. Nameroff suggests that there must be at least 2 switches capable of shutting down mitosis in the myoblast. Drs. Holtzer asks "Why do we see satellite cells in mature muscles at all?" If they are committed to myoblastic activity and have a proclivity to fusion with the existing muscle fibers, they should quickly disappear. Why do some linger on, seemingly neither dividing nor differentiating? What would set off fusion when it occurs – or, for that matter, what molecular alterations would bring about dedifferentiation of muscle nuclei?

As our questions and techniques become more elaborate, the quest for answers becomes more exciting. We can thank the Muscular Dystrophy Associations of America for providing this opportunity to compare and challenge our individual thinking and send us each off in new spirit to the work ahead.

REFERENCES

ABBOTT, J. and HOLTZER, H. (1966): The reversible behavior of chondrocytes in primary cultures. *J. Cell Biol.*, *28*, 473.

ABERCROMBIE, M. (1946): Estimation of nuclear population from microtome sections. *Anat. Rec.*, *94*, 239.

ADAMS, R. D., DENNY-BROWN, D. and PEARSON, C. M. (1962*a*): In: *Diseases of Muscle, A Study in Pathology, 2nd ed.* Harper and Row, New York, N.Y.

ADAMS, R. D., DENNY-BROWN, D. and PEARSON, C. M. (1962*b*): Embryology and histology of skeletal muscle. In: *Diseases of Muscle, A Study in Pathology, 2nd ed.*, pp. 3-61. Harper and Row, New York, N.Y.

ALLBROOK, D. B. (1962): An electron microscopic study of regenerating skeletal muscle. *J. Anat. (Lond.)*, *96*, 137.

ALLBROOK D. B. and AITKEN, J. T. (1951): Reinnervation of striated muscle after acute ischaemia. *J. Anat. (Lond.)*, *85*, 376.

ALLBROOK, D. B., BAKER, W. DE C. and KIRKALDY-WILLIS, W. H. (1966): Muscle regeneration in experimental animals and in man. *J. Bone Jt Surg. (Edinburgh)*, *48B*, 153.

ALLEN, R. and PEPE, F. (1965): Ultrastructure of developing muscle cells in the chick embryo. *Amer. J. Anat.*, *116*, 115.

BADER, J. P. (1966): Metabolic requirements for infection by Rous sarcoma virus. In: *Subviral Carcinogenesis. Proceedings, First International Symposium on Tumor Viruses*, pp. 144-155. Editor: Y. Ito. Research Institute, Aichi Cancer Center, Nagoya.

BARER, R. (1948): The structure of the striated muscle fibre. *Biol. Rev.*, *23*, 159.

BASSLEER, R., COLLIGNON, P. and MATAGNE-DHOOSCHE, F. (1963): Effets cytologiques du Myleran sur le muscle strié de l'embryon de poulet in ovo et de la queue d'Axolotl en régénération. *Arch. Biol. (Liège)*, *74*, 79.

BATESON, R. G. (1968): *The Origin and Early Development of Skeletal Muscle in the Embryo, in Tissue Culture and, after Injury, in vivo, in the Chick, Mouse and Man.* Thesis, London.

BATESON, R. G., WOODROW, D. F. and SLOPER, J. C. (1967): Circulating cell as a source of myoblasts in regenerating injured mammalian skeletal muscle. *Nature (Lond.)*, *213*, 1035.

BENCOSME, S. A., STONE, R. S., LATTA, H. and MADDEN, S. C. (1959): A rapid method for localization of tissue structures or lesions for electron microscopy. *J. biophys. biochem. Cytol.*, *5*, 508.

BENNETT, H. S. (1958): The cytology of striped muscle. In: *Frontiers in Cytology*, p. 343. Editor: S. L. Palay. Yale University Press, New Haven, Conn.

BERGMAN, R. A. (1962): Observations on the morphogenesis of rat skeletal muscle. *Bull. Johns Hopk. Hosp.*, *110*, 187.

BERKOWITZ, L. R., FIORELLO, O., KRUGER, L. and MAXWELL, D. S. (1968): Selective staining of nervous tissue for light microscopy following preparation for electron microscopy. *J. Histochem. Cytochem.*, *16*, 808.

BETZ, E. H., FIRKET, H. and REZNIK, M. (1966): Some aspects of muscle regeneration. *Int. Rev. Cytol.*, *19*, 203.

BETZ, E. H. and REZNIK, M. (1964): La régénération des fibres musculaires striées dans différentes conditions expérimentales. *Arch. Biol. (Liège)*, *75*, 567.

BINTLIFF, S. and WALKER, B. E. (1960): Radiographic study of skeletal muscle regeneration. *Amer. J. Anat.*, *106*, 235.

BIRKS, R. I., KATZ, B. and MILEDI, R. (1959): Dissociation of the 'surface membrane complex' in atrophic muscle fibres. *Nature (Lond.)*, *184*, 1507.
BISCHOFF, R. and HOLTZER, H. (1968): The effect of mitotic inhibitors on myogenesis in vitro. *J. Cell Biol.*, *36*, 111.
BISCHOFF, R. and HOLTZER, H. (1969): Mitosis and the processes of differentiation of myogenic cells in vitro. *J. Cell Biol.*, *41*, 188.
BISCHOFF, R. and HOLTZER, H. (1970): Inhibition of myoblast fusion after one round of DNA synthesis in 5-bromodeoxyuridine. *J. Cell Biol.*, *44*, 134.
BOWMAN, W. (1840): On the minute structure and movements of voluntary muscle. *Phil. Trans. B*, *130*, 457.
BOYD, J. D. (1960): Development of striated muscle. In: *Structure and Function of Muscle, Vol. I*, Chapter 3, pp. 63-85. Editor: G. H. Bourne. Academic Press, New York, N.Y. Quoting Godlewski (1902) and Tello (1917, 1922).
BUTLER, E. G. (1933): The effects of X-radiation on the regeneration of the forelimbs of Amblystoma larvae. *J. exp. Zool.*, *65*, 271.
BUTLER, E. G. (1935): Studies on limb regeneration in X-rayed Amblystoma larvae. *Anat. Rec.*, *62*, 295.
BUTLER, E. G. and O'BRIEN, J. P. (1942): Effects of localized X-irradiation on regeneration of the urodele limb. *Anat. Rec.*, *84*, 407.

CARDINO, S. S. and STEINER, J. W. (1965): Improvements of staining techniques for thin sections of epoxy-embedded tissue. *Amer. J. clin. Path.*, *43*, 1.
CARLSON, B. M. (1968*a*): Regeneration of the completely excised gastrocnemius muscle in the frog and rat from minced muscle fragments. *J. Morphol.*, *125*, 447.
CARLSON, B. M. (1968*b*): Regeneration of the completely removed rat gastrocnemius muscle from transplanted fragments (Abstract). *Anat. Rec.*, *160*, 327.
CARLSON, B. M. (1970*a*): Regeneration of the rat gastrocnemius muscle from sibling and non-sibling muscle fragments. *Amer. J. Anat.*, *128*, 21.
CARLSON, B. M. (1970*b*): The regeneration of a limb muscle in the axolotl from minced fragments. *Anat. Rec.*, *166*, 423.
CARLSON, B. M. (1970*c*): The relationship between the tissue and epimorphic regeneration of muscles. *Amer. Zool.*, in press.
CAVANAGH, J. B. (1968): Effects of previous X-irradiation on the cellular response of nervous tissue to injury. *Nature (Lond.)*, *219*, 626.
CHACKO, S., ABBOTT, J., HOLTZER, S. and HOLTZER, H. (1969): Behavior of the progeny of a single chondrocyte. *J. exp. Med.*, *130*, 417.
CHALKLEY, D. W. (1954): A quantitative histological analysis of forelimb regeneration in Triturus viridescens. *J. Morphol.*, *94*, 21.
CHIAKULAS, J. J. and PAULY, J. E. (1965): A study of postnatal growth of skeletal muscle in the rat. *Anat. Rec.*, *152*, 55.
CHURCH, J. C. T. (1968): *The Satellite Cell of Skeletal Muscle.* Thesis, Cambridge.
CHURCH, J. C. T. (1969): Satellite cells and myogenesis; a study in the fruit bat web. *J. Anat. (Lond.)*, *105*, 419.
CHURCH, J. C. T. (1970): Cell populations in skeletal muscle after regeneration. *J. Embryol. exp. Morphol.*, *23*, 531.
CHURCH, J. C. T. and GRIFFEN, E. R. (1968): Mycobacterium buruli lesions in the fruit bat web. *J. Path. Bact.*, *96*, 508.
CHURCH, J. C. T. and NORONHA, R. F. X. (1965): The use of the fruit bat in surgical research. *E. Afr. med. J.*, *42*, 348.
CHURCH, J. C. T., NORONHA, R. F. X. and ALLBROOK, D. B. (1966): Satellite cells and skeletal muscle regeneration. *Brit. J. Surg.*, *53*, 638.
CLARK, W. E. L. (1946): An experimental study of the regeneration of mammalian striped muscle. *J. Anat. (Lond.)*, *80*, 24.
COLEMAN, J. R. and COLEMAN, A. W. (1968): Muscle differentiation and macromolecular synthesis. *J. cell. Physiol.*, *72*, *Suppl. 1*, 19.
COLEMAN, J. R., COLEMAN, A. W. and HARTLINE, E. J. H. (1969): A clonal study of the reversible inhibition of muscle differentiation by the halogenated thymidine analog, 5-bromodeoxyuridine. *Develop. Biol.*, *19*, 527.

CONEN, P. E., BALIS, J. U. and BELL, C. D. (1967): Problems in myogenesis terminology in electron microscopy. In: *Proceedings, Twenty-Fifth Anniversary Meeting, Electron Microscopy Society of America, Chicago, Ill.*, p. 38. Editor: C. J. Arceneaux. Claitor's Book Store, Baton Rouge, La.

COX, P. G. (1968): In vitro myogenesis of promuscle cells from the regenerating tail of the lizard, Anolis carolinensis. *J. Morphol.*, *126*, 1.

CUPPY, W. (1935): *How to Attract the Wombat*, pp. 101-103. Rinehart & Co., Inc., New York, N.Y.

DECK, D. (1968): Cytological changes in newt limb muscle after limb amputation. *Anat. Rec.*, *160*, 338.

DE RECONDO, J., FARDEAU, M. and LAPRESLE, J. (1966): Etude au microscope électronique des lésions musculaires d'atrophie neurogène par atteinte de la corne antérieure. *Rev. neurol.*, *114*, 169.

DE RÉNYI, G. S. and HOGUE, M. J. (1934): Studies on skeletal muscle growth in tissue cultures. *Arch. exp. Zellforsch.*, *16*, 167.

DRAPER, M. H. and HODGE, A. J. (1949): Studies on muscle with the electron microscope. I. The ultrastructure of toad striated muscle. *Aust. J. exp. Biol. med. Sci.*, *27*, 465.

DUESBERG, P. H. (1968): Physical properties of Rous sarcoma virus RNA. *Proc. nat. Acad. Sci. USA*, *60*, 1511.

DULBECCO, R., HARTWELL, L. H. and VOGT, M. (1965): Induction of cellular DNA synthesis by polyoma virus. *Proc. nat. Acad. Sci. USA*, *53*, 403.

ENDERS, A. C. (1965): Formation of syncytium from cytotrophoblast in the human placenta. *Obstet. and Gynec.*, *25*, 378.

ENESCO, M. and PUDDY, D. (1964): Increase in the number of nuclei and weight in skeletal muscles of rats of various ages. *Amer. J. Anat.*, *114*, 235.

ERIKSON, R. L. (1969): Studies on the RNA from avian myeloblastosis virus. *Virology*, *37*, 124.

EZERMAN, E. B. and ISHIKAWA, H. (1967): Differentiation of the sarcoplasmic reticulum and T system in developing chick skeletal muscle in vitro. *J. Cell Biol.*, *35*, 405.

FAWCETT, D. W. (1965): Surface specializations of absorbing cells. *J. Histochem. Cytochem.*, *13*, 75.

FIELD, E. J. (1960): Muscle regeneration and repair. In: *Structure and Function of Muscle, Vol. III*, Chapter 5, pp. 139-170. Editor: G. H. Bourne. Academic Press, New York, N.Y.

FIRKET, H. (1958): Recherches sur la synthèse des acides désoxyribonucléïques et la préparation à la mitose dans des cellules cultivées in vitro. Etude cytophotométrique et autoradiographique. *Arch. Biol. (Liège)*, *69*, 1.

FIRKET, H. (1966): Polyester sheeting (Melinex C), a tissue culture support easily separable from epoxy resins after flat face embedding. *Stain Technol.*, *41*, 189.

FIRKET, H. (1967): Ultrastructural aspects of myofibril formation in cultured skeletal muscle. *Z. Zellforsch.*, *78*, 313.

FISCHMAN, D. A. (1967): An electron microscope study of myofibril formation in embryonic chick skeletal muscle. *J. Cell Biol.*, *32*, 557.

FISCHMAN, D. A. and HAY, E. D. (1962): The origin of osteoclasts from mononuclear leucocytes in regenerating newt limbs. *Anat. Rec.*, *143*, 329.

FLINKER, D. and SIDMAN, R. (1953): The influence of infusions of colchicine on the nerve and the early regenerate of the forelimb of the adult newt, Triturus. *Anat. Rec.*, *117*, 577.

FLOOD, P. R. (1964): Myosatellite cells in myxine and axolotl. In: *Proceedings, III European Regional Conference on Electron Microscopy*, p. 575. Czechoslovak Academy of Sciences, Prague.

FOGEL, M. and DEFENDI, V. (1967): Infection of muscle cultures from various species with oncogenic DNA viruses (SV-40 and polyoma). *Proc. nat. Acad. Sci. USA*, *58*, 967.

FRANZINI-ARMSTRONG, C. and PORTER, K. R. (1964): Sarcolemmal invaginations and the T-system in fish skeletal muscle. *Nature (Lond.)*, *202*, 355.

FREEMAN, A. I. and JOHNSON, W. W. (1968): A comparative study of childhood rhabdomyosarcoma and virus-induced rhabdomyosarcoma in mice. *Cancer Res.*, *28*, 1490.

GALASSI, L. (1967): Reutilisation of maternal material by embryonic and trophoblastic cells in the rat for the synthesis of deoxyribonucleic acid. *J. Histochem. Cytochem.*, *15*, 573.

GALL, J. G. and PARDUE, M. L. (1969): Formation and detection of RNA-DNA hybrid molecules in cytological preparations. *Genetics*, *63*, 378.

GALLUCCI, V., NOVELLO, F., MARGRETH, A. and ALOISI, M. (1966): Biochemical correlates of discontinuous muscle regeneration in the rat. *Brit. J. exp. Path.*, *47*, 215.

GEIGER, R. S. and GARVIN, J. S. (1957): Pattern of regeneration of muscle from progressive muscular dystrophy patients cultivated in vitro as compared to normal human skeletal muscle. *J. Neuropath. exp. Neurol.*, *16*, 532.

GLADE, R. W. (1963): Effects of tail skin, epidermis, and dermis on limb regeneration in Triturus viridescens and Siredon mexicanum. *J. exp. Zool.*, *152*, 169.

GODLEWSKI, E. (1902): Die Entwicklung des Skelett- und Herzmuskelgewebes der Säugetiere. *Arch. mikr. Anat.*, *60*, 111.

GODMAN, G. C. (1957): On the regeneration and redifferentiation of mammalian striated muscle. *J. Morphol.*, *100*, 27.

GODMAN, G. C. (1958): Cell transformation and differentiation in regenerating striated muscle. In: *Frontiers in Cytology*, p. 381. Editor: S. L. Palay. Yale University Press, New Haven, Conn.

GOLDSPINK, G. (1962): Studies on postembryonic growth and development of skeletal muscle. 1. Evidence of 2 phases in which striated muscle fibers are able to exist. *Proc. roy. Irish Acad.*, *62B*, 135.

GOSS, R. J. (1969): *Principals of Regeneration.* Academic Press, New York, N.Y.

GRILLO, H. C., LAPIERE, C. M., DRESDEN, M. H. and GROSS, J. (1968): Collagenolytic activity in regenerating forelimbs of the adult newt (Triturus viridescens). *J. Cell Biol.*, *5*, 571.

GRIMLEY, P. M., ALBRECHT, J. M. and MICHELITCH, H. J. (1965): Preparation of large epoxy sections for light microscopy as an adjunct to fine-structure studies. *Stain Technol.*, *40*, 357.

HAREL, J., HAREL, L., GOLDÉ, A. and VIGIER, P. (1966*a*): Homologie entre génome du virus du sarcoma de Rous (RSV) et génome cellulaire. *C.R. Acad. Sci. (Paris)*, *263*, 745.

HAREL, L., HAREL, J. and HUPPERT, J. (1967): Partial homology between RNA from Rauscher mouse leukemia virus and cellular DNA. *Biochem. biophys. Res. Commun.*, *28*, 44.

HAREL, L., HAREL, J., LACOUR, F. and HUPPERT, J. (1966*b*): Homologie entre génome du virus de la myéloblastose aviaire (AMV) et génome cellulaire. *C. R. Acad. Sci. (Paris)*, *263*, 616.

HARNDEN, D. G. (1968): Human genetics. In: *A Companion to Medical Studies*, p. 383. Editors: R. Passmore and J. S. Robson. Blackwell Scientific Publications, Edinburgh.

HAUSCHKA, S. D. (1968): Clonal aspects of muscle development and the stability of the differentiated state. In: *The Stability of the Differentiated State*, p. 37. Editor: H. Ursprung. Springer-Verlag, Berlin.

HAUSCHKA, S. D. and KONIGSBERG, I. R. (1966): The influence of collagen on the development of muscle clones. *Proc. nat. Acad. Sci. USA*, *55*, 119.

HAY, E. D. (1952): The role of epithelium in amphibian limb regeneration, studied by haploid and triploid transplants. *Amer. J. Anat.*, *91*, 447.

HAY, E. D. (1959): Electron microscopic observations of muscle dedifferentiation in regenerating Amblystoma limbs. *Develop. Biol.*, *1*, 555.

HAY, E. D. (1962): Cytological studies of dedifferentiation and differentiation in regenerating amphibian limbs. In: *Regeneration*, pp. 177-210. Editor: D. Rudnick. The Ronald Press Co., New York, N.Y.

HAY, E. D. (1963): The fine structure of differentiating muscle in the salamander tail. *Z. Zellforsch.*, *59*, 6.

HAY, E. D. (1965): In: *Organogenesis.* Editors: R. De Haan and H. Ursprung. Holt, Rinehart and Winston, Inc., New York, N.Y.

HAY, E. D. (1966): *Regeneration*, 148 pp. Holt, Rinehart and Winston, Inc., New York, N.Y.

HAY, E. D. (1968): Dedifferentiation and metaplasia in vertebrate and invertebrate regeneration. In: *The Stability of the Differentiated State*, pp. 85-108. Editor: H. Ursprung. Springer-Verlag, Berlin.

HAY, E. D. and DOYLE, C. M. (1970): A statistical analysis of the cell population of mature newt limbs. I. Absence of satellite cells in non-regenerating muscle and the significance of pericytes. In preparation.

HAY, E. D. and FISCHMAN, D. A. (1961): Origin of the blastema in regenerating limbs of the newt, Triturus viridescens. An autoradiographic study using tritiated thymidine to follow cell proliferation and migration. *Develop. Biol.*, *3*, 26.

HELLMUTH, A. E., HAN, M. F. and ALLBROOK, D. B. (1970): The skeletal muscle satellite cell population in relation to postnatal age. In press.

HEUSON-STIENNON, J. A. (1965): Morphogenèse de la cellule musculaire striée au microscope électronique. *J. Microsc.*, *4*, 657.

HOLTZER, H. (1959): The development of mesodermal axial structures in regeneration and embryogenesis. In: *Regeneration in Vertebrates*, pp. 15-33. Editor: C. S. Thornton. University of Chicago Press, Chicago, Ill.

HOLTZER, H. (1966): Mutually exclusive activities during myogenesis. In: *Exploratory Concepts in Muscular Dystrophy*, p. 57. Editor: A. T. Milhorat. Excerpta Medica, Amsterdam.

HOLTZER, H. (1968): Induction of chondrogenesis: a concept in quest of mechanisms. In: *Epithelial-Mesenchymal Interactions*. Editor: R. Fleishmajer. Williams and Wilkins Co., Baltimore, Md.

HOLTZER, H. and ABBOTT, J. (1968): Oscillations in the chondrogenic phenotype. In: *Stability of the Differentiated State*. Editor: H. Ursprung. Springer-Verlag, Berlin.

HOLTZER, H., ABBOTT, J. and LASH, J. (1958): On the formation of multinucleated myotubes. *Anat. Rec.*, *131*, 567.

HOLTZER, H., AVERY, G. and HOLTZER, S. (1954): Some properties of the regenerating limb blastema cells of salamanders. *Biol. Bull.*, *107*, 313.

HOLTZER, H. and BISCHOFF, R. (1970): Mitosis and myogenesis. In: *The Physiology and Biochemistry of Muscle as a Food*. Editors: E. J. Brinskey, K. G. Cassens and J. C. Trautman. University of Wisconsin Press, Madison, Wis. In press.

HOLTZER, H., BISCHOFF, R. and CHACKO, S. (1969): Activities of the cell surface during myogenesis and chrondrogenesis. In: *Cellular Recognition*. Editors: R. T. Smith and R. A. Good. Appleton-Century-Crofts, New York, N.Y.

HOLTZER, H., MARSHALL, J. and FINCK, H. (1957): An analysis of myogenesis by the use of fluorescent antimyosin. *J. biophys. biochem. Cytol.*, *3*, 705.

HOLTZER, S. (1956): The role of the spinal cord in urodele tail regeneration. *J. Morphol.*, *99*, 1.

HUGHES, W. L., BOND, V. P., BRECHER, G., CRONKITE, E. D., PAINTER, R. B., QUASTLER, H. and SHERMAN, F. G. (1958): Cellular proliferation in the mouse as revealed by autoradiography with tritiated thymidine. *Proc. nat. Acad. Sci. USA*, *44*, 476.

HUMBLE, J. G., JAYNE, W. H. W. and PULVERTAFT, R. J. V. (1956): Biological interaction between lymphocytes and other cells. *Brit. J. Haemat.*, *2*, 283.

IGNATIEVA, Z. P. (1960): Development of minced muscle tissue under conditions of subcutaneous transplantation in dogs (Russian). *Arkh. Anat. Gistol. Embriol.*, *39/12*, 33.

ILLINGWORTH, C. and DICK, B. M. (1963): Diseases of muscle tendon, sheaths and bursae. In: *A Textbook of Surgical Pathology*, *9th ed.*, pp. 196-206. J. and A. Churchill, London.

ISHIKAWA, H. (1966): Electron microscopic observations of satellite cells with special reference to the development of mammalian skeletal muscle. *Z. Anat. Entwickl.-Gesch.*, *125*, 43.

ISHIKAWA, H., BISCHOFF, R. and HOLTZER, H. (1968): Mitosis and intermediate-sized filaments in developing skeletal muscle. *J. Cell Biol.*, *38*, 538.

KAIGHN, M. E., EBERT, J. D. and STOTT, P. M. (1966): The susceptibility of differentiating muscle clones to Rous sarcoma virus. *Proc. nat. Acad. Sci. USA*, *56*, 133.

KARLSSON, U., ANDERSSON-CEDERGREN, E. and OTTOSON, D. (1966): Cellular organisation of the frog muscle spindle as revealed by serial sections for electron microscopy. *J. Ultrastruct. Res.*, *14*, 1.

KARNOVSKY, M. J. (1965): A formaldehyde-glutaraldehyde fixative of high osmolality for use in electron microscopy. *J. Cell Biol.*, *27*, 137A.

KATZ, B. (1961): The terminations of the afferent nerve fibre in the muscle spindle of the frog. *Phil. Trans. B*, *343*, 221.

KELLY, A. M. and ZACKS, S. I. (1969): The histogenesis of rat intercostal muscle. *J. Cell Biol.*, *42*, 135.

KELLY, D. E. (1969): Myofibrillogenesis and Z-band differentiation. *Anat. Rec.*, *163*, 403.

KITIYAKARA, A. and ANGEVINE, D. M. (1960): Regeneration of injured and transplanted striated voluntary muscle in rats. *Amer. J. Path.*, *37*, 613.

KITIYAKARA, A. and ANGEVINE, D. M. (1963): A study of the pattern of postembryonic growth of M. gracilis in mice. *Develop. Biol.*, *8*, 322.

KLINKERFUSS, G. H. (1967): An electron microscopic study of myotonic dystrophy. *Arch. Neurol. (Chic.)*, *16*, 181.

KOCHAKIAN, C. D., HILL, J. and HARRISON, D. G. (1964): Regulation of nucleic acids of muscles and accessory sex organs of guinea pigs by androgens. *Endocrinology*, *74*, 635.

KONIGSBERG, I. R. (1963): Clonal analysis of myogenesis. *Science*, *140*, 1273.
KONIGSBERG, I. R. (1965): Aspects of cytodifferentiation of skeletal muscle. In: *Organogenesis*, pp. 337-358. Editors: R. L. De Haan and H. Ursprung. Holt, Rinehart and Winston, Inc., New York, N.Y.

LAGUENS, R. (1963): Satellite cells of skeletal muscle fibres in human progressive muscular dystrophy. *Virchows Arch. path. Anat.*, *336*, 564.
LAIRD, J. L. and TIMMER, R. F. (1965): Homotransplantation of dystrophic and normal muscle. *Arch. Path.*, *80*, 442.
LAIRD, J. L. and TIMMER, R. F. (1966): Transplantation of skeletal muscle into a host with muscular dystrophy. *Tex. Rep. Biol. Med.*, *24*, 169.
LANDON, D. N. (1966): Electron microscopy of muscle spindles. In: *Control and Innervation of Skeletal Muscle*, pp. 96-110. Editor: B. L. Andrew. E. and S. Livingstone, Ltd., Edinburgh-London.
LANE, B. P. and EUROPA, D. L. (1965): Differential staining of ultrathin sections of Epon-embedded tissues for light microscopy. *J. Histochem. Cytochem.*, *13*, 579.
LASH, J. W., HOLTZER, H. and SWIFT, H. (1957): Regeneration of mature skeletal muscle. *Anat. Rec.*, *128*, 679.
LEBLOND, C. P. (1964): Classification of cell populations on the basis of their proliferative behavior. *Nat. Cancer Inst. Monogr.*, *14*, 119.
LEE, H. H., KAIGHN, M. E. and EBERT, J. D. (1968): Induction of thymidine-H^3 incorporation in multinucleated myotubes by Rous sarcoma virus. *Int. J. Cancer*, *3*, 126.
LEE, J. C. (1965): Electron microscope observations on myogenic free cells of denervated skeletal muscle. *Exp. Neurol.*, *12*, 123.
LE GROS CLARK, W. E. (1946): An experimental study of the regeneration of mammalian striped muscle. *J. Anat. (Lond.)*, *80*, 24.
LE GROS CLARK, W. E. and BLOMFIELD, L. B. (1945): The efficiency of intramuscular anastomoses, with observations on the regeneration of devascularized muscle. *J. Anat. (Lond.)*, *79*, 15.
LENTZ, T. L. (1969): Cytological studies of muscle dedifferentiation and differentiation during limb regeneration of the newt Triturus. *Amer. J. Anat.*, *124*, 447.
LEONG, C. F., PESSOTTI, R. L. and BRAUER, R. W. (1959): Liver function in regenerating rat liver. $CrPO_4$ colloid uptake and bile flow. *Amer. J. Physiol.*, *197*, 880.
LITVER, G. M. (1955): Importance of vitamin A and vitamin B complex on regeneration of skeletal muscles of mammals (Russian). *Scient. Writings Dept. gen. Biol., 1st Leningrad med. Sch.*, *2*, 12.
LITVER, G. M. (1962): On the effect of external factors and internal conditions of organism on reparative regeneration of skeletal muscle tissue. *Rev. canad. Biol.*, *21*, 367.
LITVER, G. M. and DAMPEL, N. N. (1959): On the possibility of whole muscle regeneration from transplanted minced muscular tissue in the rat (Russian). *Arkh. Anat. Gistol. Embriol.*, *37*, 54.
LOEFFLER, C. A. (1969): Evidence for the fusion of myoblasts in amphibian embryos. I. Homoplastic transplantations of somitic material labeled with tritiated thymidine. *J. Morphol.*, *128*, 403.
LYNN, J. A. (1965): Rapid toluidine blue staining of Epon-embedded and mounted "adjacent" sections. *Amer. J. clin. Path.*, *44*, 57.

MACCONNACHIE, H. F., ENESCO, M. and LEBLOND, C. P. (1964): The mode of increase in the number of skeletal muscle nuclei in the postnatal rat. *Amer. J. Anat.*, *114*, 245.
Quoting: Birks, R. I. (1962), personal communication.
MANASEK, F. J. (1968): Mitosis in developing cardiac muscle. *J. Cell Biol.*, *37*, 191.
MARCHOK, A. C. (1966): An autoradiographic observation of nucleoside incorporation in developing muscle cells. *Exp. Cell Res.*, *43*, 214.
MARCHOK, A. C. and HERRMANN, H. (1967): Studies of muscle development. I. Changes in cell proliferation. *Develop. Biol.*, *15*, 129.
MARTIN, J. H., LYNN, J. A. and NICKEY, W. M. (1966): A rapid polychrome stain for epoxy-embedded tissue. *Amer. J. clin. Path.*, *46*, 250.
MAURO, A. (1961): Satellite cell of skeletal muscle fibers. *J. biophys. biochem. Cytol.*, *9*, 493.
Quoting: Palade, G. E. (1961), personal communication.
MAURO, A. and ADAMS, W. R. (1961): The structure of the sarcolemma of the frog skeletal muscle fiber. *J. biophys. biochem. Cytol.*, *10*, 177.

MEIGS, E. B. (1932): Striated and smooth muscle. In: *Special Cytology, Vol. II*, pp. 1089-1125. Editor: E. V. Cowdry. Paul B. Hoeber, Inc., New York, N.Y.

MIDZUKAMI, M. (1964): Electron microscopic studies of satellite cells in the cardiac muscle of brachyura. *Okajimas Folia anat. jap., 40*, 173.

MILLAR, W. G. (1934): Regeneration of skeletal muscle in young rabbits. *J. Path. Bact., 38*, 145.

MILLER, N. R. and WOLFE, H. J. (1968): The nature and localization of acid phosphatase during the early phases of urodele limb regeneration. *Develop. Biol., 4*, 447.

MOORE, A. E. and HLINKA, J. (1969): Activities of lymphocytes inside fibroblasts – emperipolesis. *Cytobios, 1B*, 73.

MOSS, F. P. (1968): The relationship between the dimensions of the fibers and the number of nuclei during normal growth of skeletal muscle in the domestic fowl. *Amer. J. Anat., 122*, 555.

MOSS, F. P. (1969): Nature of dividing nuclei in skeletal muscles of growing rats. *Anat. Rec., 163*, 233.

MOSS, F. P. and LEBLOND, C. P. (1970): Nature of dividing nuclei in skeletal muscle of growing rats. *J. Cell Biol., 44*, 459.

MUIR, A. R. (1965): Further observations on the cellular structure of cardiac muscle. *J. Anat. (Lond.), 99*, 27.

MUIR, A. R. (1970): Normal and regenerating skeletal muscle fibres in Pietrain pigs. *J. comp. Path., 80*, 137.

MUIR, A. R., KANJI, A. H. M. and ALLBROOK, D. B. (1965): The structure of the satellite cells in skeletal muscle. *J. Anat. (Lond.), 99*, 435.

MUSCATELLO, U., MARGRETH, A. and ALOISI, M. (1965): On the differential response of sarcoplasm and myoplasm to denervation in frog muscle. *J. Cell Biol., 27*, 1.

NAMENWIRTH, M. R. (1968): Is cell differentiation inherited? *Amer. Zool., 8*, 785.

NAMENWIRTH, M. R. (1970*a*): The inheritance of cell differentiation during limb regeneration in the axolotl. *Proc. nat. Acad. Sci. USA*, in press.

NAMENWIRTH, M. R. (1970*b*): Regeneration of X-irradiated axolotl limbs after non-irradiated tissues are introduced by grafting. *J. exp. Zool.*, in press.

NAMEROFF, M. (1968): Interference with myogenesis: a mirror image of an embryonic induction system (Abstract). *Anat. Rec., 160*, 399.

NAMEROFF, M. and HOLTZER, H. (1967): The loss of phenotypic traits by differentiated cells. IV. Changes in polysaccharides produced by dividing chondrocytes. *Develop. Biol., 16*, 250.

NAMEROFF, M. and HOLTZER, H. (1969): Contact-mediated reversible suppression of myogenesis. *Develop. Biol., 19*, 380.

NAMEROFF, M., REZNIK, M., ANDERSON, P. and HANSEN, J. L. (1970): Differentiation and control of mitosis in a skeletal muscle tumor. *Cancer Res., 30*, 596.

NICHOLAS, J. (1955): In: *Analysis of Development*. Editors: B. Willier, P. Weiss and V. Hamburger. W. B. Saunders Co., Philadelphia, Pa.

NILAUSEN, K. and GREEN, H. (1965): Reversible arrest of growth of an established fibroblast line (3T3). *Exp. Cell Res., 40*, 166.

NORMAN, W. P. and SCHMIDT, A. J. (1967): The fine structure of tissues in the adult newt, Diemictylus viridescens. *J. Morphol., 123*, 271.

OKAZAKI, K. and HOLTZER, H. (1965): An analysis of myogenesis in vitro using fluorescein-labeled antimyosin. *J. Histochem. Cytochem., 13*, 726.

OKAZAKI, K. and HOLTZER, H. (1966): Myogenesis: Fusion, myosin synthesis, and the mitotic cycle. *Proc. nat. Acad. Sci. USA, 56*, 1484.

PARAKKAL, P. F. (1969): Ultrastructural changes of the basal lamina during the hair growth cycle. *J. Cell Biol., 40*, 561.

PEARCE, G. W. (1963): Tissue culture in the study of muscular dystrophy. In: *Muscular Dystrophy in Man and Animals*, pp. 171-191. Editors: G. H. Bourne and M. A. N. Golarz. S. Karger, Basel.

PIETSCH, P. (1961*a*): The effects of colchicine on regeneration of mouse skeletal muscle. *Anat. Rec., 139*, 167.

PIETSCH, P. (1961*b*): Differentiation in regeneration. I. The development of muscle and cartilage following deplantation of regenerating limb blastemata of Amblystoma larvae. *Develop. Biol., 3*, 255.

POLEZHAEV, L. V. (1946): The loss and restoration of regenerative capacity in the limbs of tailless amphibia. *Biol. Rev.*, *21*, 141.
POOL, C. R. (1969): Hematoxylin-eosin staining of OsO_4-fixed Epon-embedded tissue; prestaining oxidation by acidified H_2O_2. *Stain Technol.*, *44*, 75.
POPOGEFF, I. A. and MURRAY, M. R. (1946): Form and behaviour of adult mammalian skeletal muscle in vitro. *Anat. Rec.*, *95*, 321.
PRICE, H. M., HOWES Jr., E. L. and BLUMBERG, J. M. (1964): Ultrastructural alterations in skeletal muscle fibers injured by cold. II. Cells of the sarcolemmal tube: observations on "discontinuous" regeneration and myofibril formation. *Lab. Invest.*, *13*, 1279.
PRINCE, A. M. (1966): Characteristics of rapidly sedimenting ribonucleic acid from Harris strain of Rous sarcoma virus and infected cells. *Virology*, *30*, 151.
PRZYBYLSKI, R. J. and BLUMBERG, J. M. (1966): Ultrastructural aspects of myogenesis in the chick. *Lab. Invest.*, *15*, 836.

QUASTLER, H. and SHERMAN, F. G. (1959): Cell population kinetics in the intestinal epithelium of the mouse. *Exp. Cell Res.*, *17*, 420.

REGER, J. F. and CRAIG, A. S. (1968): Studies on the fine structure of muscle fibers and associated satellite cells in hypertrophic human deltoid muscles. *Anat. Rec.*, *162*, 483.
REYNOLDS, E. S. (1963): The use of lead citrate at high pH as an electron-opaque stain in electron microscopy. *J. Cell Biol.*, *17*, 208.
REZNIK, M. (1967*a*): Aspect ultrastructural de la dégénérescence du muscle strié ischémié. *Ann. Anat. path.*, *12*, 209.
REZNIK, M. (1967*b*): Influence d'une irradiation locale unique sur le muscle strié en régénération. *Arch. Biol. (Liège)*, *78*, 91.
REZNIK, M. (1968*a*): Possible origin of myoblasts during muscle regeneration. *J. Cell Biol.*, *39*, 110a.
REZNIK, M. (1968*b*): Etude autoradiographique du muscle strié en régénération in vivo. *J. Embryol. exp. Morphol.*, *19*, 283.
REZNIK, M. (1969*a*): Thymidine-^{3}H uptake by satellite cells of regenerating skeletal muscle. *J. Cell Biol.*, *40*, 568.
REZNIK, M. (1969*b*): Origin of myoblasts during skeletal muscle regeneration. Electron microscopic observations. *Lab. Invest.*, *20*, 353.
REZNIK, M. and BETZ, E. H. (1965): Régénération du muscle ischémié préalablement irradié. *C.R. Soc. Biol. (Paris)*, *159*, 2078.
REZNIK, M. and BETZ, E. H. (1967): Influence de l'irradiation locale préalable sur les capacités régénératrices du muscle strié squelettique. *Path. europ. (Brux.)*, *2*, 69.
REZNIK, M., NAMEROFF, M. and HANSEN, J. L. (1970): Ultrastructure of a transplantable murine rhabdomyosarcoma. *Cancer Res.*, *30*, 601.
ROBERTSON, J. D. (1960): The molecular structure and contact relationships of cell membranes. *Progr. Biophys.*, *10*, 343.
ROCKWELL, A. F., NORTON, P., CAULFIELD, J. B. and ROTH, S. I. (1966): A silicone rubber mold for embedding tissue in epoxy resins. *Science Tools*, *13*, 9.
ROSE, S. M. (1945): The effect of NaCl in stimulating regeneration of limbs of frogs. *J. Morphol.*, *77*, 119.
ROSE, S. M. (1964): In: *Physiology of the Amphibian.* Editor: J. Moore. Academic Press, New York, N.Y.
RZEHAK, K. and SINGER, M. (1966): The number of nerve fibers in the limb of the mouse and its relation to regenerative capacity. *Anat. Rec.*, *155*, 537.

SCHANTZ, A. and SCHECTER, A. (1965): Iron-hematoxylin and safranin O as a polychrome stain for Epon sections. *Stain Technol.*, *40*, 279.
SCHIAFFINO, S. and MARGRETH, A. (1969): Coordinated development of the sarcoplasmic reticulum and T system during postnatal differentiation of rat skeletal muscle. *J. Cell Biol.* ,*41* , 855.
SCHMIDT, A. J. (1966): *The Molecular Basis of Regeneration.* University of Illinois Press, Urbana, Ill.
SCHOTTÉ, O. E. and WILBUR, J. F. (1958): Effects of adrenal transplants upon forelimb regeneration in normal and in hypophysectomized adult frogs. *J. Embryol. exp. Morphol.*, *6*, 247.

SHAFIQ, S. A. and GORYCKI, M. A. (1965): Regeneration in skeletal muscle of mouse: some electron-microscope observations. *J. Path. Bact.*, *90*, 123.

SHAFIQ, S. A., GORYCKI, M. A. and MAURO, A. (1968): Mitosis during postnatal growth in skeletal and cardiac muscle of the rat. *J. Anat.* (*Lond.*), *103*, 135.

SHAFIQ, S. A., GORYCKI, M. A. and MILHORAT, A. T. (1967): An electron microscopic study of regeneration and satellite cells in human muscle. *Neurology* (*Minneap.*), *17*, 567.

SHIMADA, Y. (1968): Suppression of myogenesis by heterotypic and heterospecific cells in monolayer cultures. *Exp. Cell Res.*, *51*, 564.

SHIMADA, Y., FISCHMAN, D. A. and MOSCONA, A. A. (1967): The fine structure of embryonic chick skeletal muscle cells differentiated in vitro. *J. Cell Biol.*, *35*, 445.

SHIRAHAMA, T. and COHEN, A. S. (1966): A Congo red staining method for epoxy-embedded amyloid. *J. Histochem. Cytochem.*, *14*, 725.

SIMPSON, S. (1970): Studies on regeneration of the lizard tail. *Amer. Natural.*, *10*, 157.

SINGER, M. (1952): The influence of the nerve in regeneration of the amphibian extremity. *Quart. Rev. Biol.*, *27*, 169.

SINGER, M. (1954): Induction of regeneration of the forelimb of postmetamorphic frogs by augmentation of the nerve supply. *J. exp. Zool.*, *126*, 419.

SINGER, M. and SALPETER, M. M. (1961): Regeneration in vertebrates: the role of wound epithelium. In: *Growth in Living Systems*, pp. 277-311. Editor: M. X. Zarrow. Basic Books, Inc., New York, N.Y.

SJOSTRAND, F. S. (1963): A comparison of plasma membranes, cytomembranes and mitochondrial membrane elements with respect to ultrastructural features. *J. Ultrastruct. Res.*, *9*, 561.

SLOPER, J. C. and PEGRUM, G. D. (1967): Regeneration of crushed mammalian skeletal muscle and effects of steroids. *J. Path. Bact.*, *93*, 47.

SPEIDEL, C. C. (1938): Studies of living muscles. I. Growth, injury and repair of striated muscle, as revealed by prolonged observations of individual fibers in living frog tadpoles. *Amer. J. Anat.*, *62*, 179.

STEEN, T. P. (1968): Stability of chondrocyte differentiation and contribution of muscle to cartilage during limb regeneration in the axolotl (Siredon mexicanum). *J. exp. Zool.*, *167*, 49.

STEINMAN, R. M. (1968): An electron microscopic study of ciliogenesis in developing epidermis and trachea in the embryo of Xenopus laevis. *Amer. J. Anat.*, *122*, 19.

STENGER, R. J. and SPIRO, D. (1961): Ultrastructure of mammalian cardiac muscle. *J. biophys. biochem. Cytol.*, *9*, 325.

STOCKDALE, F. E. and HOLTZER, H. (1961): DNA synthesis and myogenesis. *Exp. Cell Res.*, *24*, 508.

STOCKDALE, F. E., OKAZAKI, K., NAMEROFF, M. and HOLTZER, H. (1964): 5-Bromodeoxyuridine: effect on myogenesis in vitro. *Science*, *146*, 533.

STOCUM, D. L. (1968*a*): The urodele limb regeneration blastema: a self-organizing system. I. Differentiation in vitro. *Develop. Biol.*, *18*, 441.

STOCUM, D. L. (1968*b*): The urodele limb regeneration blastema: a self-organizing system. II. Morphogenesis and differentiation of autografted whole and fractional blastemas. *Develop. Biol.*, *18*, 457.

STOKER, M. G. P. and RUBIN, H. (1967): Density-dependent inhibition of cell growth in culture. *Nature* (*Lond.*), *215*, 171.

STREHLER, B. L., KONIGSBERG, I. R. and KELLEY, J. E. (1963): Ploidy of myotube nuclei developing in vitro as determined with a recording double beam microspectrophotometer. *Exp. Cell Res.*, *32*, 232.

STRIGANOVA, A. R. (1961): *Reactivity and Restorative Potential of Denervated Muscle in Various Stages of Atrophy*, 147 pp (Russian). Nauka, Moscow.

STUDITSKY, A. N. (1952*a*): Restoration of muscle by means of transplantation of minced muscle tissue (Russian). *Dok. Akad. Nauk SSSR*, *84/2*, 389.

STUDITSKY, A. N. (1952*b*): The principles of restoration of muscle in higher vertebrates (Russian). *Arkh. Anat. Gistol. Embriol.*, *29/6*, 27.

STUDITSKY, A. N. (1954): Laws of restoration of muscle in higher vertebrates (Russian). *Trudy Inst. Morfol. zhivot.*, *11*, 225.

STUDITSKY, A. N. (1959): *Experimental Surgery of Muscles*, 338 pp (Russian). Nauka, Moscow.

STUDITSKY, A. N. (1963): Dynamics of the development of myogenic tissue under conditions of explantation and transplantation. In: *Cinemicrography in Cell Biology*, pp. 171-200. Editor: G. G. Rose. Academic Press, New York, N.Y.

STUDITSKY, A. N. and IGNATIEVA, Z. P. (1961): *Restoration of Muscle in Higher Mammals*, 192 pp (Russian). Nauka, Moscow.
STUDITSKY, A. N., ZHENEVSKAYA, R. P. and RUMYANTSEVA, O. N. (1956): Fundamentals of the technique of restoration of muscle by means of transplantation of minced muscle tissue (Russian). *Folia morphol. (Praha)*, *4*, 331.
STUDITSKY, A. N., ZHENEVSKAYA, R. P. and RUMYANTSEVA, O. N. (1963): The role of neurotrophic influences upon the restitution of structure and function of regenerating muscles. In: *The Effect of Use and Disuse on Neuromuscular Functions*, pp. 71-81. Editors: E. Gutmann and P. Hnik. Publishing House of the Czechoslovak Academy of Sciences, Prague.

TANAKA, Y. and BERSCHAUER, J. A. (1969): Application of the Perl's method for iron staining to sections embedded in epoxy resin. *Stain Technol.*, *44*, 255.
TELLO, J. F. (1917): Genesis de las terminaciones nerviosas motrices y sensitivas. I. En el sistema locomotor de los vertebrados superiores; histogenesis muscular. *Trab. Inst. Cajal Invest. biol.*, *15*, 101.
TELLO, J. F. (1922): Die Entstehung der motorischen und sensibelen Nervenendungen. I. In dem lokomotorischen Systeme der höheren Wirbeltiere; muskuläre Histogenese. *Z. Anat. Entwickl.-Gesch.*, *64*, 348.
TEMIN, H. (1963): The effects of actinomycin D on growth of Rous sarcoma virus in vitro. *Virology*, *20*, 577.
TEMIN, H. (1964): Homology between RNA from Rous sarcoma virus- and DNA from Rous sarcoma virus-infected cells. *Proc. nat. Acad. Sci. USA*, *52*, 323.
THORNTON, C. S. (1938): The histogenesis of muscle in the regenerating forelimb of larval Amblystoma punctatum. *J. Morphol.*, *62*, 17.
THORNTON, C. S. (1942): Studies on the origin of the regeneration blastema in Triturus viridescens. *J. exp. Zool.*, *89*, 375.
THORNTON, C. S. (1943): The effect of colchicine on limb regeneration in larval Amblystoma. *J. exp. Zool.*, *92*, 281.
THORNTON, C. S. (1968): Amphibian limb regeneration. *Advanc. Morphogenes.*, *7*, 205.
TODARO, G. J. and GREEN, H. (1966): Cell growth and the initiation of transformation by SV-40. *Proc. nat. Acad. Sci. USA*, *55*, 302.
TODARO, G. J., LAZAR, G. K. and GREEN, H. (1965): The initiation of cell division in a contact-inhibited mammalian cell line. *J. cell. comp. Physiol.*, *55*, 325.
TOSELLI, P. A. and PEPE, F. A. (1968): The fine structure of the ventral intersegmental abdominal muscles of the insect Rhodnius prolixus during the molting cycle. *J. Cell Biol.*, *37*, 462.
TRUMP, B. F., SMUCKLER, E. A. and BENDITT, E. P. (1961): A method for staining epoxy sections for light microscopy. *J. Ultrastruct. Res.*, *5*, 343.

VENABLE, J. H. (1966*a*): Constant cell populations in normal, testosterone-deprived and testosterone-stimulated levator ani muscles. *Amer. J. Anat.*, *119*, 263.
VENABLE, J. H. (1966*b*): Morphology of the cells of normal, testosterone-deprived and testosterone-stimulated levator ani muscles. *Amer. J. Anat.*, *119*, 271.
Quoting: Flood, P. R. (1964), personal communication.
VIGIER, P. and GOLDÉ, A. (1964): Effects of actinomycin D and of mitomycin C on the development of Rous sarcoma virus. *Virology*, *23*, 511.

WALKER, B. E. (1963): The origin of myoblasts and the problem of dedifferentiation. *Exp. Cell Res.*, *30*, 80.
WALLS, E. W. (1960): The micro-anatomy of muscle. In: *Structure and Function of Muscle, Vol. I*, p. 21. Editor: G. H. Bourne. Academic Press, New York, N.Y.
WALTON, J. N. and ADAMS, R. D. (1956): The response of the normal, the denervated and the dystrophic muscle-cell to injury. *J. Path. Bact.*, *72*, 273.
WARREN, R. H. (1968): The effect of colchicine on myogenesis in vivo in Rana pipiens and Rhodnius proxilus (Hemiptera). *J. Cell Biol.*, *39*, 329.
WEBER, O. (1867): Ueber die Neubildung quergestreifter Muskelfasern, insbesondere die regenerative Neubildung desselben nach Verletzungen. *Virchows Arch. path. Anat.*, *39*, 216.
WEIL, R., MICHEL, M. and RUSCHMANN, G. K. (1965): Induction of cellular DNA synthesis by polyoma virus. *Proc. nat. Acad. Sci. USA*, *53*, 1468.

WEINSTEIN, R. and HAY, E. D. (1970): Incorporation of H^3-thymidine into differentiated cardiac muscle cells. In preparation.

WILLIAMS, R. T. and PIETSCH, P. (1965): Time related inhibition of mammalian muscle regeneration by X-irradiation (Abstract). *Anat. Rec.*, *151*, 433.

WILSON, D. E. and BAUER, H. (1967): Hybridization of avian myeloblastosis virus RNA with DNA from chick embryo cells. *Virology*, *33*, 754.

WINOCOUR, E., KAYE, A. M. and STOLLAR, V. (1965): Synthesis and transmethylation of DNA in polyoma-infected cultures. *Virology*, *27*, 156.

YAFFE, D. (1969): Retention of differentiation potentialities during prolonged cultivation of myogenic cells. *Proc. nat. Acad. Sci. USA*, *61*, 477.

YAFFE, D. and FELDMAN, M. (1965): The formation of hybrid multinucleated muscle fibers from myoblasts of different genetic origin. *Develop. Biol.*, *11*, 300.

YAFFE, D. and GERSHON, D. (1967): Multinucleated muscle fibers: Induction of DNA synthesis and mitosis by polyoma virus infection. *Nature (Lond.)*, *215*, 421.

YOSHIKAWA-FUKADA, M. and EBERT, J. D. (1969*a*): Cell differentiation and viral susceptibility. In: *Year Book No. 67*, pp. 429-440. Carnegie Institute, Washington, D.C.

YOSHIKAWA-FUKADA, M. and EBERT, J. D. (1969*b*): Hybridization of RNA from Rous sarcoma virus with cellular and viral DNA's. *Proc. nat. Acad. Sci. USA*, *64*, 870.

ZHENEVSKAYA, R. P. (1960): The influence of de-efferentation on the regeneration of skeletal muscle. *Arkh. Anat. Gistol. Embriol.*, *39/12*, 42.

ZHENEVSKAYA, R. P. (1962): Experimental histologic investigation of striated muscle tissue. *Rev. canad. Biol.*, *21*, 457.

SUBJECT INDEX

Prepared by E. Dyckhoff, M.D., Nijmegen